Ambiances

How do ambiances shape our sensory, social, and built environments? This book offers interdisciplinary perspectives to understand the profound ways in which ambiances redefine our experiences and perceptions of ordinary spaces.

The study of ambiances is part of an important trend in the contemporary urban world, where increasing attention is being given to the sensitive and affective dimensions of space and experience. Beginning with an exploration of foundational concepts, this book delves into the sensory, philosophical, and cultural dimensions of ambiances at the intersection of sensory environments and lived experiences. It explores the role of ambiances play in architectural design, urban spaces, and daily life. It engages with questions of contemporary sensibility, the sensitive experience of architectures and urban spaces, and the dialogue between the notion of ambiances and that of affective atmospheres. This book connects diverse perspectives of architects, philosophers, anthropologists, and geographers to map the conceptual and practical territories of ambiances, from their roots in Francophone research to their broader global implications.

This book offers an essential guide to the study of ambiances, providing a foundational and forward-looking exploration that is valuable for researchers and upper level undergraduates of architecture, urban studies, urban planning, human geography, anthropology, and philosophy.

Jean-Paul Thibaud is a sociologist and a senior CNRS researcher at the CRESSON Research Lab (UMR 1563 Ambiances, Architectures, Urbanities).

Nicolas Tixier is an architect and a professor at the Grenoble National School of Architecture. He is a researcher at the CRESSON Research Lab (UMR 1563 Ambiances, Architectures, Urbanities).

David Zerbib is a philosopher. Scientific coordinator of Research at the Annecy Alpes School of Art (ESAAA), he is also a Lecturer in Philosophy of Art at the Geneva University of Art and Design (HEAD - HES-SO).

Ambiances, Atmospheres and Sensory Experiences of Spaces
Series Editors:
Rainer Kazig, *CNRS Research Laboratory*
Ambiances – Architectures – Urbanités, Grenoble, France
Damien Masson, *Université de Cergy-Pontoise, France*
Paul Simpson, *Plymouth University, UK*

Research on ambiances and atmospheres has grown significantly in recent years in a range of disciplines, including Francophone architecture and urban studies, German research related to philosophy and aesthetics, and a growing range of Anglophone research on affective atmospheres within human geography and sociology.

This series offers a forum for research that engages with questions around ambiances and atmospheres in exploring their significances in understanding social life. Each book in the series advances some combination of theoretical understandings, practical knowledges and methodological approaches. More specifically, a range of key questions which contributions to the series seek to address includes:

- In what ways do ambiances and atmospheres play a part in the unfolding of social life in a variety of settings?
- What kinds of ethical, aesthetic, and political possibilities might be opened up and cultivated through a focus on atmospheres/ambiances?
- How do actors such as planners, architects, managers, commercial interests and public authorities actively engage with ambiances and atmospheres or seek to shape them? How might these ambiances and atmospheres be reshaped towards critical ends?
- What original forms of representations can be found today to (re)present the sensory, the atmospheric, the experiential? What sort of writing, modes of expression, or vocabulary is required? What research methodologies and practices might we employ in engaging with ambiances and atmospheres?

Ambiances
A Sensitivity to Ordinary Situations
Edited by Jean-Paul Thibaud, Nicolas Tixier and David Zerbib

For more information about this series, please visit: www.routledge.com/Ambiances-Atmospheres-and-Sensory-Experiences-of-Spaces/book-series/AMB

Ambiances

A Sensitivity to Ordinary Situations

Edited by
Jean-Paul Thibaud, Nicolas Tixier and David Zerbib

With an artistic collaboration of Didier Tallagrand

Translator: Claudio Cambon

Routledge
Taylor & Francis Group
LONDON AND NEW YORK

First published 2026
by Routledge
4 Park Square, Milton Park, Abingdon, Oxon OX14 4RN

and by Routledge
605 Third Avenue, New York, NY 10158

Routledge is an imprint of the Taylor & Francis Group, an informa business

British Library Cataloguing-in-Publication Data
A catalogue record for this book is available from the British Library

ISBN: 978-1-032-65522-2 (hbk)
ISBN: 978-1-032-65525-3 (pbk)
ISBN: 978-1-032-65530-7 (ebk)

DOI: 10.4324/9781032655307

Typeset in Sabon
by codeMantra

Contents

Figures

About the editors

Jean-Paul Thibaud is a sociologist and a senior CNRS researcher at the CRESSON Research Lab (UMR 1563 Ambiances architectures urbanités). His field of research is the theory of urban ambiances, ordinary perception in the urban environment, sensitive ethnography of public spaces, anthropology of the public spaces, the anthropology of sound, qualitative methodologies in situ, and sensitivities to living environments. He founded the International Ambiances Network (www.ambiances.net) and published extensively on the theory of ambiances. Among other books are *En quête d'ambiances. éprouver la ville en passant* (2015); *Ambiances en débats* (A la croisée, 2004, with Pascal Amphoux and Grégoire Chelkoff, eds.), and *Ambiances urbaines en partage* (2013, with Cristiane Duarte, eds.). With Didier Tallagrand and Nicolas Tixier, he has co-organized the international conference *L'usage des ambiances* and co-edited the proceedings.

Nicolas Tixier is an architect. A professor at the Grenoble National School of Architecture, he also teaches at the Annecy Alpes School of Art and at the Grenoble Institute of Urban Planning and Alpine Geography. A researcher at Cresson (UMR 1563 Ambiances, architectures, urbanités), he also carries out a project activity within the collective BazarUrbain. From 2018 to 2024, he was Director of the Cresson and Deputy Director of the UMR AAU. His current work focuses on the urban transect as a field practice, a technique of representation and a project posture. His publications include *Ambiances, tomorrow* (co-edited with Nicolas Rémy), proceedings of the 3rd International Congress on Ambiances (Volos, 2016); *Traversées urbaines, villes et films en regard* (2015). With Didier Tallagrand and Jean-Paul Thibaud, he has co-organized the international conference *L'usage des ambiances* and co-edited the proceedings.

David Zerbib is a philosopher and coordinates the research unit of the Annecy Alpes School of Art. He is also a Lecturer in Philosophy of Art at the Geneva University of Art and Design (HEAD HES-SO). Attached to the Centre d'histoire des philosophies modernes de la Sorbonne (HIPHIMO

UR 1451), his research focuses on the principles of a contemporary aesthetic theory, in particular through the question of performance and format. He co-edited *Performance Studies in Motion, International Perspectives and Practices* (2014) and edited *In octavo. Des formats de l'art* (2015). Interested in the Philosophical Anthropology of Helmuth Plessner he published, together with artists Boisseau & Westermeyer, *Playing at being human. Between AI, animal and plant life: an artistic and philosophical experimentation* (2021).

Contributors

Niels Albertsen
Emeritus Aarhus School of Architecture
Research Lab1 Territories, Architecture, and Transformation

Pascal Amphoux
Member of the CRESSON Research Lab (AAU)
Urban Planning CONTREPOINT, Projets urbains, Lausanne

Jean-François Augoyard
Member of the CRESSON Research Lab (AAU)
Co-Founder of the CRESSON Research Lab

Bruce Bégout
University Bordeaux Montaigne

Céline Bonicco-Donato
ENSA Grenoble, Member of the CRESSON Research Lab (AAU)

Anne Bossé
ENSA Nantes, Member of the CRENAU Research Lab (AAU)
Chief Editor of the *Ambiances Journal*

Grégoire Chelkoff
ENSA Grenoble, Member of the CRESSON Research Lab (AAU),

Laurent Devisme
ENSA Nantes, Member of the CRENAU Research Lab (AAU)

Emmanuel Doutriaux
ENSA Paris-Val de Seine, Member of the EVCAU research lab

Olivier Gaudin
École de la Nature et du Paysage, Blois, INSA CVL,
Member of the CRESSON Research Lab (AAU)
Co-Editor of the online journal *Métropolitiques*
Chief Editor of the journal *Les Cahiers de l'École de Blois*

Tonino Griffero
University of Rome Tor Vergata

Rainer Kazig
CNRS researcher, Member of the CRESSON Research Lab (AAU)
Co-Director of the Routledge series *Ambiances, Atmospheres and Sensory Experiences of Spaces*
Co-Editor of the *Ambiances Journal*

François Laplantine
University Lumière Lyon 2

Maxime Le Calvé
Member of the Centre Georg Simmel
Co-Editor of a special issue on ambiances, *Communications* n°102, 2018

Damien Masson
CY Cergy Paris Université, Deputy Director of PLACES Laboratory, Member of CRESSON Research Lab (AAU)
Co-Director of the Routledge series *Ambiances, Atmospheres and Sensory Experiences of Spaces*
Co-Director of the International Ambiances Network

Derek McCormack
Oxford University

Yves Michaud
Member of the Institut Universitaire de France

Pauline Nadrigny
University Paris 1 Panthéon-Sorbonne

Véronique Nahoum-Grappe
Researcher at the EHESS

Nia Perivolaropoulou
University Duisburg-Essen, translator

Philippe Simay
ENSA Paris-Belleville, Member of IPRAUS research lab

Didier Tallagrand
Ecole Supérieure Art Annecy Alpes

Jean-Paul Thibaud
CNRS Senior Researcher, Member of the CRESSON Research Lab (AAU)

Nicolas Tixier
ENSA Grenoble, Member of the CRESSON Research Lab

Izabela Wieczorek
University of Reading
Member of the Urban Living Research Group

David Zerbib
Research Unit, Annecy Alpes School of Art (ESAAA)
Geneva University of Art and Design (HEAD HES-SO)

Introduction

Ambiances: A Sensitivity to Ordinary Situations

Jean-Paul Thibaud, Nicolas Tixier and David Zerbib

Thinking about ambiance forms part of a general trend of opening our thought process to the sensorial, especially in the humanities and cultural theory. In this respect, ambiance forms part of the emergence of new frameworks for considering contemporary sensitivities. By raising the issue of a certain form of sensory experience, ambiance renders us sensitive to transformations that reach past its own scope, thereby allowing us to feel and reflect on dimensions of our urban, social, and earthly situated existence that do not fit into the categories of readily definable theoretical objects. At the same time, ambiance, this new *sensitivity to the sensitive*, also refers to very real situations. Whether we wish to dwell on the tendency to aestheticize our urban spaces, the growing attention paid to social and ecological upheavals, the rapid development of sensory marketing, and the growing relevance of ambient intelligence, our mode of being sensitive to the spaces we inhabit is definitely changing. We are at a historic moment in which the sensory conditions of our existence are changing and becoming more explicit. Percepts and affects have regained legitimacy, bringing the diversity of sensory registers into play and recognizing the importance of bodily experience. Ambiance is not just a question of perceiving a landscape or measuring an environment; it also involves collectively feeling situations and experiencing the sensory contexture of social life (Tallagrand, Thibaud & Tixier 2021). The sensitive is resurfacing, making itself felt and heard, and ambiance is one of the most powerful ways in which this happens. It is also contributing to a change in thinking about cities and architectural practice (Amphoux 1998).

A sensitivity to ordinary situations

Ambiance became a veritable field of research in French-speaking countries in the 1990s amidst a very peculiar intellectual ecosystem. A set of operative concepts, investigative methods, analytical tools, and field studies developed during this period provided a theoretical and empirical basis for this nascent field of inquiry into ambiances. In search of alternatives to the major paradigms of the 1960s and 70s – Marxism, Structuralism, and Semiotics – the humanities and social sciences began to explore new avenues of inquiry and

DOI: 10.4324/9781032655307-1

engage in new theoretical explorations. A considerable diversity of propositions emerged, opening up hitherto unexplored perspectives and resulting in a "new configuration of the humanities" (Dosse 1995). It is true that a number of varyingly explicit incursions had already been made into the territory of ambiance. Consider, for example, the inaugural essay *Mélancholia* by Léon Daudet (Daudet 1928), or the intuitions of the Situationist movement in the 1960s, which based itself on the thinking of Guy Debord (*Internationale Situationniste* 1997), or even certain explorations conducted in the early 1970s at the Experimental University of Vincennes, which, in a post-structuralist vein, focused on situated experience in the humanities and in relation to urban planning. Thinking about ambiances traditionally entailed an expansion of our theoretical attention from merely analysing the interplay of the material and symbolic structures of human experience to include a primarily situated and multi-factorial approach, and whose observational and analytical tools had to be invented on-site to some extent. But it is in roughly the last 30 years – in light of this very particular scholarly environment – that this field of thought and action has emerged and taken shape.

The field of ambiances forms part of this intellectual renewal in its development of a sensitivity to ordinary situations. It has emerged along five intersecting, complementary guidelines: 1) the persistent interest in built environments serves to reconsider architectural and urban space by focusing anew on the experience of inhabiting; 2) the very specific attention paid to the ordinary aspects of life in society allows us to rethink daily life by rendering phenomena of "mediocre importance" (Simmel 1981) visible and manifest; 3) the need for a situated approach to the phenomena being studied allows us to recognize the key importance of research and the essential nature of field studies; 4) the recognition of the descriptive turn taken by the social sciences helps us to anchor and develop very real, embodied approaches; and 5) the explicit borrowing from a rigorous phenomenological posture allows us to thematize the sensory world in a new light, thereby blurring the distinction between a subject and the object of its experience.

The investigation of ambiances originated with concerns of an architectural and urban nature. It entailed taking a fresh look at inhabited space and the built environment by measuring the complexity, wealth, and depth of the experience of inhabiting. For that matter, it was during this period, starting in the 1970s, that research programmes began to look at the urban environment, public spaces, the landscape, and urban scenography. A number of urban anthropology works were also carried out at this same time that described the nuances and complexities of the experience of inhabited spaces in minute detail (consider, for example, Colette Pétonnet's beautiful book about the suburbs, published in 1979). In this respect, the notion of ambiance is one of the ways in which we can analyse, in an interdisciplinary manner, the materiality of our sensory relationship with the world, and explore the distinctions between designing and inhabiting our living spaces.

The emergence of this perspective reveals the growing interest in the realm of the senses as a way to envision how we inhabit and to reconsider architectural and urban design. This exists in counterpoint to a certain modernist history of a view of architecture and urban planning that focused on the built form and the effectiveness of the critical and programmatic reasoning that it introduced into the social space.

The development of the field of ambiances has instead involved paying particular attention to the ordinary aspects of our life as a society. Rather than studying exceptional spaces or monumental architectures, the primary goal has been to explore the sensorial content of everyday places. This is nothing less than a contribution to "a theory of 'the ordinary moment'" (Joseph 2009). It forms part of a body of work that seeks to describe everyday life, even in the context of architectural complexes whose formal rationality is contradicted by the materiality of the lives that pass through them. We can think of Henri Lefebvre's research and his critique of daily life or Erving Goffman's work on the staging of daily life, but it was the thought of Michel de Certeau and Georges Perec that has been invoked more often. Michel de Certeau's *The Practice of* Everyday Life (de Certeau 1980) was largely inspired by the rhetoric of pathways developed in *Pas-à-pas* by Jean-François Augoyard (1979), the person who launched the research of ambiances in France. In fact, these two authors share many of the same arguments: the criticism of a top-down viewpoint in favour of a pedestrian-level perspective, the recognition of ordinary skills and know-how, and the analysis of how city dwellers appropriate the urban space.

These arguments have had a lasting impact on the study of ambiances, with all the consequences that a comprehensive, embodied and situated perspective on inhabited space has yielded. People thus began to look at real-life situations and test this perspective in a wide variety of fields. It was in this context that the concept of "ways of inhabiting" (*Communications* 2002) developed, which preferred a modal, descriptive approach to a more deliberately determinist and causalist perspective. This involved accepting and letting oneself be surprised by the broad diversity and inventiveness of the ordinary, taking account of actual practices and ways of doing things, and describing modes and forms of action, i.e., the life "tactics" that divert key building strategies as well as the situations and styles that constitute forms of inhabiting. Rather than developing overly general and comprehensive theories about cities or architecture, scholarship focused on the fundamentally situated nature of experience. This entailed field studies that insisted on the contexts and circumstances that informed how lived spaces functioned and were experienced. Before the notion of ambiance came to refer to a certain type of sensory data, it emerged as a way to start thinking about the actual processes of living that modernist rationalism believed it could design according to predetermined criteria.

The idea of the "infra-ordinary", as developed by Georges Perec (1989), also helped people to focus their attention and assign words to the most ordinary, evident, and habitual of daily practices. This, too, involved shifting focus and developing a new kind of attention that prioritized the apparently insignificant and negligible lying at the very heart of the ordinary. The seemingly self-evident was scrutinized using a highly precise, meticulous style that revealed its wealth. This formed part of a descriptive turning point, as developed particularly in ethnomethodology and used in research conducted at the Ecole des Hautes Etudes en Sciences Sociales (EHESS) (Ackermann, Conein, Guigues, Quéré & Vidal 1985). As Louis Quéré said during a seminar, echoing Ludwig Wittgenstein, "At some point, we have to move from explanation to simple description" (Tome 1, p. 3). The ethnographic approach played a key role in this process by restoring the status of writing and textual reality, viewing observation and description as a unified gesture, exploring forms of narration and storytelling, and testing the usefulness of various descriptive models (Laplantine 1996).

Thus, the interest expressed in built environments and material spaces adopted various perspectives that came from the social sciences: micro-sociology, interactionism, situated action, embodied cognition, and ethnomethodology. But the study of ambiances bore another distinguishing feature: the use of phenomenology as a way to thematize the realm of the senses. The concept of ambiance arose in a phenomenological setting, the principal traces of which can be found in French and also German philosophy (as we will see later on when we consider the relationship between the notions of ambiance and atmosphere). We can cite various phenomenologists who addressed the sensitive and paid specific attention to the body, space, affect, and aesthetics in general. Obviously, there is the phenomenology of perception and the ontology of the flesh developed by Maurice Merleau-Ponty, the phenomenology of aesthetic experience and the emotional a priori of Henri Maldiney, and the metaphysics of the other and the world of the elemental of Emmanuel Lévinas. We should also mention several other French phenomenologists who have furthered this intellectual tradition in a highly original and fruitful manner: from the pragmatist phenomenology of Nathalie Depraz to the phenomenological cosmology of Renaud Barbaras, by way of the eco-phenomenology of Bruce Bégout and, in his more pragmatist exploration of these questions, the hyper-aesthetic of Yves Michaud, to name just a few. The term ambiance is barely present in these philosophies, with the relatively late exception of Bruce Bégout (2020), who has proposed its becoming a veritable concept, and Yves Michaud (2021), who views it as a symptom of our contemporary sensibility. The explicit reference to these philosophies in publications discussing ambiances has nevertheless remained fairly limited. It is as if these modes of thinking left their mark on the study of ambiances in only a discreet, implicit, and subterranean manner. They represent a kind of philosophical backdrop that helps us to ask new questions about the sensory and which supports a theoretical structure for the notion of ambiance, often in more applied directions, particularly in the fields of architecture and urban planning.

A laboratory for ambiances

A very singular moment in French architectural and urban research is worth mentioning in this regard: the convergence in the 1970s and 80s of scholars in the humanities and social sciences in Grenoble who were all interested in the experience of habitation, urban life, and theories of the imaginary. During a period that saw major transformations to this city and several original social experiments, two research laboratories developed new lines of inquiry: the Centre de Recherche sur l'Imaginaire (CRI), directed by Gilbert Durand, author of a major work on the anthropological structures of the imaginary (Durand 1960), and the Centre de Sociologie Urbaine (CSU), directed by Pierre Sansot, author of *The Poetics of the* City (Sansot 2011), a work that has proved particularly stimulating and original. Inspired by both the aesthetic phenomenology of Mikel Dufrenne (Dufrenne 1953) and the poetics of reverie of Gaston Bachelard (Bachelard 1968), Pierre Sansot gave a detailed description of the sensory forms of social life and opened the way to an urban phenomenology that paid as much attention as possible to lived situations. Although he did not explicitly invoke the notion of ambiance, he contributed significantly to its emergence by formulating a sensory approach that was rooted in the materiality of the city and its experiences and that also concerned the ordinariness of urban spaces and ordinary gestures. Pierre Sansot's urban poetics provided an inestimable contribution to the field of ambiances through their use of a descriptive and figurative sociology. The object-based approach to the city that he inaugurated, which consisted of "going from places to people" (p. 13), made it possible to articulate the sensory, social, and imaginary dimensions of the urban experience all at once.

On the strength of these research contributions and as an extension of these scientific perspectives, a new laboratory was created in 1979 at the Grenoble School of Architecture: CRESSON – Centre de Recherche sur l'Espace Sonore (https://aau.archi.fr/cresson/). Guided by two scholars and professors, Jean-François Augoyard and Jean-Jacques Delétré, this new team focused explicitly on the world of sound, using a deliberately interdisciplinary approach that combined the social sciences, spatial sciences, and engineering. This decentring of experience through listening and sound laid the foundation for an original approach to sensory environments. This led to the integration of the many arguments regarding the attention to ordinary situations, the possibilities created by a well-understood phenomenology, and the great inventiveness of the various research projects carried out within the CRI and the CSU. This undoubtedly represented the very beginnings of a thinking about ambiances that would develop significantly in the 1990s, as foreshadowed by Jean-François Augoyard already in 1979 with the publication of his doctoral dissertation in book form. At the risk of overstating the case, we will venture to say that the approach to the world of sound initiated in the 1980s at CRESSON constituted a nascent paradigm of the perspective that was subsequently developed in relation to ambiance (CRESSON would then concern itself with other sensory modes, especially light, smell, and motion). Among other things, sound makes it possible to question the strict dichotomy of figure and background derived from visual thinking, to take real account of the dynamics of time, and to introduce the idea of resonance as constituting the experience of an ambiance.

This passage from sound to ambiance was in fact made with the creation in the 1990s of the Centre National de la Recherche Scientifique (CNRS) research laboratory AAU (Ambiances Architecturales et Urbaines) (https://aau.archi.fr/laboratoire-aau/), in close collaboration with the CERMA (Centre de Recherche Méthodologique d'Architecture), a research team directed at the time by Jean-Pierre Peneau at the Nantes School of Architecture, which focused specifically on the modelling of thermal and aeraulic environments. The CERMA's work furthered the development of the interdisciplinary approach to ambiances and integrated the thermal and aeraulic aspects of sensory experience. The ever-present interest in the manufacturing of the built environment and urban space constituted a common thread that ran through this work. We should not forget that this field developed within French-speaking architecture schools through their exploration of urban projects and the design of inhabited spaces. Since then, this laboratory has continued to develop this field of research and action, and since 2008, to lead the International Ambiances Network (ambiances.net).

A first approximation of ambiance

Although ambiance refers primarily to the realm of the senses and sensory experience, it cannot be reduced to visual perception, which benefits from an aesthetic and we could even say a civilizational privilege in aesthetic thought. An ambiance instead summons all modes of perception and raises the difficult question of inter-sensoriality. We can of course talk about acoustic, olfactory, or lighting ambiances – and develop specialties regarding these various modes – but an ambiance fundamentally concerns the sensorium in its entirety. To this end, the notion of *medium* is of primary importance, but this cannot be reduced to an exclusive use of the term, as is the case, for example, in certain ways of thinking about the arts and the specificity of languages tied to meanings that are isolated from one another. In the context of ambiances, the medium is not primarily an expressive language; it is a vital milieu that first summons the air, this barely perceptible medium that conditions all organic life. We must remember that an ambiance is not what one perceives; it is not an object of perception. It forms the basis for our perception, enables perception, and gives rise to perception (Thibaud 2011). There is no vision without light and no hearing without sound, without a medium that propagates and diffuses them. This basis of sensory experience – sound, air, smell, heat, and light – is becoming a fundamental concern in how we live and inhabit.

But again, beyond the diversity of borrowings from phenomenology that make it possible to question the appearance of the world and its sensory manifestation, as well as the phenomenality of social life, the particular attention paid to the enigma of feeling helps us to approach ambiance. It thematizes *aesthesis* in a new light (Augoyard 2004), articulates the perceptive and the affective, and introduces the key ideas of affective tonality and *Stimmung* to situated sensory experience. Recourse to Erwin Straus' distinction (1963) between feeling and perceiving, between the "pathic" moment and the "gnostic" moment of sensory experience provides a very important contribution to this end. These two ways of relating to the world make it possible to assert an original sympathy with the world, to expand a vital, immediate means of communication with an ambient milieu that cannot be reduced to a purely cognitive and objectivizing grasp of things. One of the singular contributions of ambiance is its emphasis on the pathic dimension of our experience, as it lets us feel the emotional tones of situations and experience the ways in which everyday life is in and out of tune. In being open to any kind of situation, it forms part of a sensory ecology of the spaces we inhabit.

Equally importantly, the field of ambiances is not an isolated, self-contained domain independent of social practices. Quite to the contrary, any ambiance is ensconced in ways of living with others. It presupposes inhabitant practices that use and actualize the resources of the built environment and the forms of living together. Specific attention must therefore be paid to the backdrop of ordinary experience, from which an ambiance can never be separated. What seems self-evident in the sensory conditions of an experience, and which usually goes unnoticed, constitutes the raw materials of an ambiance. A way of moving or talking, a certain quality of light or air, a slight change in season or variation in climate, the height of a staircase, or the paving of a sidewalk all help compose an ambiance. Careful, though, an ambiance is not the result of just one or several of these components; it cannot be reduced to a simple sum of its parts. An ambiance instead consists of the relationship between these various elements, the way they hold together as they confer a single tonality to every aspect of an appearance. Ambiance has the power to create cohesiveness and atmosphere; consequently, ambiance leads to a shared, embodied, affected, enacted, and situated sensitivity.

We must also note that the notion of ambiance has elicited the interest of many disciplines, from one branch of science to another. A nomadic notion par excellence (Stengers 1987), it migrates, moves, morphs, and translates. Far from being the prerogative of a single scientific discipline or field of action, it propagates and proliferates across regions as diverse as sensory architecture, existential psychopathology, social anthropology, experiential marketing, cultural geography, phenomenological philosophy, literary studies, urban studies, the world of organizations and consumption, and the arts, among others. In so doing, it dissolves the boundaries between disciplines and itself changes each time it comes into contact with them. It thereby acquires a thickness of meaning, tests its heuristic scope, and explores its operative power. Analysing ambiance involves equipping oneself with the means to act *on* and *by means of* ambiance. And identifying its power to condition experience entails an undeniable political dimension, which makes it all the more crucial to identify its constituent parameters and the dynamics of the social experience that it helps to determine.

Ambiances as a subject of inquiry

Even though the field of ambiances developed against a phenomenological and then a pragmatist backdrop, in no case can it be reduced to a strictly philosophical endeavour. Very much to the contrary, one of the most remarkable aspects of this field is its recourse to empirical data and original investigative tools (Grosjean & Thibaud 2001). Very early on, the field of ambiances led to investigations of inhabiting, and it has never ceased to explore avenues of inquiry into the urban experience. (For an overview of the large number of field projects that have been carried out, please refer to the acts of the four international conferences held by the International Ambiances Network and available on the network's website, ambiances.net.) The empirical study of architectural and urban ambiances has led to the exploration of a series of methods based on on-site, often multi-disciplinary approaches. We should recall that these research methods remain highly flexible and subject to change, depending on the place being studied. Furthermore, these tools are located at the intersection of urban research and practice. They can be used to conduct research and carry out a project, to analyse the sensory experience of a city (heuristic power) and the ways of shaping it (operative power). Three major arguments govern the exploration and implementation of these efforts.

First is the attention paid to the sensory materiality of the situations and environments being studied. Very often, the research techniques use acoustic, photographic, and videographic data to document an area, which also represents essential sources of knowledge and can be used as prompts in interviews. It constitutes a kind of sensory ethnography. In addition to the sensory experience of the researchers in the field and the great importance assigned to the testimonies of inhabitants that are often gathered on-site, the work on the sensory material itself is fundamental. And it often lends itself to experimentation in terms of editing, expression, and forms of presentation.

Second, walking also often forms a key part of the means of inquiry, as this allows for the physical exploration and on-site description of an area under study together with its inhabitants. The sensory experience and bodily mobilization of the inhabitants and the researchers are thus put to the test of local specificities. These walks, be they solitary, in twos, or in larger groups, use a broad diversity of protocols, ranging from freely wandering to transect subject to constraints and from guided itineraries to commented, directed pathways.

Third, the research techniques employ indirect approaches to collect the experiences and statements of inhabitants. Rather than conducting classic, formal interviews as in the social sciences, they create original research situations that use a third-party actor. This third party may be of many different kinds, but its main function is to prompt people to speak. Therefore, more than a series of questions and responses between interviewer and interviewee, this involves a third-party entity that creates a shared situation in which residents feel free to express themselves. In this way, familiar spaces can be seen or heard using sensory documents produced on-site. Sound recordings from the field can be made available for listening and comment as part of inquiries involving sound reactivation. In such a case, the sound itself prompts and encourages the inhabitant/listener to speak. Videos can also perform this role in the case of recurrent observation. Other third-party media can also be produced and used, such as records of routes previously compiled by the residents themselves, which are then subject to comment and discussion as part of a storytelling process. And, of course, the walk itself also plays this third-party role, as the challenge lies precisely in putting words into motion. The space being traversed summons the spoken word and thereby gives rise to situations and sensory transitions that elicit comments as well as recollections.

These research methods also increasingly rely on groups of researchers who share their field experiences and involvement in the ambiances under study. The testimonies of the inhabitants collected during these inquiries and their analysis are often presented to the inhabitants as well (Amphoux & Tixier 2017). This produces a give-and-take relationship between the team of researchers and the local residents. In addition to these various research tools, a number of multi-disciplinary analytical tools have been developed to describe the sensory content of an ambiance as precisely as possible: sound, patterns, figures, ambient objects, sensory formants, visual perspectives, aromatic effects, among others. Without getting into the details of each one, we believe that they each contribute to the articulation of the spatial, social, and sensory dimensions of an ambiance. Together, they develop a kind of generative grammar for ambiances that recognizes the full complexity of sensory experiences to be had at a site. One of the specificities of the field of ambiances is its development over time of a veritable toolbox that has allowed for a theoretical elaboration as well as an empirical testing open to a wide range of areas of study.

On ambiances and atmospheres

The question of kinship, similarities, and differences between the notions of ambiance and atmosphere has become recurrent in the scholarly domain. Clearly, there are a number of affinities between these two notions, even though they are not interchangeable. Our aim is not to explore this matter at length – although it is one of the main themes running through this book – rather to provide several indications to help to clarify it. Two principal points are worth mentioning in this regard.

We must dwell on the linguistic dimension of this question, as it involves a problem of translation. While *atmosphère* and "atmosphere" are readily used in French and English, respectively, "ambiance" is a French term that is hard to translate into English. There are various options, depending on the context and semantic field: it can be translated as "ambience", "atmosphere", or even as "mood", "feel" or "vibe". But each one of these translations is rife with connotations that alter, add to, or subtract from the French term's scope of meaning, and without covering it entirely, for that matter. The notion of ambiance indeed represents what we could call an "untranslatable" (Cassin 2004).

We should also note the multitude of models of intelligibility on which the notions of ambiance and atmosphere are based. Each one offers a perceptibly different version of this domain of thought and orients it in a different direction. Said another way, these two notions should be related not just to their linguistic environment, but also to the scientific paradigms and philosophical positions they invoke.

On a merely exploratory basis, we can schematically distinguish five main frameworks of thought according to which theoretical positions on this subject are organized.

1 We can begin with the *Situationist movement* and its critique of daily life in the 1950s and 1960s (*Internationale Situationniste 1997)*. In its close relationship to Lettrism, Dada, and Surrealism, this protest movement sought to arrive at a definition of art that removed the boundaries between art and life. It strove to overturn daily routines and to intensify the emotional potential of urban situations. Experimenting with deviations and diversions in urban environments as well as with the techniques of psychogeography led to a unitary urbanism and a deliberately socially committed, creative, and subversive attitude. Drawing on the central notion of situation and the theory of moments in the work of Henri Lefebvre, ambiance is understood here as a set of impressions that determine the quality of a moment and inject desire into the very heart of ordinary experience. In keeping with the thought of Guy Debord, ambiance here is located at the intersection of the material, social, artistic, urban, and political world.

2 We can adopt a *pragmatic stance* that tests ambiance by means of inquiry (Thibaud 2015). The experience and manufacturing of architectural and urban ambiances give rise to a variety of forms of experimentation, methodological instruments, and multi-disciplinary analytical tools (cf. supra). The potential for the ambiance's action and transformation emerges through a multi-disciplinary investigation that always concerns itself with the experience of precise, actual situations. This entails an explicit empiricism that proposes declining ambiance in the plural (several ambiances instead of a single Ambiance), focusing on its concreteness and its fundamentally social nature, and on the simultaneous articulation of the level of action and feeling. While preserving its strong phenomenological character, ambiance is thus viewed in terms of its effects, of what it

produces, enacts, and accomplishes on the basis of its heuristic potential and its operative power. Rather than trying to define a notion of ambiance, this instead entails reflecting on what an ambiance allows us to do and perceive, feel, share, and transform.

3 We can use the German *new phenomenology* by again applying a philosophy of bodily experience and emotional tonalities. "Atmosphere" (or *Atmosphäre* in German) is the term that is used mainly in this case. Emphasis here is given to our capacity to be affected by our ambient milieu and to the body's experience of sensory spaces. Such a philosophical framework leads to the development of an aesthetic of atmospheres that pays great attention to the nuances of human sensitivity. Thus, several German phenomenologists adopted an approach to atmospheres that was quite similar to the pragmatist and eco-phenomenological approach to ambiances developed by Bruce Bégout. A veritable tradition of thought developed within Germanophone phenomenology, from Hermann Schmitz's new phenomenology (2019, 2023) to Gernöt Böhme's aesthetics of atmospheres (2017a, 2017b), by way of Hubertus Tellenbach's phenomenology of the oral sense (1968, 1981). Tonino Griffero (2014, 2017), an Italian Germanophone phenomenologist, is currently pursuing this phenomenology of atmospheres, which he has renewed and developed extensively. In other words, the approach to ambiances first developed in a uniquely Francophone context, but in the 1990s, it came into contact with another phenomenological tradition that had developed in parallel to it and with which it shared many of the same lines of inquiry.

4 According to an entirely different perspective, we may take inspiration from the non-representational theory currently in use by Anglophone geographers. Inspired by Deleuze and the Post-structuralists, this vitalist perspective seeks to free itself of human subjectivity to consider the dynamics of the ambient world. Atmospheres are thus explored in the light of pre-individual affects and in terms of intensities, forces, elements, and events (Anderson 2014, McCormack 2018). Even though this stance shares many attributes with the phenomenological approach, it remains fundamentally different, to the point of explicitly criticizing it. It questions the privilege accorded to human intentionality in phenomenological thought, refusing to reduce atmospheres to their strictly corporeal and anthropocentric dimension and instead approaching them more in terms of forces. This is undoubtedly an entirely different approach to atmospheres that emphasizes the complexities of the "more-than-human" and its essentially political significance.

5 Lastly, a path has been opened up by Peter Sloterdijk's *philosophy of spheres* (2011, 2014, 2016), according to which we inhabit climatic envelopes that protect us. These atmospheres that shape us also consist of immunity systems that allow us to live. This leads to a theory of climatizations that assigns primary importance to air, and which explores the way in which atmospheric

> systems are embedded in forms of life and technical mechanisms. This perspective allows us to think anew about the material dimensions of existence and leads us to think of human beings as designers of atmospheres. This conception of the spherical constitution of human existence entails a thorough shift in perspective: an ontology of the thing that gives primacy to substance is replaced by the categories of relation and accident.

As we can see, these lines of demarcation are not fully rigid. They simply reflect the fact that the theory of ambiances is rooted in different traditions of thought, and they argue for a plurality of approaches. Among the many points of discussion in these approaches are two important ones worth mentioning. A basic theoretical problem concerns the role assigned to the subject and to human subjectivity in an ambiance. If emotions and bodies play a key role, the question remains as to how we should understand them. Should human experience be the measure of ambiance, or should ambiances be open to the non-human, the more-than-human, and the pre-individual? Can we thus opt for a phenomenological anchoring of the notion of ambiance, or should we also draw inspiration from a post-phenomenological thinking that proposes decentring the subject? And in such a case, what nuance does the notion of ambiance introduce in relation to the notion of atmosphere, as it has so often been associated with the idea of affective tonality and been even more deeply rooted in the limits of subjective experience than atmosphere? If we base ourselves on the genealogy of Francophone scholarship on ambiance, should we hypothesize that ambiance entails thinking about contemporary anthropological and ecological decentrings primarily at the scale of everyday existence on this earth? Clearly, there is an essential debate to be had on the various positions on the world's atmospheric existence, and these questions run throughout this book. What is at stake is the ability of these approaches to come to terms with the empirical world, everyday experiences, and changes in contemporary environments. The challenge lies in giving a concrete account of the situations in which we find ourselves, without losing a sense of their wealth and specificity. In other words, we should not be too quick to equate ambiance and environment, or to confuse ambiance with landscape. So, what should we say about ambiance when we relate it to a metropolitan context and to actual urban situations? How can we comprehend a field that is above all diffuse, discreet, and minute? How can we explore and test new forms of inquiry, and how can the world of art help us do that?

Articulations of ambiance

This volume consists of six chapters that explore these issues and define the large variety of avenues of inquiry that ambiances open up for us. Each one articulates ambiance from a particular perspective and explores in greater depth the way in which ambiances subject ordinary situations to the test of our sensitivities.

This work derives in part from an international colloquium held in Cerisy, France in 2018. Titled *The use of ambiances: a sensitive experience of situations*, this colloquium brought together speakers from a diverse range of disciplines, geographies, and languages. This volume reproduces a large share of these lectures and supplements them with other essential chapters to outline more broadly the debates that currently occupy the domain of ambiances.

1 *Ambiances, sources, and resources* introduces the Anglophone reader to several seminal texts that have grounded the contemporary approach to ambiances. The field is introduced using the interdisciplinary perspective initiated in France, especially at CRESSON. For that matter, the three authors in this part are members of this research laboratory within the Grenoble School of Architecture who have worked on ways of designing inhabited spaces. Special attention is paid to the sensory environment, situated experience, and the material and social frameworks of perception. These chapters are intended as a form of guideposts to this vast field that provide an original way to articulate the differences between sensory, built, and social space. Jean-Paul Thibaud's chapter explores how various fields of research initiated and nurtured the study of ambiances. Without making any claim to comprehensiveness, it outlines the main lines of inquiry that characterize this field, and it shows how ambiance can anchor the realm of the senses in experience and the act of inhabiting. Pondered to equal extents in the fields of historical semantics, existential psychopathology, and phenomenological aesthetics, ambiance thus presents itself as a nomadic notion that encourages the circulation of knowledge, feeds on it from within, and asserts its irreducibility to the notions of milieu, environment, and landscape. Jean-François Augoyard's chapter is the landmark text that gave rise to this nascent field of research. It was the first to propose a model for understanding the phenomena of ambiances by adopting an interdisciplinary, context-based, modal, and procedural approach. This led to the formation of a complex epistemology that challenged the primacy of the visual and offered a salutary critique of physical substantialism. Ambiance was thus given an original frame of reference that was supplemented with the most recent advances in the history of science. This integrative modelled specifically to a reinterpretation of the field of aesthetics and the domain of architectural and urban construction. Pascal Amphoux's chapter is both literary and experimental, expressing ambiance in words in an original way. Composed in a ternary fashion, the refrain uses rhythm as a way to think about ambiances, not as one aspect among many, rather as itself constituting an ambiance. The writing style suggests the immersion we feel when we experience an ambiance, thus raising an issue that continues to be debated, namely the expressive force of an ambiance, which cannot be reduced to the domain of representation.

2 *Ambiances, perspectives, and positionings* contains more recent texts that demonstrate the great variety and wealth of the possible ways of considering ambiance. This part explains how this field of thought has permeated the humanities and social sciences. These more deliberately philosophical and anthropological chapters were penned by Francophone authors who have expanded and explored the initial thinking about ambiances in greater depth, thereby opening up new avenues of thought in relation to experience. Véronique Nahoum-Grappe, an anthropologist affiliated with EHESS, uses ambiance to introduce the potential of an ethno-phenomenology. In this respect, time plays a key role, in that it is the thickness of the present that enables an ambiance to arise in a given situation. This identifies a dual epistemological difficulty: on the one hand, ambiance can only be grasped holistically, as it involves too many enmeshed, unstable layers of meaning for it to be reproduced, and on the other, most often in a non-verbal manner, perceiving an ambiance presupposes a specific, almost immediate, collective cognitive skill. François Laplantine, anthropologist at Université Lumière Lyon 2, discusses an anthropology of the sensory by focusing on the fundamentally corporeal and collective aspect of an ambiance. Using Japanese culture and language as an example, he applies a decentring to our customary way of envisioning ambiance to offer a non-anthropocentric, non-logocentric perspective. This passage to Japan introduces the notion of an impersonal and a-substantive dimension of experience. Aesthetics are thus reconsidered to allow for a relational view of ambiance that assigns great importance to passivity, variability, and singularity. Bruce Bégout, philosopher at the Université de Bordeaux Montaigne, has developed an eco-phenomenological approach to ambiance and seeks to explain this concept by examining how an ambiance is and how it appears. In considering the vast literature on this topic, of which he provides a meticulous phenomenological critique, he places a perspective that insists on the subjective aspect of ambiance alongside another that, to the contrary, insists on its objective qualities. For Bégout, ambiance has a third mode of being that is fundamentally affective, and which challenges the dichotomy between subject and object. Yves Michaud, a philosopher specializing in the world of art, uses ambiance as a way to explore contemporary aesthetic experience. He hypothesizes the existence a general trend towards the atmosphericization of our living environments that enshrines the realm of the hyper-aesthetic. Ambiance is a symptom of this revolution of sensitivity that testifies to the emergence of the world as a gaseous state. He provides a conceptual clarification to distinguish and delineate the contiguous notions of ambiance, atmosphere, and environment.

3 *Ambiances, dialogues, and resonances* addresses the open debate between the study of ambiances and of atmospheres. The affinities and specificities of these two domains are explored not by viewing them as fundamentally opposed to one another, rather by bringing them closer together in terms of their mutual contributions. This complex dialogue features authors

from very diverse disciplinary, linguistic, and scientific backgrounds. Olivier Gaudin, a philosopher at the School of Nature and Landscape Architecture in Blois, and Maxime le Calvé, anthropologist at Humboldt University in Berlin, identify various historical and literary sources for analysing the complex conceptual relationship between these two notions. They demonstrate that the social sciences can study the mechanisms that engender ambiances, thereby making ambiance a reliable and operative descriptive concept for research and distinguishing it from atmosphere. In this respect, anthropology shows that conducting research on ambiances requires us to experience them directly, through our bodies. This chapter drives research to embrace an exacting plurality of perspectives. Derek McCormack, a geographer at Oxford University, uses an entirely different perspective of a post-phenomenological nature to explain the differences between atmosphere, ambiance, and the ambient by focusing on their respective affective tonalities. The remarkable originality of this approach consists of prioritizing how a concept sounds and to its varying scope. Thus, in an Anglophone context, the term atmosphere appears to have a broader scope than ambiance, as it more readily opens itself up to a world that is not exclusively human. Rainer Kazig, a geographer and researcher at CRESSON, and Damien Masson, an urban planner teaching at CY Cergy Paris University and a researcher at CRESSON, explore the role of the notion of ambiance in Francophone cultural geography. They demonstrate how the notion of atmosphere has come to situate itself in Anglophone and Germanophone cultural geography, whereas the concept of ambiance has been cultivated mainly in the Francophone world, in the field of architectural and urban research. They propose three main perspectives to further integrate ambiance into cultural geography: the role of material conditions in the emergence of spatial representations, the existence of cultural mechanisms that are differentiated in terms of their lived experience, and the difference between the micro and macro scales of the phenomena being studied. Tonino Griffero, a philosopher at the University of Rome Tor Vergata, explores the affective power of atmosphere using the new phenomenology developed by Hermann Schmitz. He engages in a very precise, philosophical thought process that specifies the notion of atmosphere from within, ultimately to propose a pathic aesthetic able to distinguish various ideotypes of atmospheres and to account for the variety of atmospheric sentiments. The concept of atmospheric competence is also developed in detail, which is used to de-axiologize the notion of atmosphere.

4 *Ambiances, sensitivity, and culture* considers the approach to ambiances by exploring how it is embedded in modern and contemporary sensitive culture. This part takes a fresh look at the issue of sensitivity in daily urban experience. Philippe Simay, a philosopher and Lecturer at the Paris-Belleville Architecture School, and a researcher at IPRAUS, opens the discussion by arguing that ambiances have played a therapeutic role

in response to the shock provoked by the development of modernist cities in Europe. To this end, he distinguishes between two paradigms: that of the "urbaphobe" Jean-Jacques Rousseau, who sees the ambiance of certain natural environments as a balm to help treat the negative effects of the city, and that of thinkers such as Georg Simmel, Siegfried Kracauer, and Walter Benjamin, for whom a critique of urban modernity is coupled with the need to acclimatize our sensitivities, for which certain ambiances, such as cinema, may constitute the proper medium. The role of ambiances in the transformation of sensitivities has taken on new meaning in the context of our contemporary ecological crisis. In her contribution, Anne Bossé, architect and geographer, Lecturer at the Nantes School of Architecture, a researcher on the CRENAU team at the AAU laboratory, and editor-in-chief of the online review *Ambiances,* discusses the multiple scales at which our ways of inhabiting the world will play out in the future, from nanoparticles penetrating our skin to vaster geological and cosmological scales. At each level, the body is affected, placed into a relation, and decentred. And ambiance is the prism through which to consider our sensitivity to these changes. Pauline Nadrigny, philosopher and Lecturer at University of Paris 1 Panthéon-Sorbonne, analyses a very eloquent case in her piece concerning the link between a given environment and a form of thought, specifically through the aesthetic forms generated by this connection. Her text looks at the resonances produced by the ambiance of the forest of Concord, Massachusetts on the thought of philosopher and poet Henry David Thoreau and on the music of Charles Ives, thereby asking us to consider the difference between thinking in terms of systems and in terms of and about ambiances. In returning to the question of the sensory experience of urban modernity that opened this chapter, Nia Perivolaropoulou, Professor of Comparative Literature at the University of Duisburg-Essen, rereads Kracauer's writings on cinema and the city to identify the emergence of an aesthetics of atmosphere as a mutation of the Romantic *Stimmung* and a dissolution of the frameworks used to define the notion of "landscape". Indissociable from a cinematic way of thinking and capturing intensities amidst a fragmented reality, this new sensitivity feeds on a form of indeterminacy to articulate different spatial scales as well as multiple temporal layers.

5 *Ambiances, architecture, and place* focuses on the place of architecture within the field of ambiances. It presents several new ways of looking at architectural design. The first explorations of architectural ambiances focused on their physicality (acoustics, lighting, sun exposure, thermal and aeraulic conditions, among others), in which technical expertise is invoked to better understand, model, and exploit interactions between the physical aspects of the senses and the built space. This research then quickly evolved to look at the design of ambiances, in which the knowledge base and know-how applied to a project are used to transform our quality of life. It also began to study how people experience ambiances,

using methods involving the observation of inhabitant practices, the collection of perceptions, and the expression of social representations. But thinking about architecture in terms of ambiances extends well past a multi-disciplinary mastery of its effects; it can also mean a significant paradigm shift in how we understand and design any inhabited space. This is the issue broached by the four scholars whose chapters form this fifth part. Grégoire Chelkoff, architect, Professor Emeritus at the Grenoble School of Architecture and researcher at CRESSON, has developed a modal approach that explores how things are experienced, rather than attempting to explain the causes of their appearance. For him, thinking of architecture as an ambiance invites us to envision and design built forms not only in terms of spatial configurations, real or figurative, but also in terms of the sensitive formants of perception and action through which they emerge. Izabela Wieczorek is an architect and Assistant Professor at the University of Reading and member of the Urban Living Research Group. For Wieczorek, the idea of "Atmospheres in Formation" carries a twofold meaning. It invites us to examine heterogeneous constituents, conditions, properties, and processes from which spatial ambiances emanate. It also suggests tracing how the concept of ambiances has formed in architectural discourse, exploring a challenging yet inspiring plurality of ambiances, as well as revealing the earlier aspects of its history. To explore ambiances in the formation means adopting a quasi-archaeological approach and searching for both material and discursive traces. This serves to illustrate the rise of a particular sensitivity to ambiances and its translation into design procedures and processes as a way to elicit new conceptions, perceptions, and experiences of space. In so doing, despite the established assumption that the history of architecture has largely neglected ambiances, this chapter situates architecture in a long tradition of building ambiances, thus offering an alternative history of architecture. Emmanuel Doutriaux, architect, Lecturer at the Paris – Val de Seine Architecture School and a researcher at EVCAU, considers the notions of affordances of ambiances and architectures of the commons. For Doutriaux, if we accept the failure of the notion of space while noting the concomitant success of the notion of ambiance, architecture provides us with a formidable experiential opening. Thus, we can consider the benefits (potentially) offered by ambiances, as, sheltered by an overhang of a certain size, they can bring the exercise of certain idea of the commons into play. And while the first three chapters in this part were authored by architect-researchers, the fourth is by a philosopher, Céline Bonicco-Donato, Professor at the Grenoble School of Architecture and researcher at CRESSON. Her contribution examines the affinities between the notion of ambiance and that of atmosphere by situating the latter within the history of German language philosophy. She discusses the original way in which Böhme developed his concept by intertwining two traditions often presented as antagonistic, namely phenomenology and critical theory. An examination of these archives demonstrates the capacity of ambiances understood in the broadest sense to challenge aesthetic philosophy and its approach to architecture.

6 *Ambiances, city, and urban space* deals specifically with the question of the city and its ambiances. It seeks to show how ambiances contribute to a renewal of our view of the urban experience and to a new and original way of designing cities. In its focus on urban issues, this final part presents recent work by three scholars from different disciplines (an architect, a political scientist, and a sociologist) who invoke and consider the notion of ambiance in the context of the project, as well as the notion of the project when seen through the prism of ambiance. Each in their own way engages in a critical discussion of the notion of the project in general and the urban project in particular as it has evolved from the second half of the twentieth century to the present day. Nicolas Tixier, architect, Professor at the Grenoble School of Architecture and researcher at CRESSON, has developed an approach to ambiances through the use of the transect for both research and operative projects, in an echo of older bodies of work, such as those of the Scottish biologist and urban planner Patrick Geddes in the early twentieth century. He uses the open field of ecology (ecology of perception, social ecology, and environmental ecology) and the field of the urban project (urban design and urban history), which is by its very nature both prospective and retrospective, to develop methodological tools. His chapter focuses on the urban transect as a field practice, a technique of representation, and a way to position a project. He demonstrates how the transect offers an original approach to territories and their fabrication through ambiances, and how it encourages us to reveal the legacies of places while designing their potential futures through the use of fiction. Niels Albertsen earned his degree in political science and is a Professor Emeritus at the Aarhus School of Architecture in Denmark. His research lies at the intersection of social theory, architecture, design theory, the sociology of the architectural profession, and the sociology and philosophy of art. His chapter explores the connection between Metapolitan diversification, anthropogenic processes, and atmospheric change. He pays special attention to the anthropogenic "call" for an interdisciplinary understanding of relations between "atmosphere, i.e., in the meteorological and the aesthetic sense, in general and in the Metapolitan context. The chapter closes with some remarks on atmospheric aspects of Metapolitan climate politics. Laurent Devisme is a sociologist and Professor at the Nantes School of Architecture. He conducts his research as a member of the CRENAU team at the AAU laboratory. He brings an ethnographic approach to the building of cities and explorations of urban expressiveness. His chapter starts with the finding that there are currently competing visions of urban planning that invoke ambiances to varying degrees of explicitness to support an often fantastical vision of a living city, as the opposite of a standard city. In such a context, we should consider representations as stakeholders in reality, take the apparent banalization of urban language seriously, and pay attention to

the discrepancies between emotional tonalities and ambiances as they have been planned. Significant paradoxes and questions emerge for any urban scholar considering the passages between languages and practices.

In a final "opening" inscribed in the field of aesthetics, David Zerbib, philosopher and lecturer at the Annecy Alpes School of Art and the Geneva University of Art and Design, reflects on the form of thought involved in the experience of ambiances. Drawing upon the concept of "ambiance of thought" he defends, through examples in art, literature and philosophy, the heuristic value of ambiances, adding to the classical analysis of the sensible knowledge the qualities of "diffuse ideas", which bring into play a specific type of situated indeterminacy. For the editors, artistic practices have a lot to teach us about ambiances. Art may intensify the existence of ambiances. It has the ability to generate an ambiance, heighten its reality, and to give it an existence. Art is thus able to vary our relationship with even the most ordinary ambient environments. Art has also the ability to shift boundaries. It questions the not overly strict dichotomies of emotion and reason, body and spirit, and reality and fiction. It brings a fictional universe, gesturality, corporality, and musicality to the world of ambiances, thereby contributing to the creation of new ways of thinking about them. Finally, artistic practices are an occasion to experiment not only by means of language, but within language, thereby helping us to modify our thought patterns and to delve into the nuances of experience.

References

Ackermann, Werner, Bernard Conein, Christiane Guigues, Louis Quéré, Daniel Vidal (eds.). *Décrire: un impératif?* Paris, EHESS, 1985. 2 tomes.

Amphoux, Pascal (ed.). *La notion d'ambiance. Une mutation de la pensée urbaine et de la pratique architecturale.* Paris, PUCA, 1998.

Amphoux, Pascal and Nicolas Tixier. "Paroles données, paroles rendues. La marche collective comme écriture du projet urbain", *Europe. Revue littéraire mensuelle*, No. 1055 (2017): 196–215.

Anderson, Ben. *Encountering Affect. Capacities, Apparatuses, Conditions.* Ashgate, Farnham, 2014.

Augoyard, Jean-François. *Pas-à-pas. Essai sur le cheminement en milieu urbain.* Paris, Seuil, 1979.

Augoyard, Jean-François. "Vers une esthétique des ambiances". In *Ambiances en débats.* Pascal Amphoux, Jean-Paul Thibaud, Grégoire Chelkoff (eds.), À la Croisée, 2004, 17–30.

Bachelard, Gaston. *Poétique de la rêverie.* Paris, PUF, 1968.

Bégout, Bruce. *Le concept d'ambiance.* Paris, Seuil, 2020.

Böhme, Gernot. *The Aesthetics of Atmospheres.* Abingdon, Routledge, 2017a.

Böhme, Gernot. *Atmospheric Architectures: The Aesthetics of Felt Spaces.* London, Bloomsbury, 2017b.

Bonnin, Philippe (ed.). *Communications No. 73, "Manières d'habiter".* Paris, Seuil, 2002.

Cassin, Barbara (ed.). *Le Vocabulaire européen des philosophies: Dictionnaire des intraduisibles.* Paris, Le Seuil, 2004.

Daudet, Léon. *Mélancholia.* Paris, Grasset, 1928.

De Certeau, Michel. *L'invention du quotidien*. Paris, Gallimard, 1980.
Dosse, François. *L'Empire du sens. L'humanisation des sciences humaines*. Paris, La Découverte, 1995.
Dufrenne, Mikel. *Phénoménologie de l'expérience esthétique*. Paris, PUF, 1953.
Durand, Gilbert. *Les Structures anthropologiques de l'imaginaire*. Paris, PUF, 1960.
Griffero, Tonino. *Atmospheres. Aesthetics of Emotional Spaces*. Ashgate, Farnham, 2014.
Griffero, Tonino. *Quasi-Things. The Paradigm of Atmospheres*. New York, SUNY Press, 2017.
Grosjean, Michel and Jean-Paul Thibaud (ed.). *L'espace urbain en méthodes*. Marseilles, Parenthèses, 2001.
Internationale Situationniste. Paris, Librairie Arthème Fayard, 1997.
Joseph, Isaac. *Erving Goffman et la microsociologie*. Paris, PUF, 2009.
Laplantine, François. *La description ethnographique*. Paris, Éditions Nathan, 1996.
McCormack, Derek. *Atmospheric Things. On the Allure of Elemental Envelopment*. Durham, Duke University Press, 2018
Michaud, Yves. *L'art, c'est bien fini. Essai sur l'hyper-esthétique et les atmosphères*. Paris, Gallimard, 2021.
Pétonnet, Colette. *On est tous dans le brouillard. Ethnologie des banlieues*. Paris, Éditions Galilée, 1979.
Perec, Georges. *L'infra-ordinaire*. Paris, Seuil, 1989.
Sansot, Pierre. *Poetics of the City*. Translated by Willard Wood. Lewiston, Edwin Mellen Press, 2011.
Schmitz, Hermann. *New Phenomenology. A Brief Introduction*. London, Mimesis International, 2019.
Schmitz, Hermann. *Atmospheres*. London, Mimesis International, 2023.
Simmel, Georg. *Sociologie et épistémologie*. Paris, PUF, 1981.
Sloterdijk, Peter. *Bubbles: Spheres Volume I: Microspherology*. Translated by Wieland Hoban. Los Angeles, Semiotext(e), 2011.
Sloterdijk, Peter. *Globes: Spheres Volume II: Macrospherology*. Translated by Wieland Hoban. Los Angeles, Semiotext(e), 2014.
Sloterdijk, Peter. *Foams: Spheres Volume III: Plural Spherology*. Translated by Wieland Hoban. Los Angeles, Semiotext(e), 2016.
Stengers, Isabelle (ed.). *D'une science à l'autre. Des concepts nomades*. Paris, Seuil, 1987.
Straus, Erwin. *The Primary World of Senses: A Vindication of Sensory Experience*. Translated by Jacob Needleman. Glencoe, Free Press of Glencoe, 1963.
Tallagrand, Didier, Jean-Paul Thibaud, Nicolas Tixier (eds.). *L'usage des ambiances. Une épreuve sensible des situations*. Paris, Hermann, 2021.
Tellenbach, Hubertus. *Geschmack und Atmosphäre*. Salzburg, Otto Müller, 1968.
Tellenbach, Hubertus. "Tasting and Smelling-Taste and Atmosphere-Atmosphere and Trust". *Journal of Phenomenological Psychology*. Vol. 12, No. 2 (1981): 221–230.
Thibaud, Jean-Paul. "The Sensory Fabric of Urban Ambiances". *Senses and Society*. Vol. 6, No. 2 (2011): 203–215.
Thibaud, Jean-Paul. *En quête d'ambiances. Éprouver la ville en passant*. Geneva, MétisPresses, 2015.

Notes from the translator

In one of the foundation myths of India's Parsi community, the first Zoroastrians arrived in the Subcontinent as early as the eighth century CE, in the wake of the fall of the Sassanid Empire. When they disembarked near Sanjan, in what is now the state of Gujarat, they sent a message to the local Hindu King, Jadi Rana, requesting permission to settle. Too embarrassed to refuse them directly, the King sent his regrets with an envoy instead, who presented the refugees with a glass of milk that was full to the brim – a symbolic indication that there was no room for anyone new. The leader of the Zoroastrians took the glass and stirred a spoonful of sugar into the milk until it dissolved, without spilling a single drop, to say that they would enrich the local culture without disrupting it. Recognizing the graciousness of this gesture, the King allowed them to settle and practice their religion freely. The Parsis thrived, and India with them.

Translation also involves such a gesture of both humility and self-assuredness. Let us consider the French word *ambiance*, which lacks a direct cognate in English and instead only has analogous, partially corresponding terms. It too asks to be admitted as it is into English, as "ambiance", and not to be renamed. And it too promises to bring something special to English by remaining itself, expanding the reader's sense of the world without provoking discord or discomfort.

In turn, the translation also offers hospitality, welcoming the reader into a world largely from elsewhere and now made more familiar. In this case, the invitation is twofold, because at their best, ambiances themselves strive to be accommodating to one and all. We could say that the translation is itself a unique textual ambiance, one that kindly requests acceptance, only to offer much in return. Through this medium, writer and reader thus welcome each other.

My thanks to all the contributors to this volume, especially to its editors, for their many suggestions on their essays, which greatly helped the undersigned translator arrive at a result that he hopes will engage and please the reader, much like a spoonful of sugar dissolved in a glass of fresh milk.

Claudio Cambon
Le Héron, September 2024

Photographic Intermezzos

The six black and white photographic images in this series, made in Sao Paulo and Rio de Janeiro, were not conceived as an illustration of the idea of ambiance.

Far from the impressionist or pictorialist approaches most often associated with the notion of atmosphere, the images instead serve as cross-sections of a dense urban fabric.

These tightly cropped and contrasting situations provide an overview. The viewer is placed at a certain distance, with points of view that are sometimes at ground level. Everything seems crushed, flattened, and yet, the viewer is able to penetrate these saturated, even choked architectural spaces through a handful of small details and micro-scenes that can be grasped from a distance. Their visual acuity lets us discover and perhaps objectify ordinary urban situations that are not immersive. They instead cut through the living fabric of the city like a series of sensitive, contrasting, and moving spaces. By means of what our vision alone cannot fully render, they help us grasp the ambiance from a distance, as if it were a possible space.

Didier Tallagrand, March 2024
(with thanks to David Zerbib for his constructive remarks).

[ADG-83]
VIVA SÃO PAULO
VIVA NO CENTRO.
5582.7709
Sta. Ifigênia
Bradesco
Seguros

Part 1

Ambiances, sources, and resources

1 A brief archaeology of the notion of ambiance

Jean-Paul Thibaud

As the sensory field begins to assert itself in the social sciences and humanities, the notion of ambiance has taken centre stage and appears to be thriving. This is surely no coincidence given that ambiance is the *basso continuo* of the sensory world, the background against which our perceptions and sensations are renewed. Paradoxically, while much research has explicitly identified itself with the topic of ambiance, very little work has tried to pinpoint the primary sources and deep roots of the notion itself. However, such an undertaking is indispensable if we wish to take the notion seriously, treat it as more than just a portmanteau word, and use it to further knowledge of the sensory ecology of cities.

Here we present three perspectives at the very root of the notion of ambiance, thus revealing the scope and diversity of the issues that it allows us to tackle.[1] The goal is not so much to provide a formal definition of the notion as to point out the types of issues for which it is useful and the heuristic potential it harbours. We shall begin by examining an approach rooted in *historical semantics*, then look at a second based on *existential psychopathology*, followed by a third approach driven by *phenomenological aesthetics*.

The semantics of ambiances

Several linguistic studies in the 1940s and 1950s attempted to update the semantics of the term *ambiance*. They adopted two main approaches: an etymological approach that aimed to identify the origin, underlying conditions, and possible roots of the word "ambiance", and a comparative approach that focused on translations and semantic discrepancies with its cognates in other languages.

Although the adjective *ambient* appeared in scientific and technical spheres in the 16th century, it was only three centuries later – around 1890 – that *ambiance* became a noun in French: in 1885, in one of de Villiers de l'Isle-Adam's *Cruel Tales* called "Sublime Love", and in 1890 in André Gide's *Traité du Narcisse*. The nominative use of *ambiance* also appeared in 1891 in a passage in the *Journal* of the Goncourt brothers, to whom it is also often attributed. The term was initially nurtured in the symbolist and impressionist

DOI: 10.4324/9781032655307-5

literary sphere. The symbolist school made great use of the (French) suffix –*ance* during this period, thus forming the word *ambiance*, among others. For Paul Adam, one of the representatives of this movement, "*ance* particularly attenuates the primitive meaning, which becomes less determined, vaguer and takes a more nuanced perspective".[2] In his inaugural research into the word *ambiance*, Karl Michaëlsson built on this argument in an attempt to show that the neologism went against dominant Cartesian reasoning in the French language:

> While French continues to maintain the fundamental traits of its former clarity, in its traditional sense, it also resorts increasingly to intuition, expressiveness, half-light and terms which suggest more than they mean, to vague words without precise boundaries. Among such suggestive and imprecise terms, the word ambiance holds centre stage due to its use and meaning, as well as for its current popularity.[3]

Despite the appeal of this idea, not all linguists agreed. Leo Spitzer's research later clearly showed the extent to which this reasoning was based on a somewhat hasty oversimplification. Indeed, while the term *ambiance* quickly spread into everyday language with a very broad and open meaning, the adjective *ambient* was subject to numerous technical and scientific uses that were much more rigorous and precise.

In a remarkable essay in 1942,[4] Leo Spitzer outlined a veritable theory of knowledge based on the term *ambiance*. While he paid tribute to Michaëlsson's early research, he moved beyond it in many respects by relying on the historical semantics method (of which he is the founder). He argued that the mere use of dictionaries is insufficient for providing information about the origin and evolution of a term. This means of information provides access to "petrified sediment" that is only the visible surface of the field lines and tensions that underpin any given word. Spitzer's approach was much more ambitious: it aimed to provide a history of ideas, from ancient Greece and Rome up to the contemporary period. With this approach, Spitzer was able to show how the word *ambiance* is embedded in different ways of understanding the world. It is by looking back on these visions of the world that it becomes possible to understand how the term *ambiance* progressively gained semantic, philosophical, and scientific substance.

It would be off-topic here to summarize the conclusions of Spitzer's essay on historical semantics, but we will point out a few of its main ideas. The central argument is that the term *ambiance* is closely connected to the term *milieu* and, consequently, that it is impossible to work on the former without also examining the latter. These two words help clarify each other because they both refer to "what surrounds men and things" and are often used together (e.g., when we refer to an *ambient milieu*). In other words, research into different meanings and uses of the word *milieu* provides precious insight into the history of ideas in which the word *ambiance* is rooted.

In long passages of great erudition, Spitzer attempted to identify the shifts, changes, and challenges to the notion of ambiance throughout the history of Western thought.

While it is now recognized that the word *ambiance* derives from the Latin verb *ambire*, Spitzer pointed out that the prefix *amb-* did not originally mean "around" or "what surrounds", but rather "on both sides" (left and right). This observation is by no means insignificant; indeed, it reveals the protective connotation associated with the verb *ambire*. The latter was actually a reference to the act of using both arms in a warm embrace. This sense of protection and benevolence, of a protective milieu that is kind to mankind, was an extension of the Ancient Greek vision of the world; it was, however, challenged by the advent of modern science, which sought out determinants external to mankind. Newton's idea of an ambient medium was a notable turning point in this respect. An environment regulated by a series of laws in which mankind was no longer the measure for all things came to replace the protective and benevolent nature of ambiance. There was a shift away from a "warm" notion connected to the corporeal and benevolent relationship of mankind with his milieu towards a "cold" and deterministic, abstract idea which made mankind an isolated entity confronted with an ensemble of forces past its control.[5]

Despite this shift, the shared idea of an active environment that affects the body, spirit, and behaviour of individuals endured in different ways in different periods: in the ideas of Hippocrates, for whom climate and atmosphere affect the human constitution; in Montesquieu's *The Spirit of Laws*; in Newton's notion of ambient milieu; and in Taine's theory of milieu. In short, if the notion of ambiance is underpinned by the idea of an active force influencing mankind, this force took on either a protective or threatening connotation, one that existed in symbiosis with, or external to mankind, depending on the worldview of the time.

Aside from etymological research, other methods have been used to try to elucidate the term "ambiance", namely looking for its equivalent in other languages, researching the semantic discrepancies between similar words in French (i.e., milieu, climate, atmosphere, and environment), and connecting the term ambiance to its accompanying qualifiers in everyday language. Again, these approaches challenged the prevailing use of dictionaries:

> Often, the precise meaning of a word cannot be expressed in a basic dictionary definition: its use and values can ultimately be expressed only through examples and contexts – the latter actually constitute the very ambiance of words and are indispensable for a precise understanding of their nature.[6]

Without going into too much detail, let us note a few points.

First, the semantic paternity of ambiance, milieu, and climate derives from the fact that all three refer to things that surround, envelop, and influence us.

But the specificity of ambiance compared to the other two is that it directly targets the affective dimension:

> *Milieu* is a sober, neutral, strict term which is always more or less bound to its scientific past. *Ambiance* and *climate* are colourful, evocative terms – *climate* because of the metaphor it constantly evokes and *ambiance* due to the feelings it suggests.[7]

If climate is defined as "the set of atmospheric conditions to which a region is subject", the only affective connotation possible is metaphorical or figurative. Ambiance, on the other hand, constantly and immediately conveys a "moral" or affective aspect. Second, unlike milieu or climate, the term ambiance often has a positive connotation and is associated with a pleasant situation or place. As Hans Nilsson-Ehle has argued, while the word "ambiance" can be used as a promotional argument for the advantages of a hotel or restaurant, "it is hard to imagine a hotel advertisement reading **Milieu*! Refined Cuisine! or **Climate*! Entertainment!, etc.". Lastly, while the adjectives that qualify *milieu* express a relationship, those associated with *ambiance* generally describe a quality. For example, when we talk about an "artistic milieu", we simply mean a "milieu of artists" that identifies a specific group within society. But when we talk about "artistic ambiance", it implies "an interior designed with artistic taste, containing beautiful works of art, giving a charming impression of subtle forms and colours".[8] In short, ambiance inherently involves qualitative thinking.

Let us retain two points in respect of the semantics of ambiances. One, while the meaning of the term *ambiance* has evolved over time, it is because our vision of the world itself has changed. Far from a neutral term devoid of theoretical importance, ambiance has slowly gained a semantic layer that is tied to different conceptions of science. Although ambiance is now becoming a research topic, this challenges the scientific paradigms we use to study it. Two, the term ambiance takes on meaning only within the language registers of which it is a part. Rather than talking about the possibility of a formal and unequivocal definition, it is more appropriate to point out the various types of discourse it elicits.

The psychopathology of ambiances

In the first half of the 20th century, several German psychiatrists became interested in phenomenology and laid the foundations of existential psychopathology. Despite their numerous differences, the research of Ludwig Binswanger, Erwin Straus, and Eugene Minkowski[9] constituted a radical critique of behaviourism and the objectivist ideology of the period to offer a new way of thinking about mankind's relationship with the world. Often relying on clinical case studies and their own practice as therapists, these authors engaged in rich and uncompromising dialogue with Husserlian philosophy, Heideggerian existential analysis, and Freudian psychoanalysis.

While their main goal was to further an understanding of mental disorders, the process also involved a fundamental analysis of the world in which we live. In the most famous of his essays, Binswanger[10] adapted and highlighted Kierkegaard's proposal: "Above all, we must keep in mind what it means to be a human being".

While this indeed constituted psychopathology, we cannot ignore the anthropological focus of the approach.[11] Dissatisfied with the distinction between psychic and somatic processes and sceptical of strictly functional and neurophysiological explanations, existential psychopathology examined the forms and structures of human existence. Consequently, the study of clinical cases could not be strictly introspective; it instead needed to be open to the different ways of being in the world, regardless of whether they were "normal" or "pathological". In other words, mental pathologies – and psychosis in particular – were seen as particularly relevant analysers for describing ways of existing.

There are two central arguments behind this idea. First, the detailed description of a specific clinical case need not be limited to the description of a unique experience; it also needs to point out the unchanging traits upon which it is based. As Minkowski noted:

> It is also important to keep psychopathological phenomenology completely *autonomous* from both clinical psychology and psychology in the usual sense of the term. It is not so much with the individual as a unique case that we compare the common and the general; rather we compare the shared, variable, and inconstant characteristics of individuals with the *essence* that simultaneously goes beyond them and carries them along.[12]

In short, the aim was to get away from the subjective to discover the structure underpinning a way of being in the world. This anthropology was not focused on the subject itself, but rather on the morphology of the world in which individuals exist. In other words, mankind's relationship with the world and how humans exist in it constitute the common thread that runs through the analysis:

> There is no existence that is not an existence in the world, or existence is nothing more than being in the world. Consequently: 1) the perpetual opposition between subject and object is overcome: subjects exist only to the extent that they are in the world; 2) psychology focused on a subject that is detached from the world gives way to anthropology, i.e., to the study of human beings necessarily connected to the world in which they exist.[13]

It is around this clinical anthropology that the topic of ambiances emerged and acquired its full meaning. We can first note that the notion of ambiance allowed those interested in the subject to identify a few basic characteristics of human existence. Even more than ambiance itself, it was the attitude towards ambiance that was being examined. As such, following on from Bleuler, Minkowski identified two vital principles that regulate our mental equilibrium: syntony and schizoidism.[14] Syntony designates the principle that allows us to "vibrate in unison with the environment". It ensures vital contact with reality and is related to feeling in harmony with the world. Conversely, schizoidism designates the ability to "[detach] ourselves from that environment". It is connected to personal *elan*, allowing the "me" to be affirmed and meaning and direction to be given to the future. These are two complementary functions of human life, and it is only when one of them hypertrophies takes the upper hand that pathological troubles arise; an excess of schizoidism might encourage the development of schizophrenia, while an excess of syntony might instead lead to manic depression.

Taking a more boldly Heideggerian approach, Binswanger also described clinical cases that led to fairly identical conclusions. The case of Suzanne Urban[15] involved a patient who lost her ability to maintain perspective and connect with the world in which she lived after suffering a traumatic experience. When in direct contact with an ambiance, without the ability to move away or withdraw, she became submerged in a homogenous atmosphere that was terrifying for her. This "atmospherization"[16] of the world made her feel dispossessed of her being and caused her to fall into a delirium in which everything felt threatening and disturbing. Cases of manic-depressive psychosis work completely differently.[17] Regardless of the events that occur around those afflicted, nothing appears to affect or touch them. Sufferers undergo mood swings and opposing thoughts and flit from one thing to another as if nothing were of any real importance. The world as such loses its depth and perspective, and contact with surrounding ambiances appears to be cut off. For both Minkowski and Binswanger, our way of being in the world seemed intrinsically tied to the way in which we experience ambiance.

To understand how psychopathology defined the notion of ambiance, we most likely need to return to the distinctions that have been made between different types of space. Whether we refer to the distinctions between Binswanger's "oriented space" and "attuned space",[18] Minkowski's "light space" and "dark space",[19] or Straus's "geographical space" and "landscape space",[20] it is through space that we can shed light on the notion of ambiance. While these distinctions are not identical, they are nonetheless closely related and they all attempt to highlight what we may refer to as "ambient space". So we have pragmatic and finalized space, on the one hand, that of our actions and perceptions, of objects and affiliated activities. "Oriented", "light", and "geographical" spaces all designate the functional spaces of everyday life. On the other hand, we have affective and qualified space, that of our moods and feelings, of the atmospheres that surround us and how they make us feel. "Attuned", "dark", and "landscape" spaces refer to the ambient world and to our ways of being in the world. It is the former way of experiencing space that is usually given precedence and studied from a scientific perspective;

the latter is generally not considered or even described. And yet, it concerns the physiognomy of the world in which we live, its more or less hospitable or familiar nature, and our direct interaction with surrounding phenomena. The distinction between these two types of spatiality should not mislead us, however. Both are always simultaneously present in everyday experiences. The world is experienced both as a world of objects in relation to which we act and position ourselves and as a world of qualities that affect us make us tick and to which we attune ourselves.

Let us elaborate further on the ambient world. While up to now we have dwelt on its spatial aspect, we need to understand what this encompasses. The focus here is not on the world of representation but rather on that of presence; rather than on the "what" of the surrounding world, on the "how" of our being in the world. In other words – and this is what is both interesting and challenging here – it is a matter of grasping what eludes all objectification and characterization and of understanding what stems from a pre-predicative experience of the world.[21] As mentioned above, the idea of ambient space refers to the anthropological structures of mankind's experience in the world, which can never be reduced to a simple topological position.

While Minkowski, Binswanger, and Straus agreed on numerous things with regard to ambiance, they did not adhere to the same thought process in their respective approaches to the issue. One means of differentiating between these authors is to examine the role that sensoriality plays in each of their approaches. Having spent part of his life in France, Minkowski was the only one of the three to explicitly refer to the term "ambiance". The central concept of *vital contact with reality* implies that it is not so much ambiance itself which is important as the interaction between the individual and ambiance. The term "vital" is fundamental here as – aside from the sensory contact with an ambiance – it is the driving force underlying this contact that is targeted. Consequently, ambiance needs to be considered in terms of becoming:

> Ambiance here must not be assimilated with what we call the 'outside world' when we refer to perceptions, nor with a result broken down into its supposed elements. It must be taken as a vast and living whole, in all its primitive force, in which human beings may only subsequently – with help from the analytical processes at their disposal – manage to distinguish between living beings, objects, and even physiological stimulants. An ambiance, in all its uniqueness, is not a static given; it is forged through contact with mankind, and mankind through contact with it. Crudely speaking, ambiance is a moving ocean. It is the becoming itself.[22]

This way of thinking does not exclude the sensorial world from ambiances. On the contrary, it involves freeing up the vital scope of our senses by showing how each of them is the product of opening up to the specific world in question. As such, for each sensory sphere there is a specific corresponding human attitude towards ambient life: resound (hearing), spreading (smell), tasting (taste), and touching (touch).[23] For Minkowski, metaphors express the overarching role that each sense plays in the general texture of life. In short, it is through ambient outcomes that the phenomenal is connected to the psychic and the material world to the spiritual world.

For Binswanger, attuned space is based on three "essential radicals" that are complementary and inextricably bound to one another: spatial-temporality, affectivity, and corporeality (to which we could add the relationship to oneself and others[24]). The founder of *Daseinanalysis* thus aimed to identify and describe the complex force that connects these three aspects. Starting from the assumption that the meaning of our relationship with the world is already inscribed in language, he suggested the notion of "sense-direction" (*Bedeutungsrichtung*) to describe this.[25] In some respects, the goal was to restore complete substance and meaning to everyday expressions to reveal the basic structures and forms of being in the world. Let us look at his paradigmatic example of the fall:

> When we are brutally disappointed and 'fall back to earth', we do indeed fall, but it is neither a purely physical fall, nor a fall which imitates (metaphorically or analogically speaking) or is derived from one; more precisely, the essence of brutal deception and fear mean that the harmony of the ambient and common world, which carried us along until then, suddenly receives a blow and stumbles. In this instant, our existence is indeed damaged, torn from its bearings and thrown back upon itself. Until we find a new mooring in the world, our total *Dasein* will be in the direction of stumbling, collapsing, and falling. We call this type of general sense-direction naked terror, and it allows us to see that the two are actually *one*.[26]

As such, the phenomenon of "the fall" cannot be reduced simply to one of its accepted significations, be it physical, moral, affective, or spiritual. The fall is bound up with a sense-direction to the extent that it creates unity between a type of space-time experience, a style of bodily movement, and a dominant affective tonality. From this perspective, sense-direction describes a way of being in a specific world.[27] Binswanger not only placed particular emphasis on the fall, but he also pointed out other sense-directions such as ascent, narrowness, fullness, and distance. If attuned space – as described and defined by sense-direction – is a space loaded with qualities that harness the subject's corporeality and sensoriality, it also refers primarily to the sphere of vital feelings, and not only to that of sensory phenomena. In other words, sense-direction targets the atmospheric or climatic dimension of being-in-the-world. Like Minkowski who used the senses to point out their vital potential, Binswanger greatly emphasized the vital character of affective tonalities. Here again, the subject's sensoriality is not defined in and for itself; rather, it takes on its full meaning only when connected to an emotional state or mood.

It is most likely Erwin Straus who truly developed the aesthetic angle on our presence in the world. Starting from an in-depth critique of behaviourism, he developed the notion of "sensing" to designate the original type of relationship we have with the world. Like Minkowski and Binswanger, he too was interested in living matter and the means of human existence, but he placed particular emphasis on the sensory dimension of lived experience. As Renaud Barbaras has noted about Straus: "Living matter is not the subject constituted by sensing; on the contrary, it is developed through sensing".[28] We must, however, be careful not to misinterpret what Straus actually meant by *sensing*. He chose this term primarily to differentiate it from the notion of sensation, understood as localized and occasional excitement resulting from a purely physiological process. Conversely, sensing corresponds to a means of global and immediate communication with the world; it is through sensing that living beings form a whole with the world. This involves both the animal and human worlds. Further, Straus sought to distinguish between sensing and perceiving. While perceiving is already a form of knowledge involving a "gnostic moment", sensing designates above all a feeling that involves a "pathic moment", free from all objectification or thematic distancing. With sensing, there is an empathetic relationship and corporeal attachment to the world. For Straus, sensing cannot be dissociated from "moving" since the sensory world cannot be reduced to a state of things or to characteristics that are independent from the subject; rather it requires a vector and affective mechanisms. In sum, it is by attaining a specific physiognomy that things appear to us and exert their affective power over us:

> Sensing is oriented towards the physiognomic character of what is frightening or terrifying. (…) Empathy is the broadest concept which encompasses both the act of separating and uniting, that of fleeing and following, attraction and repulsion, etc., which thus also includes both what is pleasant and unpleasant.[29]

It is by emphasizing the way things appear and how we approach them that Straus managed to make the sensory sphere the primary vector of our presence in the world. Landscape space and dance movements are in this respect the two paradigmatic cases upon which the pathic moment unique to sensing is presented. In sum, while Strauss most likely overestimated the distinction between the pathic and the gnostic, he is one of the people who managed to get closest to the phenomenality of the ambient world.[30]

We should retain three key ideas regarding the psychopathology of ambiances. First, the field of ambiances involves a vital, non-thematic level of experience. Ambiance is more than just a relationship of knowledge, objectification, or representation of the world; it is very much bound up with our presence in the world. It involves both how we feel in the world and the way we experience it. And yet, we must not exaggerate the divide between sensing and perceiving. Instead, we should look for ways of connecting these two dimensions.[31] Second, ambiance primarily involves affective tonalities; however, these are far from independent of the other components of experience, and they make sense only when connected to spatial-temporal forms, types of movement, and sensory data. Finally, ambiance can be broken down based on modal logic. It does not designate the "what" of an experience, rather the "how". Connecting the different components thus allows us to identify specific ways of being that involve both the subject itself and the world in which the subject exists. In short, ambiance is nothing less than a way to identify, describe, and distinguish between "styles of existence".[32]

The aesthetics of ambiances

Having addressed the semantics and psychopathology of ambiances, let us now turn to the aesthetic approach. Aesthetics are central to the notion of ambiance since they explicitly address the issue of human sensoriality. There are two elements that frame aesthetics in terms of ambiances. The first is that ambiance conjures an "environmental aesthetic" that cannot be reduced to a fine arts aesthetic. It is not a work of art that is at the centre of this aesthetic but rather nature, understood in the broadest sense of the term (including the city and urban sphere). And yet, while this aesthetic has both a narrative and ambient facet, the former has until now largely dominated the latter.[33] The second element is that ambiance brings us back to the original meaning of aesthetics, understood as a theory of sensory perception. The goal is to restore an understanding of *aesthesis* that goes beyond taste judgements, rehabilitates the place of the body, sensitivity, and emotion, and relativizes the weight of semiotics and the primacy of language.[34] Ambiance as such encourages us to rethink both the object of aesthetics and the discipline of aesthetics itself.

We shall limit ourselves here to the aesthetics of ambiances as developed by Jean-François Augoyard and Gernot Böhme.[35] While the former does indeed use the term "ambiance", the latter employs the term "atmosphere" (a German speaker, Böhme uses the word *Atmosphäre*[36]). Although these two conceptions were formed independently of each other, they nonetheless have several things in common. Beyond the fact that they seek a return to a theory of sensory perception, they both underscore more generally the architectural component and the spatial and material dimension of ambiances.[37] In both cases, the goal is to produce an aesthetic of ambiances that can be applied to the built space. However, the aesthetics of ambiances does not refer only to monumental architecture or buildings of reference; it broadly encompasses all types of everyday situations. This approach is opposed to a "museum-like" view of art and refuses an amalgam between artistic and aesthetic experiences; it provides an outlet for analysing the most ordinary situations.

The importance placed on the contextual nature of ambiances encourages the idea of ecological aesthetics. For Augoyard, the goal is to develop *in situ* approaches and to build interdisciplinary tools that use "perceptible physical signs and the full range of norms, rules, codes and references, as well as the instruments, features and uses assigned to built forms".[38] For Böhme, this involves analysing the connections between environmental qualities and human sensitivity to better understand "how one feels in an environment".[39] While the terms used by the two are not the same, both tackle the complexity of situations by suggesting that heterogeneous contextual components be connected to one another.

Indeed, this logic of articulation exists at all levels of their thinking. Everything occurs as though ambiance is the very starting point from which various components or polarities need to be connected or harnessed to a single dynamic. Böhme constantly refers to atmosphere as an "in-between". For clarity, I shall distinguish here between three primary levels of articulation.

The first level concerns the relationship between subject and object. As Augoyard has noted,[40] we can look at the dictionary definition of the term ambiance: "the moral and material atmosphere surrounding a place or person..." As basic as it is, this definition highlights the connection between the "material" and "moral" dimensions. Far from insignificant, this connection is problematic and requires that attention be paid to the "quasi-beings" and "semi-things" that lead to an aesthetic of intermediaries.[41] For his part, Böhme insists on the fact that the atmosphere is both objective, since it emanates from things and can be produced from material arrangements, and subjective because it cannot be defined independently of the people who experience it. In short, atmosphere is regarded as "the characteristic manifestation of the co-presence of subject and object".[42]

The second level involves the relationship between the senses. Here Augoyard distinguishes between ambiance (singular) and ambiances (plural). While science and technology tend to dissociate the different sensory bases to better study an environment, questions remain regarding its ability to convey inter-sensory and amodal phenomena. "While architects produce *an ambiance*, we ask them to align different *ambiances*: a thermal ambiance, an acoustic ambiance, etc. And yet, can we actually scientifically define *a single* architectural ambiance?"[43] Similarly, Böhme challenges the tendency to dissociate the senses initially, only to reunite them later on. In taking a new approach to the issue of synaesthesia, he conversely shows that the sensory unity of a situation actually precedes differentiation between the senses. It is on the basis of atmosphere – understood as a global and indivisible backdrop – that the identification of detail and the distinction between different sensory processes seem to occur.[44]

The third level concerns the relationship between reception and creation. Ambiance involves both what can be perceived and what can be produced. Better still, it often challenges such distinctions since what is perceivable is itself an action. Just as architects and scenographers physically construct sensory forms, users configure their milieu through their actions. Augoyard contends that "our relationship with the formal and sensory environment should be understood as an interaction, a constructive movement between what is given and configured, what is felt and acted, perceptible and representable".[45] Böhme has expressed the same idea in his own manner when arguing that the production of an atmosphere is not simply the result of designing an object but also, simultaneously, "creating the conditions for its appearance".[46] From this perspective, the way the world appears is the product of both the designer's work and the inhabitant's activity.

While there is a great affinity between these two actors – particularly in their use of phenomenology – each has a number of unique features. Let us first note that the material and objective aspects of ambiance are not addressed in exactly the same terms. For Augoyard, they are connected to both physical signals and constructed space, while for Böhme they are more closely tied to the world of things. Based on the idea that all physical signals necessarily include aspects of the built environment in which they are produced, Augoyard attempts to develop a contextual physics:

> A specifically situated sound traverses a space of propagation that gives the sound a certain "here and now" quality: for example, in terms of sound, the time of its reverberation and shaping of its tone; for light, particular reflections, changes in temperature and colour, and the organization of shadows; and specific air turbulence around certain architectural configurations, and the volatility of odours that varies with the wind speed. The signal that can be physically isolated after the fact only exists in this spatial and temporal context that is entirely dependent on the site's morphological and material characteristics.[47]

This type of approach reconnects with "controlling ambiances" as taught in schools of architecture by reintroducing the contribution of spatial context into meteorological procedures. The aim is nothing less than to rethink the complementarity of the quantitative and the qualitative. And yet, if physical signals exist only when connected to a built environment, they really only take on meaning when they are perceived or filtered by the codes, norms, representations, and social interactions at work in a given place. As Augoyard has noted, a signal-forming line with multiple contextual features (spatial-temporal, perceptive, cultural, and social) "leads to defining a central object which is no longer a signal but a phenomenon".[48] In other words, the issue is not so much the physical signal itself or the ambiance in general, as the pivotal place created by an "ambiance phenomenon".

Böhme, for his part, addresses the material dimension of ambiance based on the world of things. His main argument revolves around stating the "ecstatic" nature of things. He questions the traditional ontology of things that considers its qualities as determinations.[49] In other words, according to this traditional view, the characteristics of a thing (its shape, colour, odour, etc.) are what distinguish it from other things. Each thing has internal unity and is self-contained, distinct from others and independent of its surrounding environment. For Böhme however, things have the power to reveal themselves, to interact, and to disseminate on their own. As painting illustrates, the colour of an object can be modified by another nearby object, and the shape of a thing can create field lines and suggestions of movement. One of the examples given by Böhme to illustrate his argument deserves our attention:

> If we say for example: a cup is blue, then we think of a thing which is determined by the colour blue, which distinguishes it from other things. This colour is something that the cup 'has'. In addition to its blueness, we can also ask whether such a cup exists. Its existence is then determined through a localization in time and space. The blueness of the cup, however, can be thought of in quite another way, namely as the way, or better, a way, in which the cup is present in space and makes its presence perceptible. The blueness of the cup is then thought of not as something which is restricted in some way to the cup and adheres to it, but on the contrary as something which radiates out of the environment of the cup, colouring or 'tincturing' in a certain way this environment, as Jakob Böhme would say.[50]

This is how each thing can affect its environment through its presence.

And yet, in pointing up the ecstatic dimension of things, Böhme introduces a third term that is generally left out of theories of perception: the medium. As he has argued, "The structure of perception is not 'I see something' but rather 'in the medium, the presence of things is perceptible'".[51] This is most certainly the key element of Böhme's thinking about atmosphere. Introducing medium as the third term leads to a twofold development. For one, it is on this basis that it becomes possible to move beyond the idea that perception can be reduced to the simple fact of identifying or noticing objects. Indeed, the medium is the basis upon which the world acquires a certain physiognomy and places the perceiving subject in a certain corporeal and affective state. Further, with regard to experience, atmosphere is nothing other than the medium – or more precisely the state of the medium – in a given situation. In other words:

> The primary 'object' of perception is atmosphere. What is first and immediately perceived is neither sensations nor shapes or objects or their constellations, as Gestalt psychology thought, but atmospheres, against whose background the analytic regard distinguishes such things as objects, forms, colours, etc.[52]

Augoyard and Böhme draw upon markedly different categories. For the former, the idea of ambiance as a phenomenon is primordial, with its six complementary elements (physical signals, spatial-temporal forms, percepts, representations, codes and norms, and social interactions). For the latter, atmosphere is instead constructed around the notion of presence, with three main components (the thing, the medium, and the senses).

To summarize, we can retain three main contributions from the aesthetics of ambiances. First, it provides a new way of thinking about sensory perception by pointing to its eminently contextual nature, by including it as a part of everyday social life, by opening it to all of the senses, and by restoring the full importance of physical and material factors. In other words, the focus on the phenomenality of built space helps develop the idea *of urban and architectural ambiances*. Second, such phenomenological aesthetics help underscore the complexity of ambiances by encouraging a modal and interdisciplinary approach. From this perspective, rather than dissociating the terms of an ambiance and isolating the elements that compose it, the goal is to connect and get them to bind together. Although each author takes a different path, the aim is to establish connections and continuity to give substance to an ambiance. Finally, the aesthetics of ambiances involves a dynamic approach. Rather than taking ambiances as a given or as a state, the goal is to think of them as a process in action that involves both the activity of inhabitants and that of their designer.

The scope of the notion of ambiance

The notion of ambiance has a broad heuristic and operational scope, open to a diversity of approaches and uses. It takes on its full meaning and strength when implemented from a particular perspective. It is by submitting ambiance to specific questions that we can best understand the extent of its consequences and implications. Beyond the three areas addressed here, others – ranging from environmental psychology to sensory architecture, human geography to artistic creation, and pedagogy to dramaturgy – could also have been explored. Clearly, one original aspect of the notion of ambiance is that it encourages interaction between different types of knowledge and a connection between disciplines. As a nomadic and multi-disciplinary notion, ambiance offers an original alternative to traditional object/subject, sensory/intelligible, and active/passive dualism. It most certainly helps to sensitize thought.

However, let us not mislead ourselves: our difficulty in grasping such a field of thought must not lead us to commit common mistakes. Ambiance is certainly not the same as environment. While there are similarities between the two terms, ambiance is bound up with sensoriality, emotion, and lived experience, and it cannot tie in with an overly objectivist or positivist approach. This is due to the epistemological implications of the notion of ambiance. And yet, ambiance is not purely subjective either. We have seen that it cannot do without the materiality of built and planned space (which is why we then refer to architectural and urban ambiances) and that it also draws on an anthropological and collective dimension that cannot be reduced to an individual experience. Finally, it is patently obvious how much a well-understood ambiance exceeds from the merely cosmetic sphere. Far from a simple excess of luxury or comfort, it helps us think about the existential aspect of human experience. Ambiance places the sensorial world at the very centre of living space and constitutes a *condition of possibility*.

Notes

1 This is of course an extremely limited selection of authors and approaches selected from the vast literature on the subject. For an overview of the domain of ambiance, see the presentation of the *International Ambiances Network*: https://www.ambiances.net/. Two main edition series directly address this topic: *Ambiances, Atmospheres and Sensory Experiences of Spaces* (Routledge, edited by Rainer Kazig, Damien Masson, and Paul Simpson) and *Atmospheric Spaces* (Mimesis International, edited by Tonino Griffero and Giampiero Moretti). For a more developed presentation of the theory of ambiances, see Jean-Paul Thibaud, *En quête d'ambiances. Eprouver la ville en passant* (Geneva: MetisPresses, 2015).

2 Paul Adam quoted by Alexis François: "Suffixe littéraire –ance", Vox Romanica. Vol IV, Bern (1939): 20–34. As Leo Spitzer has shown, a more in-depth study of this suffix reveals that it points more to the perpetuation or subsistence of a state of being, something that lasts and is prolonged over time (*Le français moderne*, Vol. VII, 1939, p. 276; my translation).

3 Karl Michaëlsson, "Ambiance", *Studia Neophilologica*, Vol. XII (1939): 91–119, my translation.

4 Leo Spitzer, "Milieu and Ambiance: An Essay in Historical Semantics", *Philosophy and Phenomenological* Research, Vol. III (1942): 1–42 and 169–218.

5 It is worth noting that certain thinkers such as Goethe and the Goncourt brothers reacted against this modern interpretation and attempted to renew the lessons of Antiquity.

6 Hans Nilsson-Ehle, "Ambiance, Milieu et Climat", *Studia Neophilologica*, Vol. XXIX, No. 2 (1957): 181–191, my translation.

7 Hans Nilsson-Ehle, "Ambiance, Milieu et Climat", *Studia Neophilologica*, Vol. XXIX, No. 2 (1957): 181–191, my translation.

8 Hans Nilsson-Ehle, "Ambiance, Milieu et Climat", *Studia Neophilologica*, Vol. XXIX, No. 2 (1957): 181–191, my translation.

9 Other major thinkers – directly connected to existential analysis – also deserve at least a mention: Viktor Von Weizsacker, Frederik Jacobus Johannes Buytendijk, and Henri Maldiney, just to name a few.

10 Ludwig Binswanger and Michel Foucault, *Dream and Existence*, edited by Keith Hoeller (Atlantic Highlands, NJ: Humanities Press, 1993).

11 It is specifically this anthropological angle that interested Michel Foucault in his introduction to Binswanger's *Dream and Existence*.

12 Eugène Minkowski, "Phénoménologie et analyse existentielle en psychopathologie", in *Ecrits cliniques*, (Ramonville Saint-Agne: Erès, 2002) 95–138, my translation.

13 Eugène Minkowski, "Phénoménologie et analyse existentielle en psychopathologie", in *Ecrits cliniques*, (Ramonville Saint-Agne: Erès, 2002) 95–138, my translation.

14 Eugène Minkowski, *Le temps vécu* (Paris: PUF, 1995). Note the great similarity between this distinction made by Minkowski, and the one made by Gilbert Durand between the nocturnal and diurnal realms of the imagination in *Les structures anthropologiques de l'imaginaire* (Paris: Bordas, 1969).

15 Ludwig Binswanger, *Der Fall Suzanne Urban*, Zurich, *Schweizer Archiv für Neurologie und Psychiatrie*, Nos. 69, 70, and 71 (1952–1953).
16 With regard to the "atmospheric" notion from a psychopathological perspective and the importance of oral meaning in the atmospherization of the world, see the seminal work of Hubertus Tellenbach, *Geschmack und Atmosphäre. Medien menschlichen Elementarkontaktes* (Salzburg: Otto Müller Verlag, 1968).
17 Ludwig Binswanger, *Über Ideenflucht* (New York: Garland, 1980).
18 Ludwig Binswanger, *Das Raumproblem in der Psychopathologie* (Berlin: Springer Verlag, 1933).
19 Eugène Minkowski, "Vers une psychopathologie de l'espace vécu", in *Le temps vécu* (Paris: PUF, 1995), 366–398.
20 Erwin Straus, "Die Formen des Räumlichen. Ihre Bedeutung für die Motorik und die Wahrnehmung", in *Nervenartz*, 3rd year, Edition No. 11, (Berlin: Julius Springer 1930), 142–178.
21 It is worth noting that the history of philosophy has been marked by this preoccupation, whether in terms of Plato's khôra, Husserl's passive syntheses, or Kant's reflective judgement.
22 Eugène Minkowski, "Constitution et conflit", in *Ecrits cliniques,* 67–79, my translation.
23 Eugène Minkowski, *Vers une cosmologie* (Paris: Payot, 1999).
24 I am thinking particularly of the importance he placed on the phenomenon of the encounter.
25 For a recent overview of this notion, see Jeanine Chamond (ed.), *Les directions de sens. Phénoménologie et psychopathologie de l'espace vécu* (Argenteuil: Le Cercle herméneutique, 2004).
26 Ludwig Binswanger, *Dream and Existence,* my translation.
27 As Henri Maldiney has noted: "Its meaning is above or rather beyond all local acceptations. And it must be called *sensorial direction*, i.e., the alliance of words that make the unity of meaning-direction and meaning-signification sensitive – which is precisely that of the existential sketch and the unveiling of the being at work within it" in *Regard, Parole, Espace* (Paris: L'Age d'Homme, 1973), my translation.
28 Renaud Barbaras, "Affectivité et mouvement: le sens du Sentir chez E. Straus", *Alter. Revue de Phénoménologie*, No. 7 (1999) 15–29, my translation.
29 Erwin Straus, *Du sens des sens* (Grenoble: Jérôme Million, 1989), my translation.
30 As Henri Maldiney (*op.cit.*, my translation) has noted, Straus's thinking "begins where Husserl's intentional analysis leaves off, at the hyletic he named but was unable to construct". For a brief overview of this topic, see Michèle Gennart, "Une phénoménologie des données hylétiques est-elle possible?", *Etudes Phénoménologiques*, No. 4 (1986): 19–46.
31 As such, Renaud Barbaras (*op.cit.*, my translation) has aptly noted: "It is a given that we do not live exclusively in this pathic dimension in which the animal is itself locked up; we have always gone beyond the strictly empathetic relationship in favour of a thematic distancing which understands expressions as determinants of a thing. We live in the universe of perception". Stating the same idea somewhat differently, Maria Villela-Petit has written: "As human beings we do not know how to remain within a strictly 'sensing' dimension for fear of not having access to thought, but we never leave this dimension of our being in the world either, except in extreme cases of pathological rupture of our pathic communication with the world". See "Espace, temps, mouvement chez Erwin Straus", in Jean-François Courtine (ed.), *Figures… op.cit.*, 51–69, my translation.
32 Henri Maldiney, *op.cit.*

33 Cheryl Foster has distinguished between the "narrative" approach, which involves reading and writing the environment as a history using the semiotics of indexicality, and an "ambient" approach, which insists on sensitivity and resists a language-based formulation. See "The Narrative and the Ambient in Environmental Aesthetics", *The Journal of Aesthetics and Art Criticism*, Vol. 56, No. 2 (1998): 127–137.

34 It is through such arguments that Gernot Böhme constructed his project for a "new aesthetics": "Aesthetics Knowledge of Nature", *Issues in Contemporary Culture and Aesthetics*, No. 5 (1997): 27–37.

35 Other well-known authors have also addressed the aesthetics of ambiances, including Martin Seel, Yuriko Saito, Pierre Sansot, Henri Maldiney, Herman Schmitz, and Michael Hauskeller. Further research into their approaches would obviously be of interest.

36 Gernot Böhme has borrowed the notion of "atmosphere" from Herman Schmitz's philosophy of the body. While he acknowledges this contribution, Böhme provides a critique of Schmitz's thought and has suggested an original and personal interpretation of the notion. For Böhme, Schmitz's thinking is limited in terms of the construction of an aesthetic of ambiances for two main reasons: Schmitz does not completely free himself of a traditional interpretation of the aesthetic that restricts the notion to the artistic field; he only develops the subjective angle of atmosphere and does not sufficiently consider its material and objective aspects. See "Atmosphere as the Fundamental Concept of a New Aesthetics", *Thesis Eleven*, No. 36 (1993): 113–126.

37 Several issues of architectural journals have confirmed – if confirmation were needed – the inclusion of ambiances within the discipline of architecture. See for example "Constructing Atmospheres", *Daidalos*, No. 68 (1998); "Ambiances architecturales et urbaines", *Les Cahiers de la Recherche Architecturale*, No. 42/43 (1998); and "Atmosphère", *Faces. Journal d'architecture*, No. 67 (2010).

38 Jean-François Augoyard, "Eléments pour une théorie des ambiances architecturales et urbaines", *Les Cahiers de la Recherche Architecturale*, No. 42/43 (1998), 13–23, my translation.

39 Gernot Böhme, "Acoustic Atmospheres. A Contribution to the Study of Ecological Aesthetics", *Soundscape. The Journal of Acoustic Ecology*, Vol. I, No. 1 (2000): 14–18.

40 Jean-François Augoyard, "Eléments pour une théorie…".

41 Jean-François Augoyard, "Faire une ambiance?" in Jean-François Augoyard (ed.), *Faire une ambiance* (Bernin: A la Croisée, 2011), 17–35.

42 Gernot Böhme, "Atmosphere as An Aesthetic Concept", *Daidalos*, No. 42/43 (1998): 112–115.

43 Jean-François Augoyard, "Eléments pour une théorie…" my translation.

44 Gernot Böhme, "Über Synästhesien / On Synaesthesiae", *Daidalos*, No. 15 (1991): 26–37.

45 Jean-François Augoyard, "Eléments pour une théorie…", my translation.

46 "It is never purely a question of designing an object but always, at the same time, of creating the conditions for its appearance", Gernot Böhme, "On Synaesthesiae…".

47 Jean-François Augoyard, "L'environnement sensible et les ambiances architecturales", *L'espace Géographique*, No. 4 (1995): 302–318, my translation.

48 Jean-François Augoyard, "L'environnement sensible et les ambiances architecturales", *L'espace Géographique*, No. 4 (1995): 302–318, my translation.

49 The argument below is particularly based on Gernot Böhme, "Atmosphere as the Fundamental...".
50 The argument below is particularly based on Gernot Böhme, "Atmosphere as the Fundamental...", 121.
51 Gernot Böhme, "An Aesthetics Theory of Nature: An Interim Report", *Thesis Eleven*, No. 32 (1992): 90–102.
52 Gernot Böhme, "Atmosphere as the Fundamental...".

2 The sensory environment and architectural ambiances

Jean-François Augoyard

When ecology meets the city

In their remarkable growth from the 1960s on, environmental sciences have not failed to address the phenomenon of the city, broadly and in their own way. After all, don't cities generate a major portion of the disturbances that we experience and try to eliminate? It was inevitable that architecture and urban research could no longer afford to ignore the emergence of an increasingly worrisome aspect of the built space, along with the concomitant growth of theoretical and methodological tools that had joined forces to "fight disturbances", as demanded by so many.

Within the small and studious milieu of French architectural research, which had long been searching for its ideal institutional biotope, the environmental approach seemed to burst onto the scene somewhat abruptly, by taking a technicist shortcut and sounding the cry of urgency. In fact, starting in the 1970s, research into spatial design had slowly and carefully formed as a field of knowledge around three core problems that made abundant use of several theoretical frameworks. The question of architectural meaning and identity called for an epistemological or semiological approach.[1] The question of the specificity of the methods for describing and representing the architectural object called for an edification of typo-morphologies as well as a search in terms of modelling.[2] Lastly, the question of the intelligibility of the corpus of architectural references called for a long-term, historical approach.

And then, one after the other, various unforeseen ecological "crises" occurred that directly concerned the built space. People expected research to yield rapid solutions; hence, the urgency of thermal insulation in the wake of the 1973 energy crisis and the pressing need to improve acoustic insulation in response to considerable social demand. We wondered whether the research wouldn't become hypertelic. And we feared that a systematic environmentalism, supported both by a spike in mortgage loans and a blurring of notions (any architectural and urban situation being an *environment* in the broad sense of the term), might penetrate and vitiate the theories and methods of research on the built space.

DOI: 10.4324/9781032655307-6

Twenty years of both environmental research and research into the built space have shown us that nothing was that simple. The built, inhabited space is not just any "terrain". Its complexity holds many surprises for any ecologist or environmentalist too much in a hurry. To the contrary, research into architectural environments does not stop at statements and representations. It instead pays great attention to the sensory and physical nature of things, thereby contributing to our general knowledge of the urban phenomenon.

Based on my participation in this interdisciplinary trend, I would like to describe the positive dialectical relationship yielded by the convergence of environmental sciences and urban research. How has the built urban space changed the scientific approach that viewed this space as a new topic of study? What new contributions does research on the sensory environment make to architectural and urban research?

The paradigm of noise research

Environmental sciences have found a remarkably high concentration of existing forms of disturbances within the urban milieu: air, water, noise, and light pollution, and multiple forms of disturbances and exposures. However, it is this very milieu, treated as a field of observation, which has generated the largest epistemological obstacles. For a careful observer who leaves the laboratory to visit a site, the city appears as a tightly woven fabric of highly interdependent contextual variables that resist the reductions common in the experimental method. To provide a brief illustration of this problematic encounter between a method stemming from the physical sciences and a complex, interdisciplinary entity such as a city, it may be helpful to consider in broad strokes the history of France's national research programmes into noise, which were conducted between 1975 and 1990.

Until the early 1970s, research in the fields of acoustics and psychoacoustics had focused mainly on fundamental physics, audiology, phonology, musical acoustics, and the acoustics of auditoriums. But then, it was confronted with the topic of the urban environment. Society had begun to demand solutions to the problems of road noise and noise in the workplace. There were three periods of national research programmes between 1970 and 1990, all sponsored by the Ministry of the Environment.

During the first phase, which was heavily polarized because of a healthcare emergency, people first sought to characterize the auditory and extra-auditory effects of noise pollution, to create norms for them (such as tolerance thresholds and disturbance levels), and to test protective mechanisms. At the core of these efforts was a postulate directly inspired by the in vitro approach: the need to distinguish between action taken on causes and action taken on effects. The remedies for noise pollution thus entailed either modifying the sound at the source of the problem or providing the individuals victimized by the noise with soundproofing insulation. People also expected the abundant technical norms and legal regulations crafted during this period to modify our conducts and the sources of the pollution decisively.[3] And yet, empirical observation and a review of case law show that regulations are often far removed from the actual experience of the disturbance and that the application of norms does not necessarily produce a satisfactory result for those who complain of the problem.[4]

During the second phase, the research tried to understand the "human" factors better. Emphasis was placed on refining the concept of "disturbance" from both a physiological and a psychological perspective,[5] but the results were disappointing.[6] The average reactions were too imprecise, and the most revealing variables were often embedded within the history of the topic. At the same time, the first, truly interdisciplinary works[7] began to show the multiple, complex nature of the norms, that the architectural and urban context is not neutral in the context of noise-related *problems*, and finally that noise pollution is a remarkable case that must be studied within the context of a more general anthropology of the sound environment. Positive or negative, noises are also instruments of collective relations. This has two consequences: norms must be relativized to the experiential conditions on which they are based, and sound insulation can no longer be the only curative solution to recommend in all cases.

Beginning in 1983, the third phase worked more systematically on the effects of noise on site. In this way, the spatial and acoustic context of the urban milieu became a primary topic of inquiry. The work explored the *human* dimensions that had been overlooked: the psychology of people who make noise, the role of representations around noise (such as silence), the interdependence between the people complaining of noise in their neighbourhood, and the management of noise problems by local authorities and social actors. Based on the previous work, a programme was launched in 1990 titled "Management of the urban sound environment". It provided perspectives on work that focused on just one topic (noise from transportation, work, and facilities) and highlighted the complex relations between the physical and social characteristics of the urban milieu. Another significant aspect was that, in looking past the merely regulatory and defensive goals, this programme researched tools for helping plan and design cities.

Complementarily, other research programmes that focused more on the question of the city, especially those conducted by the Ministry of Public Facilities, Urban Planning, and Housing, investigated the environmental or sensory aspects[8]. To recap, we could say that the common trend in the 1990s has been to study the contextual aspects and to try to exploit the qualitative variables as best as possible.

The experience of being on site

Among the perceptible factors[9] of the architectural and urban environment that are being studied (i.e., light, sound, tactile materials, and air in terms of its temperature, speed, and smell), noise is certainly the most developed, complex example of resituating an object within its context. There are four reasons for this: the difficulty in finding fully satisfactory norms and physical mechanisms for current sound situations,[10] the significant influence of psychological and sociological connotations on the perception of sound, the persistent social demand that remains vigorous for being dissatisfied with the proposed technical and legal solutions, and, finally, the individual profile of the entities and people heading research in this sector.

It should be noted that the other aspects of urban ambiance have not been as extensively developed, past an open interdisciplinarity, especially in the field of human sciences. For example, the rare and interesting works on the individual and social perception of thermal phenomena were abandoned several years ago. Currently there is just one architectural research laboratory that is developing an awareness of usage values between the technical aspects and those characterizing the architectural project.[11] Thus research into the perception of light in cities is not yet able to distance itself from the functional or visual ergonomics of roadways. Proof of this comes when we read the lighting recommendations common in municipal technical service departments. Despite an overabundance of studies on vision, we know very little about the set of components that make up a pedestrian's visual attitude at night.[12] Lastly, with regard to the other senses with which we perceive the urban environment, despite a few isolated studies,[13] there is still no interdisciplinary field of knowledge in this area[14]. Thus, the example of noise shows most clearly what can happen to environmental sciences when they consider the question of the city.

What happens when a researcher leaves the laboratory to observe the complexity of phenomena in the urban environment directly on site? The more the information, the more one grasps observable interdependencies. At the same time, certain limits to the in vitro method that one had hoped to reuse on site emerge more clearly.

What are these limits? We can list at least five very remarkable ones: (1) the delocalization of the *problems* to be dealt with in favour of universal pronouncements; (2) the orientation of the goals towards the production of universal norms and technical mechanisms that are expected to provide effective remediation; (3) the limitation of observing the *human* element to either the neurological receptor or type of subject; (4) the reuse of the model of intelligibility from the physical sciences to deal with the entire chain of effects caused by signals[15]; and, lastly, (5) the choice of a defensive axiological attitude that can be summed up using the following dual interlocked view: the environment should be studied when it becomes a disturbance, and this disturbance becomes all the more evident when it has a measurable impact on people's health.[16]

A historic and typological analysis of the scientific committees that oversaw research into the environment would certainly reveal a remarkable divergence, regardless of their fields of origin, between scientists working in vitro and those who, for various reasons (applied feasibility research, the practical approval of tools and results, or investigations of populations in a certain area) assiduously frequented the urban environment. The discovery of the structural complexity of cities inevitably leads to a slow evolution in ways of imagining the environment.

What is therefore called into question is less the validity of the main scientific models used, and more the postulate of an intangible hierarchy of excellence among these models. One discovers that the built environment is more than a collection of measurable signals and technical systems. One discovers the adjacent reasons why a law is not applied or observed, or why a system for defending against disturbances is diverted or subverted. One discovers that a physical disturbance may have positive social effects (community organizations, solidarity, and an encouragement to participate directly in one's democracy). Thus, one discovers that cities observed on site demand a pluralism of hypotheses, methods, and theoretical developments.

To sum up the remarkable evolution that environmental research has undergone since it began to look at the urban milieu, we could say that the observed object ultimately changes the observer, meaning his/her investigative techniques and way of thinking about what he/she observes.

The environmental reasons for a city's intelligibility

Through a terrain's resistance to the in vitro logic familiar to the environmental sciences, which is dominated by physics, the city yields a number of reasons tied to the importance of context to observable phenomena and related to atypical models of intelligibility.

Recurring contextuality

Why do certain conditions for the appearance of observable phenomena that experimental physics distinguishes as belonging to the contingent context actually form an integral part of the object being observed? It is precisely when we go on-site that we don't find any *pure* phenomena. Let us first recall that, in an urban environment, all physical signals are facilitated by other physical elements. A specifically situated sound traverses a space of propagation that gives the sound a certain "here and now" quality: for example, in terms of sound, the time of its reverberation and shaping of its tone; for light, particular reflections, changes in temperature and colour, and the organization of shadows; and specific air turbulence around certain architectural configurations, and the volatility of odours that varies with the wind speed. The signal that can be physically isolated after the fact only exists in this spatial and temporal context that is entirely dependent on the site's morphological and material characteristics. For that matter, not all essential conditions can be measured; for example, there is no equivalent of the decibel for vibrations or smells. Thus, practitioners of the urban environment are pushing research to find multi-criterion modellings that include qualitative variables.

By default, the *qualitative nature* of the urban environment arises initially from the quotient of the immeasurable in an urban situation. It more generally obeys the rule that, on site, any physical signal makes sense only if it can be perceived. Signals do not exist on their own, and variability is not accidental. The *same* sound and the *same* light are perfectly capable of having different meanings and values, even antithetical ones, depending on the subject, social group, and moment of perception. Thus, signals are never exactly the *same*, neither by how they are filtered nor by their connotations. According to this variance, the same, identical composition of signals (the sound of a passing train, or a bright, lively advertising billboard) can be either a useful medium, a normative indicator, an object expressing a reactive dynamic, or an aesthetic object. In terms of perception, the salience of individual, social, cultural, and economic factors place context at the core of this phenomenon.

Would we say that everything consequently plods along in a state of confusion? It is true that cities, these compendiums of the world where the most varied objects, practices, and representations congregate, coexist, and proliferate, are particularly exposed to reveries that bring opposites together with varying degrees of success, amateur interpretations that blithely mix genres, and efforts as amnesiac as they are lacking in rigour, examples of which abound in both our past and present.

And yet, complexity does not necessarily entail a general state of chaos or equivalence. And interdependence does not prevent there being a formal distinction between the terms present. Research into architectural and urban environments that have yielded results shows that the successful coordination of various approaches is based on a distinction of competencies and a minimal independence between fields of knowledge. Under such conditions, a consideration of the qualitative does not entail the exclusion of the quantitative. Researchers working directly in the field know how much these two spheres can enrich each other, and that observable phenomena can be seen as much through ethnographic observation as through a calculation of such a measurement. On the other hand, the more complex the situation, the more patience is required in the investigations and prudence in the conclusions if one wants to identify relations and interdependencies with any clarity and rigour, and the more the study of the observable will evolve away from the exclusive imprint of the physical sciences.

The evolution of models of intelligibility

An ecological approach to the built space that seeks to comprehend its perceptible phenomena presupposes a change in models of intelligibility. We can now identify some of the traits of this ongoing change. First of all, the necessary consideration of the anthropological dimension results in a dual confrontation. Across from laboratory subjects, the recipient of stimuli, we have actual city dwellers, imbued with their own individual histories, imaginations, and unique interactions with their social networks, but who are also actors of their environment, producers of signals, if not also disturbances. And across from statistical populations, we find groups and communities tied to their environments through an indissoluble interaction between the sensory and the social, which feed and enable one another. The causal links between the different aspects of the phenomenon being studied are thus no longer univocal and linear. There is both a convergence between different systems of causes and conditions used to define the nature and existence of a localized phenomenon and an interdependence of the factors at work.

A second observable trait has begun to form part of both evidence and actual practice: interdisciplinarity. This approach, which gets various competencies to interact with one another on an equal footing, is not a given. Research into disturbances has often engaged in interdisciplinarity only as a last resort. In other words, not only does interdisciplinarity – understood as the interaction and grouping of each specialty under the umbrella of common concepts and integrating methods – not take place, but the openness of the physical model to other paradigms of understanding occurs only when this approach bumps up against the unknowable. The mystery is in such cases qualified by hypothetical social and cultural variables that muddy the clarity of the causal relationship between physical signals and its physiological effects. In such a mindset, the indications used to locate these variables clearly do not benefit from any semantic autonomy. They merely refer to the occasional deterioration of the causal laws based on the unilinear model that goes from the most known (i.e., the most quantifiable) to the least certain. The history of research into noise pollution mentioned above provides an excellent example of such a pseudo-multidisciplinary approach.

Almost 30 years ago, Pierre Schaeffer engaged in a sustained critique of the indiscriminate extrapolation of the physical model to the fields of musicology, the psychology of music, and the everyday discourse of musicians, educators, and cultivated audiences. However, the construction of the intelligibility of sound phenomena on the basis of the signal only applies within the limits of the experimental field generating this type of knowledge.[17] Other approaches to the comprehension of listening in general, and listening on site in particular, are just as legitimate. Everything depends on the perspective from which the phenomenon is observed. We cannot always say, "In the beginning, there was the signal".

This epistemological reversal opens the doors to a plurality of models of understanding whose distinction is merely formal (in terms of a field of knowledge or a discipline) and whose exploration of the whole results in a definition of a central object that is no longer the signal, instead the phenomenon. The development of knowledge of the perceptible environment can be represented using a radial diagram (Figure 2.1). A situated ambiance consists of six complementary modes: it is a physical signal, a spatio-temporal form, a percept, a representation, an element of a code or norm, and an instrument of social interaction. Each of these modes requires specialized and non-equivalent means of investigation, analytical methods, and languages. But does this use

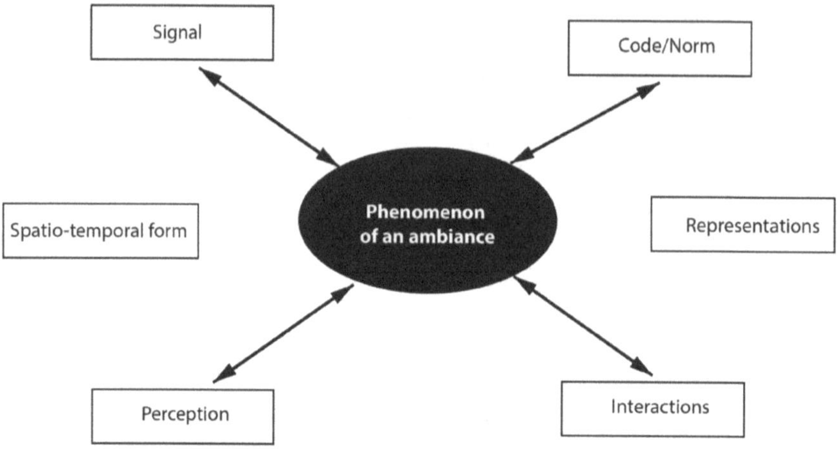

Figure 2.1 Modalities of a phenomenon of ambiance on site.

of converging approaches remain simply multidisciplinary, and does it cultivate a proximity of fields of knowledge only for speculative purposes?

Figure 2.2 represents how the diversity of two of the better-known approaches that are frequently set in opposition to one another achieves an operational and thus interdisciplinary sense of complementarity. In this approach, the specificity of paradigms and kinds of observation and experimentation are respected. The interaction sees five points of contact: on site, during the collection of the results, in the production of transversal concepts, in the development of models of intelligibility, and in the application of interdisciplinary tools.

At the common site, the interaction is expressed as the expansion of the list of observable objects, depending on what electively emerges: this surprising measure will refer to this morphological examination or that investigative question that was not provided in the initial protocol. This or that remark by a user will encourage the creation or diversification of such a measure.

The interdisciplinary comparison of results is all the more useful if it is made during each research phase, thereby permitting a progressive refinement or correction of hypotheses.

The production of transversal concepts is never easy or quick. Our initial intuition may lead us to make an extensive review of the extent and comprehension of the term that does not always arrive at the definition of a concept. It takes as long to approve the means of applying the concept in all the relevant disciplines and kinds of situations where it can apply.[18]

The development of integrating models of intelligibility may take various forms: interdisciplinary theory, procedural or declaratory modellings, a critical review of importable models, or the conception of typical methodological approaches.

Lastly, the test of *applying interdisciplinary tools* on site is particularly well-suited to the methodology of architectural and urban design, in which

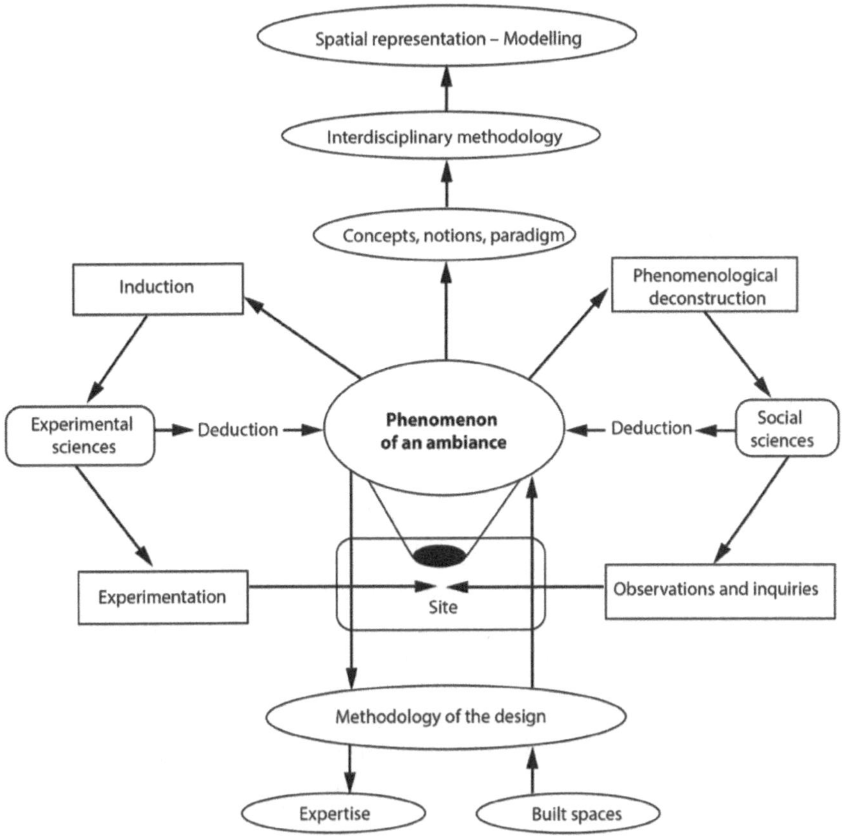

Figure 2.2 Interdisciplinary analysis of a phenomenon of ambiance on site.

the link between society and research is patently clear. The field of knowledge and expertise in urban planning and architecture is a plentiful source of questions for interdisciplinary research. At the same time, the practical valuing of the tools must be considered a social duty and a chance to increase our understanding of phenomena in both factual and operative terms.

Thinking in terms of process

The third remarkable aspect of the growth in ways of approaching the architectural and urban environment is *the gradual transition from thinking about the state of things to thinking about process*. The environmental sciences, a sector that concerns itself with urban, physical factors (in terms of air, heat, sound, and light) have, often unwittingly, privileged fixist, simplistic attitudes that are all too widespread in government technical offices, as we mentioned above ("All you have to do is insulate", "You just have to fix the glare", and so on). In this sense, the issues of the milieu that better consider the relationship of living beings to their environment would be more productive than viewing the environment as a question of prioritizing isolated, physical states, in which the measurable signal used to analyse reactions remains stable. For example, the notion of "environmental quality" that we would like to see replace that of pollution when addressing our healthcare and political concerns should itself be understood as an ongoing process of dynamic qualification produced by the competition of a variety of physical and human components. We could even say that a city, which is by its very nature a field of permanent competition between a number of dynamics, embodies what we mean by "environment" to a tee. Is it not the oldest, the most archetypal, and most complete model of a perpetually integrating control of space, of physical and biological constraints, and cultural and technical assets that humanity has to offer? In such a case, city and environment are actually two faces of the same dynamic reality.

The sensory environment and architectural ambiances

The picture we have just sketched out briefly seeks to show under what conditions we may consider the city to be an environment. But which *environment* are we talking about? From the interdisciplinary and contextualized perspective that we have constructed, the notion may expand past its current understanding, which remains heavily connoted by the main characteristics of the physical model we mentioned earlier. Coming from American scientific language, such an environment means strictly the ensemble of physical, biological, and social conditions that act on living things. It would be pointless to work our way back through the language to rehabilitate the prior definition, which in French, since the twelfth century, referred to the surroundings of phenomena, the *local context* that lies at the core of our thinking.

The particular configuration of our current notion of environment both selects aspects that can be incorporated into the model of the natural sciences and also extends to include any phenomenon of disturbance. At the same time, as it has been mediatized, it has become very polysemic and overdetermined, confusedly mixing together notions of urgency, health, global thinking, and advocacy. Thus, environmental activism, though declaring an occasional interest for the local, too often prizes this reductionism by hypostatizing the disturbance at hand, but almost always ignoring the social production of the representations that designate a disturbance as such. The few works on notions of the environment in past centuries show that the global emergence of one particular disturbance as opposed to another is not necessarily due to the physical or technical quality of the signal, rather the sensibilities of public opinion, the type of social relations, and class behaviour.[19]

As considered by architectural and urban planning research, the notion of environment thus entails a two-fold gap.[20] More broadly, its comprehension is open to several models of intelligibility. More narrowly, its scope extends to phenomena whose quality is apprehended by the senses with enough precision to enable three acts: the effective perception of variations (the dimension of individual experience), the incorporation of aesthetic and axiological representations (a contextualization of the phenomenon), and the integration into the processes of instrumentation of social cohesion (collective references and meaning of the phenomenon).[21] We will add a fourth criterion to these three that directly concerns us: the identified aspects of the environment are the ones on which architecture and urban forms have a structural impact, meaning that their perception includes the morphological modalities of the place and that these components systematically form an integral part of the process of design and planning.

There are two prominent subjects of environmental knowledge that remain peripheral for the time being: water and air pollution, neither of which meets these criteria. Thus, the perception of taste and smell of water or air remains a poorly developed indicator in our culture, one that is hard to predict on a practical level, in terms of their chemical reality and physiological effects. Thus, water pollution may generate reactions of collective claims, but it is not a sensory medium for interpersonal exchanges. The organization of the urban environment obviously has an effect on such forms of pollution, but architects and urban planners are not primarily responsible for controlling it.

What new terms can evoke these major differences? We propose two: the "sensory urban environment" and "architectural and urban ambiances". When we talk about the sensory environment, we are developing and diversifying a region, the "human dimension", which is understood within the general concept of environment as referring to a behaviourist framework.

By using the adjective "sensory", we refer more broadly to the overall domain of perceptions and usages of a certain time and place, it being understood that the configuring, active dimension impacts all of our perceptual constructions, from specifically physical data to individual, cultural, and social factors. In accordance with the on-site approach to the urban environment that we described earlier, this meaning only has one defect, namely that it works better for the interdisciplinary analysis of everyday situations, for a natured city, rather than for an articulation between architectural theory and practice.

There is another notion that is better suited to a naturing city, to the urban space as a production: the (plural) term of "architectural ambiances" or "urban ambiances". Lighting, thermal, and acoustic ambiances are both sensory and produced phenomena. In the professional and academic field of sciences applying to architecture, the concept is also closely linked to a notion of control (e.g., "controlling ambiances"). Understanding an ambiance has three parts: science and technology, knowledge of daily uses and representations, and, finally, a quotient of art.

If the first field concerns the progress of instruments and models from the world of applied physics, and if the second evolves with changing attitudes in the humanities and social sciences towards the observable, is the third part just a mystery? What role does the quotient of art play?

The production of ambiances refers to art and expertise in two ways: too often implicitly in any ordinary architectural gesture, and as a sign of expertise in the current development of the sensory aspects of urban development. In the practicing of architecture and urban planning, any overall plan, any formal composition, and any choice of material necessarily entails lighting, acoustic, thermal, and tactile effects, the whole of which is generally mastered only within the limits of regulated technical obligations. The rest, meaning anything that we cannot classify as a disturbance is almost always ignored, deemed insignificant, or shrouded in an aesthetic judgement of only what is visually apparent. The rest is supposedly unknown and unpredictable, as an ambiance's successes and failures only appear once the building has been completed.[22]

As a result of this magical process, and due to the need to manage the built structure, there is a second field for the production of ambiances in which the quotient of art is unevenly spread. In the most current practice, urban ambiances are handled and managed by diverse actors who are more reproducers of knowledge than actual creators, i.e., a company placing exterior insulation over an old building, a city employee installing lampposts, an electrician setting up a temporary sound system for a clearance sale, or a public transportation system that tolerates busking musicians – a far cry from the new actors in what we might call "urban scenography". The staging of a city in terms of its lighting and sound was initially the prerogative of land artists such as Christo or Max Neuhaus, but it has now begun to be integrated into the overall process of urban development, where it provides a sense of the imaginary and the spectacular.[23] In France, Nicolas Frize has transformed the practice of composer in residence by working directly on the space and with the population of the city hosting him. In transposing the aesthetics of his teacher in stage lighting, Henri Alekan, to the treatment of facades and of monuments, Laurent Fachard has turned cities into epiphanies of light.[24]

The notion of architectural ambiances more readily incorporates this quotient of art, of knowing how to produce an urban space, which remains foreign to the semantic field of the environment. On the other hand, the plural of this concept, *ambiances*, reveals a non-negligible disadvantage. As things currently stand, talking about an *urban ambiance* in the singular does not correspond to anything that can be analysed precisely, except as a succession of sensory registers, each sense entailing a logic of observation and specific tools. And it excludes any kind of interdisciplinary interaction. We have knowledge in the fields of thermal ambiance, acoustic ambiance, lighting ambiance, and even olfactory ambiance (in which people have begun to show an interest), but we still lack a trans-sensory theory to cover all the various disciplines that touch on perception and the variety of sensations. Nor is work possible on binary interactions as conducted in physiology and ergonomics, the famous, hypothetical theory of correspondences, or the existence of general, explanatory models that presumably apply to all – or almost all – of the senses.[25]

At the end of a thorough interaction with environmental research, architectural and urban research yields two notions that are unquestionably more precise than that of the environment. One, the "sensory environment", corresponds better to urban research, while the other, "architectural ambiances", is better suited to investigations into the knowledge of the production of built spaces. A predilection for one or the other notion depends on one's momentary research or teaching goals, but it only makes sense when one acknowledges their necessary complementarity. If certain works on cities have remained confined to a compulsive repetition of the subject of disturbances and pollution, and if the teaching of mastering ambiances in architecture has too long and too often remained strictly a matter of mechanically applying technical norms and formulas, this is precisely because of a lack of interaction between an interdisciplinary knowledge of the sensory environment and a theoretical and methodological thinking about ambiances.

The contribution of works on the sensory environment and on ambiances to architectural and urban research

Research into cities considered as a sensory milieu has diversified and fertilized a part of the vast field of environmental sciences, and the inverse is also true. What new contributions have works on the sensory environment and on ambiances made to architectural and urban research?

One of the main difficulties of research in urban planning and especially architecture has always been the distinction between a knowledge of an "architecture" architecture and of an "architecturing" architecture, between an analysis of architecture as an object of history, semiology, psychology, sociology, epistemology, aesthetics, and applied physics on the one hand and an analysis of the processes of producing the built space on the other. Among the various reasons for this impasse (which would take up too much space to analyse here) is the competition of prerogatives, which places knowledge and speculation on one side and art and expertise on the other. Though maintained by polemical tensions, this dualism is not symptomatic. It is the very nature of the endeavour that is in question, as if architecture could only access the real on the basis of a rupture, of a qualitative leap from the known to the unknown.

It seems that the above distinction between the two complementary approaches to the built environment hints at the possibility of re-establishing a dialogue between art and technique, talent and competencies, and knowledge and expertise. By limiting itself to a closely circumscribed aspect of the urban phenomenon, through a broad interdisciplinarity practiced by a community of researchers that is as small as it is open, for being such a new field, research into the sensory environment and architectural ambiances offers several means to change how we see things, overcome certain impasses, and produce transversal and useful notions and methods, both in practice and in theory.

On the other senses in architecture

Since the Renaissance, architecture, like other forms of reasoned knowledge, has remained mainly within the sphere of the visual.[26] The eye became the primary sense, thereby grounding the idea (*dessein*) within a drawn reality (*dessin*), and also fundamentally connoting the language of analysis and influencing the rhetoric of architectural design.[27] Seeing is automatic. From a plan to a model, from a drawing to a worksite, the production of built forms depends so much on our vision that we forget how hegemonic this sense is for us.

Returning to the initial state of things, in the order of perceptions, research into architectural ambiances encounters a collection of signals and *sensata* that are considered equal. Inquiring into the nature of a non-visible form, following the logic of hearing or smell to describe an inhabited area, and wondering what a thermal, olfactory, or aerodynamic landscape composition is like – nothing could be more scientifically legitimate.[28] And yet, nothing could be more unsettling for the certainties of architectural and urban thinking, as there is no necessary orthogonality in a sound space, an olfactory space, or a thermal space. Nor is there a mandatory congruence between visual, acoustic, or tactile limits, which are very different from one another. The spatial reference, as defined by Newtonian characters, loses its explanatory relevance for other sensorial processes besides sight. Moreover, we can demonstrate that what we see is not entirely subject to a Cartesian sense of space.[29] And yet, it is so difficult to imagine representations of the tactile, the thermal, or the olfactory, which would obey a logic unique to each sense! How to get an architect to admit that a sound form is first and foremost a question of time? "And how would I draw *that*?", he/she would say.

These resistances illustrate to what extent visual metaphors have permeated our manner of conceiving the built space.[30] Is it delusional to want to re-establish this *balance of the senses* that a vernacular architecture or an intuitive building not subject to a visual hegemony appears to have respected more fully?[31] We must at least jolt the attitudes of architects and urban planners who too often forget that there are other senses. The minor senses can be rehabilitated in three steps: the first concerns a critique of the stereotypes that dominate design; the second involves changing our cognitive approaches to the built space; and the third addresses architectural creative imagination.

In the practice of building, we have to demonstrate, with supporting studies in hand, the formal, social, and economic potential and advantages of preventively integrating all elements concerning our comfort, starting with the planning phase. The economic and social costs of building badly, which damage the health and well-being of users, always exceed the foreseeable and possible surcharge of construction procedures. The argument that people in charge of architectural and urban projects often make so as not to exceed the lower limits of comfort required by the regulations is based on the problem of having to increase the size or quantity of materials. The effectiveness of an intelligent organization of the space and of a design of forms that encourage a particular acoustic or thermal effect is not taken into account according to this logic.

How can we make this exercise possible? That is the task of information and especially education. It is a long-term project because it requires changing a cognitive attitude that was forged during the Renaissance and which impregnates not only architectural studies but our entire educational curriculum as well.[32] For example, if we remain on a purely theoretical level, learning about the properties of sound environments does not garner much interest. A true understanding of what we should call a hearing of the world, in the sense of *Weltanhörung*, and a consolidation of knowledge in this area are predicated on a pedagogy of listening, on reforming our listening habits, which are, most often, attitudes of non-listening. Without such a preliminary measure, we cannot expect an architect to be able to conceive of sound forms in the same way as visible forms when the morphology of a buildable space emerges.

Without incorporating this sensory experience, which must be applied to each of the senses, the acoustic, aerodynamic, olfactory, and tactile attributes of space will merely be add-ons to the visibility of the created form. And the imprint of the forgotten senses[33] can never be prevalent, nor can it contribute effectively to the form's production. The third task of research is to facilitate the ordinary exercise of a constructing imagination that is as inventive when it comes to hearing, moving, smelling, and relating to the air as it is for seeing. This intentional and desirable multisensory approach to all kinds of architecture will not derive simply from the specialized fields of expertise used in exceptional architectures such as opera houses, auditoriums, exhibition buildings, or experimental solar houses. Ordinary architecture must, according to its means, invent suitable processes and systems for creating comfort.[34] An ethno-historical inquiry remains to be carried out on the intuitive and inventive knowledge and expertise developed out of a natural as well as a cultural need to master comfort in ordinary architecture.[35]

On architectural interdisciplinarity

As a cumulative body of knowledge, architecture is multidisciplinary by its very nature, whether its various fields of knowledge simply coexist or are compared and arranged against a common epistemic background, the latter being rarer. And yet, the process of design at work in any architectural project entails a true interdisciplinarity, meaning a competitive and dialectical confrontation between different fields of knowledge and processes.

What do practitioners have to say about this? A completed project is a matter of experience, expertise, and negotiation between *constraints*, some of which are random and depend on the project's own history. The updating of architecture's epistemic background is already old. It began to emerge, bit by bit, during Romanticism, each time an author would consider the question of the essence of this atypical "art" that mixes function and beauty.[36]

The cognitive functions have been explored to a far lesser extent. What is the nature of this mysterious unity that brings together very different bodies of knowledge within the architectural act, and which involves specific areas of knowledge and expertise? Even if architecturology remains in its infancy, we know, thanks to Philippe Boudon, that we have deliberately placed the question within the field of epistemology.

One of the most important architectural questions that have thus been reformulated is knowing by means of which processes the epistemic background is able to articulate the collection of different areas of architecture-related knowledge. The environmental question that requires us to address the relationship between the human dimension and the physical dimension of contextualized phenomena resembles the architectural question in this regard. This is when we see that the descriptive and taxonomic approach applied to architectural forms under various kinds of typo-morphologies is far from able to ground architecturological knowledge by itself. Like our memory, architecture is merely a polyparium of images. Once we leave the domain of the visible, types and classes lose their relevance.[37]

The trend of French research into architectural ambiances often shows that architectural morphology is based on the structural interdependencies between the built form, the perceived form, and the represented form. It is impossible to explain the context of one of these modalities without referring to the other two. There is nothing mysterious about the three operators of this modal conjunction, which are the perceptible physical signal, the corpus of norms, rules, and codes, and the instruments, functions, and uses assigned to the built form. Current studies of the complex, interdependent functioning of these operators should help us better understand the nature of our sensory environment and, more broadly, the interdisciplinary processes involved in the act of producing a built space.

On the need-to-know usage

Knowledge of practices and uses of inhabited space has been a major topic in urban and architectural research since the 1960s. The methodological influence of these works on observing the sensory environment has been significant, as we mentioned before. And we expect new knowledge of architectural ambiances to contribute to the consolidation of knowledge of the usage of architecture, which will happen in two ways: one, by restoring the function of physical factors in our perception of space and, two, by highlighting the tools for perceptual constructions and for social cohesion on site.

We recall that French research in the second half of the twentieth century has approached the issue of perceiving the built space in three very diverging ways: one, through the development of an experimental psychology that has often sought to overcome its own behaviourist origins by looking for openings in comprehensive psycho-sociology[38]; two, through the development of a phenomenological vein cultivated with varying degrees of success by researchers in architecture and urban planning, as the use of the logical and verbal *organum* of the post-Husserl field – however seductive it may be – does not dispense one from the long process of reductions, of a meticulous attention paid to those things without which the phenomenological enterprise remains pointless[39]; and three, through certain works of urban sociology that address collective representations of things such as landscape, habitat, public space, and heritage.

Research into ambiances does not seek to reconcile these various currents, where the strength of the methods and the interest in the results are in part due to the specificity of the selected viewpoint, to a deliberate, antagonistic position. But as we have emphasized, the observer of an environmental phenomenon on site is unable to abstract either the physical signal, its treatment dictated by one's perception, or its collective component. In other words, the observer must develop problematic articulations on which other areas of research have skimped.

Thus, by rising up against the physical substantialism, caused by a reductive behaviourism that did not assign any causal value to the collective, urban sociology has rightly re-established the autonomy of social causalities. Its counterpart is the enduring mistrust of anything concerning the materiality of space. Now, this means forgetting that behavioural sciences have evolved since the 1970s. In fact, we hear more and more often in the physiology of perception that the physical environment is nothing more and nothing less than a list of properties offered to an infinity of perceptual constructs.

Once again, the development of a sociological reasoning at the heart of the urban phenomenon could produce an idealism of social cohesion. As if the order of the analysis ultimately generates the conditions of daily existence, and as if knowledge, in forgetting the senses, produced a reality devoid of sensations. And yet, social interactions would not have any effect without the sensory-motor tools that embody these very interactions. How shall we think, given that sensory markers will not have any effect on the *message* except as a hylomorphism that can no longer be sustained, and which basic observation disproves?[40]

Research into the sensory environment thus considers the physical signal not just as matter for perceiving space but also as an instrument of social cohesion within space. It develops this issue in two, complementary ways. The question "How can the social be sensorial?" is the subject of the praxeology of perception and observation (which has not been pursued to any significant extent until now) of the tools for social relationships.[41] The question "How can the sensorial be social?" has been developed by aesthetics (understood as the theory of the organization of sensations) in two ways that have not been explored to any extent until now: aesthetic experience as action and as shared know-how.

Architecture and cities are not just objects. Their sensory effects act at the core of a node that brings together perception and action, form and function, and individual and collective. Therefore, can architects and urban planners conceive of the usage of built space without knowing how this usage *reconstructs* them? We may presume that, with the progress of work on the sensory environment, the understanding of the relationship between a space's user and its professional/expert will evolve to some degree. The user's sensory environment would thus no longer be envisaged merely as an object of consumption or contemplation or of action in the sense of protest. Knowledge of the *modalities* of daily practice becomes a priority. So, how can one settle for the current set of conceptual tools and explanatory models? Updating interdependencies[42] requires paradigmatic tools such as the effect, motif, the ambient, and the transcript[43] and procedures for virtual explanation such as digital modelling.

On the rehabilitation of aesthetic knowledge as an interdisciplinary field

As the subject of many works of art history, architecture has not been the subject of much aesthetic discourse in the Baumgartenian sense of the term, namely the theory of the beautiful as grounded in an autonomous sensibility.[44] It seems easier to address the production of built forms and classify genres, styles, and schools than to analyse the *reception* of an architectural or urban work. Hegel's few, partial pages in his *Aesthetics* instead form part of a theory of the symbolic hierarchy of forms in function of their prescriptive nature. Few authors have ventured into the topic of the perception of architecture. They have done so as part of a critique of the preconceptions underlying the judgement of taste, but for the purpose of reforming representations of the city (Camillo Sitte) or monuments (Aloïs Riegl). Psychologists and philosophers of art such as Erwin Panofsky, Pierre Francastel, Sir Ernst Gombrich, and Nelson Goodman have made illuminating remarks about the link between perceptual organization and the representational structures of architectural works. In architecture itself, we have theories as famous as those of Venturi and Christian Norberg-Schulz, the phenomenological vein of which brings us closer to the sentient. Yet, an approach based on sensory modalities of perceiving the built form has not been broached any more extensively by psycho-sociological approaches that prudently rely on a semiological interpretive base of user representations rather than on the direct observation of sensory experience.[45] The distance between the aesthetic experience of the architectural work and the sensory experience of the built object is significant.

What is the cause of this difficulty? First, the fact that architecture exceeds the domain of art, understood in its strictest sense, in two ways, because, to preserve Vitruvius' triadic definition, it adds constructive technicity (*firmitas*) and functionality of usage (*commoditas*) to beauty (*venustas*), which is instead the sole subject of artistic contemplation. Therefore, it is improbable that architectural aesthetics will progress until we remove two obstacles: first, understanding the relationship between formal creation and technique and, second, understanding the relationship between built and inhabited form.

Because inhabitants are both users of functions and practitioners of sensory forms, the built environment merges two questions at the heart of contemporary aesthetic debate: knowing where the borders of the artistic lie and knowing what the spectator's role is. There are still too few tools in the field of general aesthetics to elucidate the paradoxical articulation between form and function. A strand born of nineteenth-century German aesthetics and continued by Wörringer, Bauhaus, and those who took inspiration from it provides a set of hypotheses awaiting further development. As to the role of the spectator, this topic, before its appropriation (and frequent reduction) by the sociology of art in the twentieth century, was the subject of fundamental observations regarding sensorial and cultural construction starting in the nineteenth century. This gave rise to a kind of spectator pragmatism followed more similarly to us by the Prague School, the Constance School, and the Warburg Institute.[46]

There is a significant point of convergence between these two, seemingly different topics. The analyses all point to issues concerning the contextualized organization of perceptions. What are the logics inherent to each sense? How is the inter-sensoriality by which we access the unit of form articulated? How does the directed use of the senses (by social and cultural factors) participate in constructing the sensory (and not just what is represented)? And these questions are actually debated against the background of a shared hypothesis, namely that our relationship to the sensory and formal environment should be seen as an exchange, as a constructive circulation between the given and the configured, the felt and the acted, and the perceptible and the representable.

It is in this specifically environmental direction[47] that work on architectural ambiances finds theoretical support and should be able to enrich architectural aesthetics by considering phenomena at their root, namely according to the trajectory that we have sketched out in the previous sections. The built form must be considered in four ways. It is a set of physical properties distinguished according to the nature of each kind of signal. It is a sensory configuration situated in a specific place. It is the expression of a culture of experts that merges function, art, and technique. And it is the expression-reception of a culture of inhabitant users. We have seen that the study of each one of these modalities refers to fields of study that have already been formed. The interest of a programme developed within architectural research lies in the transversal approach to these attributes of the built form.

By practicing the broad interdisciplinarity that the on-site study of forms demands, architectural aesthetics can escape the confines of morphological and stylistic taxonomies. By becoming a place where people ponder the meaning of the senses, it can become fundamental knowledge for architectural and urban research. By bringing together the variety of sectors of knowledge used to observe the sensory environment and by focusing them on the issue of form, it helps specify the meaning of a question about practice that has become essential to architecture in the present day: what exactly is the process by which built forms arrive at reality?

Notes

1 The former was first espoused by Philippe Boudon.
2 Research into architectural modelling was developed starting in the early 1970s by two pioneering laboratories: CERMA (the Nantes School of Architecture) and GAMSAU (the Marseilles School of Architecture).
3 The work performed at this time was carried out almost entirely by people in the fields of engineering and medical research.
4 There are various reasons for this: a growing sensitivity to noise, the transfer of querulousness to another object, direct victimization, and the socialization of conflict, among others. See P. Amphoux *et al.*, *Le Bruit, la plainte et le voisin*)"Noise, complaints, and neighbours"] (Grenoble: CRESSON, 1988), 2.
5 Work by the CEP (M. Perianez, F. Desbons) and by the CSTB (D. Aubree).
6 See *Douze ans de recherche sur la gêne due au bruit* ["Twelve years of research into noise-related disturbances"], C. Levy-Leboyer (ed.) (Paris: IRAP/SRETIE, 1987).
7 See the work conducted by the Centre for Research into Sound Spaces and the Urban Environment (CRESSON, URA CNRS, Grenoble School of Architecture) between 1980 and 1986.
8 This trend of investigating environmental factors derived in particular from the Urban Plan and then from the interdisciplinary programme on cities ("PIR-Ville" by the CNRS).
9 Therefore, this is not the totality of components of the urban environment, simply those whose qualities can be perceived by the senses with sufficient precision that variations 1) are conscious and can be described using words; 2) can support constructed aesthetic and axiological representations; 3) can be integrated into the tools for social cohesion; and 4) can be related, at least intuitively, to spatial morphologies (see "The Sensory Environment" below).
10 Except for specialized, single-function acoustic situations such as roadway noise, listening to performances, and listening at work, among others.
11 Works on the sociology of usage in the thermal domain were developed at the CSTB by Philippe Dard. The architectural research laboratory that, while developing a thermal modelling applied to architecture, tries to integrate usage values is CERMA (URA CNRS) at the Nantes School of Architecture, directed by J.P. Peneau.
12 A ground-breaking study was conducted in the early 1980s by the late Abraham Moles: "Des fonctions de la lumière dans la ville" ["The role of light in the city"], in *Communication on the National Days of Light, Paris, September 1980,* in LUX, Ed. No. 111 (1980), 10–25. See also the records of the conference "Les facteurs lumineux du sentiment d'insécurité" ["Light-related factors in the sense of insecurity"], J.F. Augoyard (ed.) (Grenoble: CRESSON, 1990).
13 Thus, the well-known works of Alain Corbin and of landscape historians (K. Pomian, J.D. Hundt, A. Roger, and Y. Nakamura).
14 The development of research into the senses seems closely tied to demands for a technical or scientific mastery of a situation of immediate concern; thus, alongside some rare general studies into smells, there is a large amount of grey literature into the olfactory aspects of rat behaviour. The reason for this is the optimization of the selection and control of the strains of SPF (specific pathogen-free) rats that are so useful to biology laboratories.
15 Hence, the approximating distortions inflicted on the qualitative, which needs to be reintroduced to the complex analysis of human relations. We can refer to the history of the decibel, which has not managed to be a true physical measurement, or hybrid notions of *loudness* and *noisiness*, which constitute the traces of a subjectivity invoked in the last part of treatises on applied acoustics. On the epistemological criticism of improbable scientific reuses, two works are particularly

revealing: I. Stengers, *L'invention des sciences modernes* ["The invention of the modern sciences"], (Paris: La Découverte, 1993), and G. Chatelet, *Les Enjeux du mobile, Mathématique, physique, philosophie* ["The mobile between mathematics, physics, and philosophy"] (Paris: Le Seuil, 1993).

16 Hence, the trend is to give priority automatically to medical research.

17 P. Schaeffer, *Treatise on musical objects*, translated by Christine North and John Dack (Berkeley: University of California Press, 2017).

18 For example, the notion of sound effect formulated at CRESSON in 1978 to account for the interdisciplinary nature of on-site phenomena took seven years to verify. The validation of its applications and predictive quality began in 1988 and is still underway.

19 See the works of Alain Corbin; P. Goubert *et al.*, *Du Luxe au Confort* ["From luxury to comfort"] (Paris: Belin, 1988); O. Balay, *Discours et savoire-faire sur l'aménagement de l'environnement sonor urbain au XIXe siècle* ["Discourse and knowledge of the urban acoustic environment in the nineteenth century"] (Grenoble: CRESSON, dissertation, 1992); and G. Chelkoff *et al.*, *Bien-être sonore à domicile. Architectures du logement et potentiel de confort sonore* ["Acoustic well-being at home: housing architecture and the potential for acoustic comfort"] (Grenoble: CRESSON, 1991).

20 We can see the notion of "built environment" in the milieu of architectural research. A Swiss research laboratory has even included it in its name: the "Institute of Research into the Built Environment", also known as the IREC/EPFL. More precise than just "environment", this concept, which refers to what is built and planned, could be used as a comfortable synonym; nevertheless, its inconvenience lies in its predominant connotation of construction sciences and in its prioritizing of a factual over an operative dimension.

21 These three criteria should be understood as effective potentialities. *Sensata* do not necessarily or always form conscious perceptions that can be used to form social connections and that are affected by our judgement. But they must have this potential, depending on the occasions and the rules of daily life, and this effectiveness must be directly or indirectly observable.

22 This is obviously just a defence mechanism, as none of the justifications that are usually provided, including those pertaining to a project's budget, hold up to closer scrutiny. The lack of education and dissemination of knowledge of this subject has led several research laboratories (CERMA, CRESSON, ISITEM, LASH, and CSTB) to create a department of "Architectural and urban ambiances".

23 There are rare but spectacular cases where a city has integrated lighting into its urban planning, for example. The city of Lyon's "Light Plan" implemented in the late 1980s thus marries the pleasant with the functional and the secure.

24 On the reuse by urban planners of the techniques and know-how of the performing arts and on the first analyses of this parallel between the square and the stage, please see the conference *Scenography and Public Spaces*, organized at the Arche de la Défense by the Ministry of Urban Planning and Public Facilities in 1993.

25 Thus, key concepts of gestalt theory such as figure-ground remain much more relevant to sight than to hearing.

26 We should specify that while "pan-optism", understood epistemologically, propagated quickly throughout the world of the arts, technology, and letters, as explained by such diverse authors as Rudolf Arnheim, Eugenio Garin, Erwin Panofsky, Ernst Gombrich, Marshall McLuhan, and Michael Baxandall, the cultural predominance of the visual in the West is rooted in Mediterranean Antiquity. The history of this remarkable cultural construct has become the subject of a number of works, especially on the topic of viewing landscape.

27 On this practical logic of urban and architectural design as coming out of a rhetoric, please see our analyses and annexed comparative table in *Pas à pas* ["Step by step"] (Paris: Le Seuil, 1979).

28 This committed approach presupposes a sensory immanentism of thought (see our developments of this hypothesis in J.F. Augoyard, "La sonorización antropológica del lugar" ["The anthropological sonorization of places"], in M. J. Amerlinck (ed.), *Hacia una antropología arquitectonica* ["Towards an architectural anthropology"]. (Jalisco, Mexico: University of Guadalajara, "Jornadas Académicas" series, 1995.)

29 For more extensive developments, see J.F. Augoyard, "La vue est-elle souveraine dans l'esthétique paysagère?" ["Does sight reign supreme in landscape aesthetics?"], *Le Débat*, No. 65, Paris, Gallimard (May-August 1991): 51–59.

30 Works like those of Grégoire Chelkoff show not only what happens when we analyse a public space in acoustic terms but also that we can view the public space as a configuration that escapes from the Newtonian model of space. G. Chelkoff, *Entendre les espaces publics* ["Hearing public spaces"] (Grenoble: CRESSON, 1988). G. Chelkoff, J.P. Thibaud, *Les mises en vue de l'espace public* ["The visual presentation of the public space"] (Grenoble: CRESSON, 1992).

31 This was one of the questions asked at the conference *Au-delà du paysage moderne* ["Beyond the current landscape"], organized by the research association at EHESS, under the direction of Augustin Berque, Paris, Centre Pompidou, 19–20 October 1990. See the report in Le Débat, No. 65, Paris, Gallimard (May-August 1991): 51-59.

32 We can readily see this through a structural analysis of the linguistic modalities used in the pedagogical instructions and injunctions imparted during the first years of schooling. There is a predominant visual connotation in the verbs, concepts, and descriptive adjectives used, even when talking about listening, smelling, or touching.

33 *The imprint of the senses* is the title of a report on the knowledge of the interaction between physiology and psychology, in which the author, Roger Ninio, developed the prevalence of "perceptive reason" as a major topic: Jacques Ninio, *L'empreinte des sens* (Paris: Odile Jacob/Le Seuil, 1989).

34 Worth examining in this sense is the interesting experience of the "Houses of the five senses" built by L'Effort Rémois a few years ago.

35 This is one of the major topics addressed by the international network *La Qualité Sonore des Espaces Habités* ["The Acoustic Quality of Inhabited Spaces"], which is managed by CRESSON.

36 We recall that the question of architecture taking a seat among the fine arts extends from Hegel and Schopenhauer to Christian Norberg-Schultz by way of Paul Valéry, Bauhaus, and Le Corbusier.

37 An interesting attempt was made in 1980 by an acoustician and an architect-researcher to test typo-morphological relevance in the field of sound. The application of types and relationships of scale is only valid for a closed form of habitat (like a housing block with an interior courtyard) and for insulation ratios. As regards the rest, sound phenomena do not correspond to the classification categories and boxes proposed by architectural typo-morphology. See P. Panerai, D. Tatutesco, *Diversité sonore et formes urbaines* ["Sound diversity and urban forms"], (Paris: ADRS/UP3, 1983), 2 volumes. On this critical analysis, see J.F. Augouyard *et al.*, *Sonorité, Sociabilité, Urbanité. Méthode pour l'établissement d'un repertoire des effets sonores en milieu urbain* ["Sonority, sociability, and urbanity: a method of establishing a list of acoustic effects in urban milieus"] (Grenoble: CRESSON, 1982), 155 pages + 2 cassettes (120 minutes).

38 An example of this comes from the last summary provided by the late Elizabeth Dumaurier, *Psychologie expérimentale de la perception* ["Experimental psychology of perception"] (Paris: PUF, 1992).

39 Remarkably, the most ambitious attempts at a phenomenological analysis of the perception of the built space were made in the 1960s and 1970s by authors who

took their theoretical inspiration from various sources. Lefebvre never fully forgot his Marxist influences, and Abraham Moles always remained close to experimental psychology, from which he drew some explanatory models. As to urban phenomenologists who have truly paid close attention to things and to people, we must admit that they are quite rare. Few attempts in this area can rival those made by Pierre Sansot.

40 For example, it is in the name of this immanence of the sensory in social interactions that the figure of the foreigner is defined using significations and values that vary greatly, depending on the sensory marker: sight, hearing, smell, touch, and so on.

41 This was the focus of Jean-Paul Thibaud's work in the early 1990s: J.P. Thibaud, J.P. Odion, *À l'écoute du chantier. Des productions sonores aux modes de prevention* ["Listening to worksites: from sound production to modes of prevention"] (Paris: Plan-Construction, 1990); J.P. Thibaud, *Le baladeur dans l'espace public urbain. Essai sur l'interaction sensorielle de l'interaction sociale* ["The walker in the urban public space: an essay on the sensory interaction of social interactions"], dissertation, (Grenoble: University of Grenoble 2, 1992).

42 See above, *Les Raisons écologiques de l'urbain.*

43 This development of paradigms was specifically carried out as part of the collaboration between CRESSON and CERMA (n. cit.). The notion of transcript was proposed by the architect Bernard Tschumi.

44 There are countless publications that, rightly or wrongly, describe the formal characteristics of a given architecture as works to be classified in the name of "aesthetics of architecture". This is again the issue of the prevalence of style, not of the user's aesthetic experience. In this same category are those who advocate for architectural programmes that apodictically decree the rusticity, modernity, transparency, or urbanity of buildings, whose foreseeable sensory effects on daily usage shall not be commented on.

45 See the semiotics of vernacular styles in Sylvia Ostrowetsky, *L'imaginaire batisseur* ["The constructing imaginary"], (Paris: Librairie des Méridiens, 1983) and the notion of "ordinary aesthetic competence" in the work of Marion Segaud (dissertation, University of Nanterre, 1989).

46 See Ernst Gombrich, *Art and* Illusion. Pantheon Books, 1960 (Chapter III: "The role of the spectator") and H.R. Jauss, *Pour une esthétique de la reception* ["Toward an Aesthetic of Reception"] (Paris: Gallimard, 1990).

47 The term "aesthetics" refers more specifically to the meaning given by Kant at the beginning of the *Critique of Pure Reason.* It entails an elementary aesthetics concerning any sensory organization. The bringing together of such an aesthetic with ecology, understood as a field of knowledge, has already been done by several authors (see, in particular, Ernst Gombrich, *L'Ecologie des Images*, (Paris: Flammarion, 1982), and his commentary on this topic in the preface).

3 The ambiance waltz

Pascal Amphoux

When we say the word "ambiance", we should say "ambivalence". In French, *ambiance* is construed as either a sign of youth, of having a great time, of modernity, and of plurality or as a sensation of a density that is temporal, unifying, and singular, as an "atmosphere" or a "climate". The former dresses up reality, assigning it a content that critics may claim is false or illusory, because it wears a mask of representation. The latter is instead perceived as reality stripped down to its essence; it reveals reality or constitutes its touchstone. Nevertheless, critics will allege the threat of mystification, that this too is another form of illusion.

Ambivalence is, thus, the movement from one to the other, the breaking point between reality and illusion – as the word itself indicates in its apparent hesitation between its true etymology from the Latin (*amb-ire*, to go around) and a false derivation (*ambi-*, both at once). So, let us get caught up in the word and amble with the sole ambition of going around a notion that goes around itself.

The ambiance we know, the ambiance we live, and the ambiance we sense will constitute the three components of our refrain. Three functions, three concepts, and three percepts will, respectively, describe its content of knowledge, life, and meaning. These will be the three tempos of each of these three moments: a ternary rhythm. This is the representation waltz.

Three functions

The market. Economic history. Yesterday's luxury has been replaced by today's comfort. Today's comfort will give way to tomorrow's ambiances

What at one time was the prerogative of the rich and powerful now belongs to the people, even to those living on the lowest rungs of the ladder. Who today would live without a fork, a bureau, soap, a mirror, heating, a refrigerator, or a car? We are surrounded by such technical objects; they are, quite literally, "ambient" objects. They have become ordinary, commonplace. This is technical comfort. But our standards and requirements for comfort have increased. Heating is now adjustable, the refrigerator has compartments, the

DOI: 10.4324/9781032655307-7

washing machine can be programmed, and the car has become an extension of our living space more than simply a means of transport. Technical comfort has been replaced by the comfort of control. But control is in turn slowly becoming outdated. As programming and remote controls offer their services, and the synthesis of sounds and images opens up the fields of the virtual and the unprecedented, management, security, and communication will be – so we are told – better and better. The comfort of control will be replaced by a comfort of ambiance. The popularization of a "control of our environments" is at hand.

We should be happy about this evolution, but we should also underscore its totalitarian penchant and latent terrorism. The supremacy of the market for ambiances only works at full capacity when, by entering into a symbiosis with the individual feelings of those on whom it acts, it fills them with the secret, unmentionable dread of being one step behind.

The fact remains that ambiance, now in vogue, appears to be a principal feature of the market for comfort. The primary function of ambiance is, thus, economic.

Desire. Psychology of the environment. Ambiance fascinates us, and it has to be fascinating

How so? By getting us to look at an object (i.e., the ambiance) other than the one being manipulated (which constitutes the reality of a practice). People sell you on the design of a domestic appliance, the silence of a vehicle, the softness of a chair, but they don't tell you anything about the reality of the practices they induce. You are shown a function, technique, or code, but you no longer examine the use. And the recent proffering of a supposed "expert comfort" that relies on the naïve belief in the omnipotence of telecommunication, tele-management, or tele-security no longer says anything about the content of what is being communicated, managed, or protected. You can communicate anything anywhere to anyone anytime. What could be more fascinating than that?

Second, ambiance is superlative: it's more, it's too much, and even more than that. It is always "the most" this or "the most" that. Over-qualify your interior and go beyond everything you have ever known – it's tango lighting. It exceeds anything you could imagine – and even for those who can't stand it, it remains excessive. But for everyone else, what could be more desirable? Ambiance becomes an object of desire that belongs to the domain of excess that Clément Rosset defines not as real, instead as an "addition to the real".

Lastly, in its supra-realist excess, ambiance also tends to render everything "real" as undesirable – social practices, direct contact with the material, and a body against another body. Thus, the music of one's furniture conceals the noise of the neighbourhood, ambiance lighting dematerializes one's vision of things, and the household robot distances the homemaker's hands from culinary gestures. There is nothing magical about this principle; by enchanting the individual, it renders the world disenchanting. By making itself desirable as a representation of everything that does not exist, it helps render as undesirable the reality of everything that does exist.

The other, the superlative, is the unreal. Ambiance must always catch our eyes – or ears. This is what makes it invisible. It lies at the core of the individual's desire – and that is its psychological function.

... And the spectre. A sociological hypothesis. The production of ambiances is a way of transfiguring the world and coming to terms with its unstable, fragmentary, and fungible nature

Is this not so much a reflection of an individualist society as what Marc Guillaume calls a spectral society, in which individuals tend to play different roles depending on the context where they find themselves (a spectre of attitudes), to blend several personalities, fictional and real (a ghostly spectre), and to indulge in ephemeral, random, and fungible relationships on micro-networks, the traces of which are usually invisible (a spectre of invisible lines or relationships)?

In a world that has lost its sense of communal unity and its symbolic reference points, the production of ambiances is a means of adapting one's daily practices to society's "spectral" evolution. Ambiance, as it is being marketed, semiotizes, breaks down, and materializes what it is not (light, sound, and social relations). However, we can at the same time say that it makes us lose our sense of reality, which it recomposes and dematerializes through the technologies of illusion, virtual spaces, and real time. Ambiances generate interactions that had become impossible between the functional world (which is utilitarian, programmed, and codified) and the realm of our feelings (which is imaginary, ephemeral, and informal).

A blend that is odious – or melodious. The production of new techniques and "ambiances" becomes a social issue. The sociological function.

Three concepts

The motif. The notion of domestic or urban spaces currently follows an adaptive logic: one analyses the functional relations between humans and their environment and "adapts" the space to the "function", the material to the physical body, and the software to the user's psychology, thus adhering to the tried-and-true precept attributed to the architect Louis Sullivan that "form follows function"

But when we look more closely at our domestic habits and ways of inhabiting, we see that use exceeds function, that it diverts it or reappropriates it. One blocks a vent, fills in a door, and reconfigures one's space. One takes a short-cut across the lawn, one jaywalks, and one's private sphere encroaches onto the public space. The space certainly determines some of the uses, practices, and gestures, but the opposite is also true; there is always a co-determination between use and space, a reciprocal fashioning of gesture and tool, a mysterious coalescence between daily ritual and the spirit of a place – and this mutual appropriation is what gives rise to a particular ambiance. How then should we describe this blending in anything other than mystifying terms, this incarnation of a "spirit" in the materiality of a space? We must substitute the insufficient concept of function with the ambivalent one of a motif.

This is the reciprocal gaze of a man and a woman expressed in the centred layout of their apartment, the symmetrical placement of two chairs, two coffee tables, and two family portraits, in the dual arrangement of household objects, knick-knacks, and house plants: living room versus dining room, black furniture versus white furniture, toolbox versus golden slippers, and barometer versus a framed piece of embroidery. It's the face-to-face motif.

It's the management of a family's intimate relations within a space expressed in the strategies for concealing doors and windows, in the arrangement of curtains and shades, and in the filtering of the light and the concomitant progression of order and disorder within the rooms. It's the light-and-dark motif.

And it's the emptiness of a young, expectant couple expressed in the emptiness of their living spaces, a padded cradle in a room for a child who is not there, a lone chair in a cold living room, and a few foldable chairs stacked on an empty veranda. It's the emptiness motif.

Face-to-face, light and dark, emptiness... The motif is literally what motivates the domestic ambiance, giving it form and enlivening it. The motif expresses the ambiance within the space.

The figure. The notion of contemporary domestic or urban life currently obeys a taxonomic logic. Different ages or socio-professional categories determine "lifestyles", a marketing invention from the 1970s that generates stereotypical uses, spatial configurations, and furniture layouts that we know all too well

But the behaviour of the average inhabitant cannot be reduced to any one of these categories, because he/she mixes them, intertwines them, and transforms them. The lifestyle identified for an individual determines their behaviour, but the individual also shapes this conduct and immerses it in the pool of a collective, social experience. So, we must substitute the concept of lifestyle with the notion of the figure.

It is the weight of all that solitude, indigence, and illness, embodied in behaviours of submission or resignation, in the unstructured, contradictory expression of a complaint made to no one, in the obsessive, confused tallying of a neighbour's deeds and misdeeds, in the impossibility of explaining the damage suffered, or at least, the loss of means to prove it.

> I am the only one to have stairs above my bedroom, just 1.4 meters above my head, machinery above my apartment, piano playing every day, a dog that barks. I hear everything my neighbours say in the duplex and in the 17th arrondissement, crystal-clearly. It's intolerable.

The figure of the victim.

It's the weight of an enforced bourgeois morality, well-meaning, and moderate, embodied in the habits of self-coercion and inhibition, in the rituals of anticipation, routines, obligations, and redress, and in an ironic, aggressive manner of expressing oneself that relies on euphemism and a shared ideology. "When the music goes olé olé, it's ooh la la!" "I believe we have the right to peace and quiet, and to respect. These people who live abnormally should be removed and resettled in places more suited to them". The figure of the bourgeois.

And it's the weight of the normalization and codification of a technologized society, embodied in relations that are mediatized, formal, and consciously nit-picking, in the systematic recourse to institutional third parties whenever there is a problem, in a meticulous, technical, and measured manner of expressing oneself to ensure that the construction or neighbourly relations comply with the regulations. "Madame, I would like to know why the case submitted to the prefecture is being handled by animal services instead of the noise abatement department? There is something mysterious about it that I just don't understand". The litigious figure.

Victim, bourgeois, litigious. The figure is both the dominant personality that puts a face to the place where one lives (it could also be the village idiot, the old lady on the fourth floor, or the neighbourhood baker) and the schematic form that summarizes the collective behaviour and imagination (i.e., the framework that reveals the social bond). The figure is, thus, fictional. It appears as a reconstruction of "characters" that do not exist in and of themselves, but which best demonstrate this or that type of relationship to a place or neighbourhood. Literally, it depicts a particular ambiance, in the dual sense of representing and caricaturizing it. Or rather, the figure embodies ambiance in a character.

... And the effect. The notion of domestic or urban time follows a logic of profits and losses. We are sold cars, computers, and household robots that will save us time, and we are sold ambiances and developments to enjoy them. Time that is limited or not, work time, time spent in transit, or free time – we constantly want to qualify it, fill it, arrange it, furnish it, but we absolutely can never leave it empty

But once again, on an everyday basis, the inhabitant doesn't care that much. He or she may not have any time to spare, but they don't try to save any either. It may appear as if they're managing a time budget, but this concept is insufficient. When we are taken in by an ambiance, in these significant moments that punctuate our existence, we fall under the spell or charm of a specific effect – the third, intermediate concept that, between the fact, the action, and the perception, instead qualifies the temporal dimension of a domestic or urban ambiance. The effect literally brings about the ambiance, actualizing it and being actualized by it.

It's a spatial configuration that is actualized in the nature of an acoustic environment, in the way the sound is made and reverberates, in its being heavily or slightly absorbed, clearly or poorly distinguishable, instantaneous, or persistent. Organic versus mineral: rugs and curtains versus concrete, plaster or glass walls. From a recording studio to a cathedral, an open garden to an urban courtyard, from a Haussmann-era parlour to a functionalist living room, acoustic ambiance is a marker of the place one inhabits. A reverberation effect.

It's a movement that is actualized as a sudden change in ambiance, a drop in the intensity of the light or volume of the sound that may or may not be associated with an obvious change in the spectral envelope, tonality, or reverberation. We round a street corner, enter a lobby, pass behind a screen, and everything is transformed. We have passed into another world. A cut-off effect.

And it's also a state of transition that is actualized in the slow, imperceptible disappearance of a particular ambiance: the fading light at dusk, a train that moves away from us and disappears over the horizon, an urban sound of dissipating, deep notes. Moments of delight or of uncertainty. A cross-fading effect.

A reverberation, a cut-out, a cross-fade. The motif expresses an ambiance in the space, the figure embodied it in the social imagination, and the sensory effect now actualizes it in time.

Three percepts

1 *Fashion. Ambiance is in the air – what I mean is that it has become fashionable. A keyword for developers, a catchword for hipsters, a use-all word for DIYers. Ambiance is all the rage. It is loud and thunderous, young and vibrant. We put it everywhere, we claim to create it, and we want to add more of it – to an ad, the car, to a home. It's declined in the plural. Cool, hard, high tech, hippie, Japanese, bourgeois, or natural – these ambiances have become predetermined, named, qualified, exported, sold, and resold in the markets for lighting, acoustics, and electronic gadgets. A widespread exchange.*

   As fashionable objects, ambiances are nothing more than signs of ambiance; they are emblems of one's lifestyle and home. Instruments of compensation: fill the void, mask the space, and compensate for the absence.

2 *Invisible. Ambiance is in the air – and now I'm saying it's invisible. It floats, envelops, meanders. It designates "a certain" climate – whether of familiarity or strangeness – revealed by a perfume, a light, or rather, a mixture of things. Its invisibility certainly does not prevent us from "seeing", but we can't look at it, observe it, or catalogue it, because we are immersed in its bath, be it appropriate or inappropriate, and because such an immersion prohibits any external consciousness. Everyone is taken by and lives in a deep unawareness of their interior. My home is mine because*

I no longer see it. That irresistible feeling of being in your own space. Every object is in its place, every gesture has been integrated, every event recorded in the traces of an invisible memory – a slow build-up that comes to express a specific ambiance.

As the expression of a lifestyle, ambiance is an index of a specific milieu. It is no longer an instrument of compensation, rather one of foundation: mooring the world, creating the milieu, and establishing the identity of the place and of the person who inhabits it.

3 *Melodious. Ambiance is in the air – the last thing I want to say is that it is melodious, strangely melodious. Even without a sound to kindle it, it always expresses the harmony of a world, it elicits a sense of agreement, a muffled resounding in the deepest part of our being. Ambiance is always a refrain, a rhythm that becomes independent, a unique composition – a fragile yet dynamic balance between heterogeneous components (colours, postures, and sounds) that cannot be analysed using reason. It is a sensation that defies understanding – because it has no purpose. We sense it even before we feel it. We reach for it without being able to name it, and those who have not experienced it aren't able to conjure the slightest idea of what it is. It is unnameable and inconceivable, detached from all representation. Fable-like, ineffable. Utter singularity.*

Out of fashion, the ambiance becomes an icon of the world – for those who understand it, it is the world itself. No longer a compensation nor foundation, instead a detachment: leaving representation, "making landscape", touching reality.

Conclusion

Reprise and acceleration. 1. Ambiance is fashionable: the forces of market, desire, and spectre haunt its definition. 2. Ambiance is invisible: the motif, the figure, and the effect reveal it. 3. Ambiance is melodious: the fashionable, the invisible, and the melody (!) suggest it.

Dizziness. 1. In three steps, knowledge gets reality and illusion to dance together as a couple. 2. In three steps, design gets the visible and the invisible to dance together as a couple. 3. In three steps, perception gets the odious and the melodious to dance together as a couple.

Panic. A frenzied waltz. Symbol, index, icon. One, two, three. General, particular, singular. One, two, three. We know its trend. We can imagine its invisibility. We perceive its melody.

Ambiance is in the air...

Bibliography

Amphoux, P., C. Jaccoud, and R. Perrinjaquet. "Dictionnaire critique de domotique". In *Dictionnaire de domotique, Milieux et techniques,* edited by W. Zaniewicki. Paris: Le Creusot, Eyrolles, 1990.

Amphoux P., M. Leroux et al. *Le bruit, la plainte et le voisin*, Grenoble: Cresson, Ecole d'Architecture de Grenoble, February 1989, 2 volumes.

Amphoux P., B. Galland, and F. Galley. *Habitat solaire à l'usage.* Lausanne: IREC, Ecole Polytechnique Fédérale de Lausanne, Report No. 89, February 1990.

Augoyard, J. F., ed. *Répertoire des effets sonores.* Grenoble: Cresson, Ecole d'Architecture de Grenoble, going to press.

Deleuze G. and F. Guattari. *What is philosophy?,* translated by Hugh Tomlinson and Charles Burchell. New York: Columbia University Press, 1996.

Guillaume M., *La contagion des passions, essai sur l'exotisme intérieur.* Paris: Plon, 1989.

Lyotard J.-F., *The Postmodern Explained to Children*, translated by Thomas Pefanis. Eastbourne: Gardners' Books, 1992.

Rosset, C., *L'objet singulier.* Paris: Minuit, 1979.

Part 2

Ambiances, perspectives, and positionings

4 The thickness of the present

A brief ethno-phenomenological reflection[1]

Véronique Nahoum-Grappe

"Atmosphere... it's the only word in argot that I don't understand. Atmosphere, atmosphere! Do I look like a freaking atmosphere to you?!"[2] This famous response needs to be pronounced with a Parisian accent to recreate its own atmosphere, that of a pre-war, working-class Paris where banter was used to deal with tragedy. The surprising success of this phrase, tied in part to the working-class use of a sophisticated, almost scholarly term, is interesting because it designates a non-object, the opposite of an event. And paradoxically, with this phrase, the heroine defines an atmosphere of her very own: something empty and vague that surrounds her and that her lover rejects. It is thus an offence that touches her identity, vaguely and incomprehensibly.

A non-object, a non-event

Since Jean-Paul Thibaud's pioneering work, the list of works in the humanities that have contributed to the scholarly construct of "atmosphere" has grown,[3] but the subject remains uniquely opaque and fluid, slipping through the mesh of our theoretical nets. It demands that we try to define it each time. An ambient atmosphere does not have a shape. We don't "see it coming", unlike an event, a performance, or a pleasure, which "arrives", bursting on the stage of the present and marking a break between a "before" and an "after". When we begin to sense that we are perceiving an ambient atmosphere, it's already there, enveloping events in its halo, but it is not a non-event occurring at a precise moment in time. Its diffuse mass of signs eludes us when it is over without ever "slamming the door shut" on its way out. It often remains absent from our memories and writings about an event. Without stable limits or any visible representation or intent, its level of reality is imprecise. The finer points of its social production and the conditions for the collective's ability to perceive it have yet to be elucidated. It lives and dies in the present, its very own "today", like the aroma of a wine that exists only as one drinks it.

The possibility of an *ambient atmosphere produced by a socialization of a situation* is hard to objectify as a real, collective, and social fact, despite

DOI: 10.4324/9781032655307-9

countless, magnificent literary descriptions most often read by specialists as subjective creations and textual performances, but not as sociology. Ungraspable, slipping "through our hands" and between set categories, blurry, provisional, and overarching, its theoretical "substance" is moreover confined by the metaphor of the atmosphere and clouds. It seems totally unmoored, like a balloon swept away by the wind. Initial attempts at grasping this notion often fall to the temptation of rhetorical flights into a misty sky of dubious figures. And yet, ethnologists in the field who may find themselves immersed in its powerful bath of signs and who have a hard time differentiating between their own ennui, their bad moods, and the greyness of the empty, dreary day around them may at times feel the invasive power of an ambient atmosphere creeping under their skin. "Oh is there concrete all around or is it in my head?" as David Bowie once exclaimed.[4] In David Bowie's *Rock'n'roll Suicide*,[5] one has to hear the lyrics with their music, listen intently with one's eyes closed, and perhaps lean one's head back – a sensual, symphonic listening process in which concentration and abandon converge, without which "atmosphere" as an object does not reveal itself. This becomes an entirely different version of reality, in which time, one's fingers, the cigarette, one's mouth, the surrounding walls, the rug – everything that encloses us and the "I" – become one and the same heavy, oppressive matter stuck to the walls, collapsed onto the floor, tossed about by the waves of the beat. And the chiasmus between the force of the soundscape and the density of the atmosphere seems to give birth to a subject-less gesture, that of lifting the other cigarette from one's hand to one's mouth. During the sequence, everything concerning the non-verbal communication, sounds, and forms will evade the narration's memory. The actual evidence of the ambient atmosphere will remain locked within the "extreme present"[6] of the situation. This extreme present can be transported elsewhere through a musical creation and the time it takes to listen to it, or a poem and the time it takes to read it. The extreme present of the scene forms a loop that evades linear time, a social sphere that is the only place where the ephemeral wave of an atmosphere can be experienced. This leaves successful aesthetic works – musical, pictorial, and written – which can serve as a "vehicle" of the extreme present in its entirety, that is, thickened with all the signifying power of the "social air" one breathes only for the time that this present lasts.

The extreme present

> We see that the hardest moment of our existence to know is the current one, today. The "yesterday" represented by elements intervening in their genesis can be analysed with relative ease.[7]

For Furio Jesi, who wrote in the 1960s and 1970s, the "non-currency of the elements of the structure" of a phenomenon, its "yesterday", constitutes the subject of the social sciences, which can invoke their vision of probable causes in this context. But the "today", the extreme present of a celebration, a revolt,[8] or any other scene that is in the process of happening – including the one summoned every time a certain song is sung – remains out of these potential analyses' reach. To ensure that an atmosphere is distinctive enough that one can experience it at a given moment, we must therefore put into perspective the "non-currency of the elements of its structure", i.e., the theories of its plausible "causes", as for the aroma of a good bottle of wine; the gustatory adventure of a single mouthful, both precise and astonishing, owes its aromatic characteristics in part to the "non-currency" of the earlier process of vinification, and even before that, to all the historical and technical parameters that impact this product's fabrication.

The particularity of the object "atmosphere" is that it is undoubtedly the product of an overall, heterogeneous system of intricate causalities that sometimes entail a vertiginous non-currency, a radical heterogeneity, some of which are too tenuous or fleeting to be isolated (or even perceived); this system must be represented as a whole, but its time of expression is discontinuous and mortal, confined to the present.

The aroma of the social sphere

The social sciences are required to predicate atmospheres as an objective social fact that is hard to describe, which is born of a process of the *socialization of a situation*, and not the moody projection that attempts to describe this fact. It is only grasped in the sphere of the present social situation, within its "today". It consists of what is happening, of what has not yet happened. "That which has not happened is as much a part of history as that which has".[9] The suspense engendered by a sense of plausibility is the phenomenological definition of the present. Thus, atmosphere is a drawing whose colours spill over and "leak" into this immediate future, a matrix of the possible sequence of events in a situated format that remains nailed to its extreme present. In the literal and physical sense of the term, it expresses a situation, to the point of identifying it exactly, on the condition of its almost immediate collective perception in real time.

We may find an example of an enigmatic but effective perception in a recent book by Geoffrey West, a theoretical physicist interested in the mathematics of cities. He studies the effects of demographic scales on extremely heterogeneous social forms. What is fascinating about this scholarly enterprise is its attempt to mathematicize the urban fact, which largely remains outside our area of interest, with the following exception: West calculated that a walker's pace in a city of more than one million inhabitants is twice as fast as that of inhabitants in a city of a few thousand people. This acceleration is a constant tied to a whole series of other curves situated at very heterogeneous scales and which concern sociological and economic factors of an incredible diversity "integrated with one another by and within a holistic framework of parameters into systemic interactions"[10] at different scales.

The issue here is not the author's (very precise and sophisticated) methodology, nor the universalizable validity of his overall results. I would instead like to comment solely on the scientific possibility of echoing a deep-seated cliché in our culture, namely the unsettling acceleration specific to the urban space,[11] the vague notion that "everything is faster or too fast" in the city. People in cities "dance faster than the violin", to quote an eighteenth-century French expression.[12] Careers, love lives, trends, innovations, failures, falls, divorces, consummations, excesses, differences, lack of words, illnesses, steps, and basically "everything" in a big city "goes too fast!" In Emily Brontë's *Wuthering Heights*, one of the narrators, Mr Lockwood, freshly arrived from the big city in the wild places where the novel unfolds, says that, before he left the city, he never imagined that a love could last for more than three years, but that here, in this deserted countryside, he instead saw that one love could last a lifetime. As if the spatial overcoming of differences of all kinds, which exist in close proximity within the urban space, somehow set all possible spheres of meaning into motion, spinning ever more quickly and dizzily.

Our culture envisions big cities as sites of an increasing de-socialization, either in the direction of a progress of a civilizing "enlightenment" or "vertiginously", heading down a deep slope. How many times has an ethnologist of contemporary rurality heard say that Paris "moves too fast"? This deep-seated *topos* does not constitute our subject as a cultural stereotype – whose appearance should be dated (sometime during Ancient Rome) and its extent analysed – instead taken as a sign of an exact perception of a parameter constituting the atmosphere of the big city. As such, this holistic object is physical, historical, and social. The cliché is denoted exactly by this specific pace that is perceived as "being too fast", by an entire series of established practices, conducts, and facts, to the point that, in Paris, even pedestrians walk faster without ever having decided to do so, whereas in areas of low demographic density, or in the Swiss mountains, even the red lights seem to last longer, and the tick-tock of the clocks seems slower. This worn stereotype constitutes a relevant collective knowledge when all the urban accelerations in all domains are related to (and not explained by) the demographic curve, the size of the city, and its scale – and Geoffrey West's book is in fact titled *Scale*.

Impossible to describe as an unequivocal, established fact, this sphere of parameters without any common measurement between them – other than acceleration, which spins and gravitates more as the city becomes bigger – undoubtedly contributes to the characteristic atmosphere of a large, capital city. This entails the fabrication of a non-visible but perceptible specific trait of an urban atmosphere, an objective, global, and holistic fact produced by the entire "systemic framework" that Geoffrey West's mathematical calculations are able to represent, and which is powerful enough to influence the pace of pedestrians and, at the other end of the spectrum, produce a deeply held stereotype that ties the big city to speed. The term "frame" connotes an internal articulation, not just what surrounds a picture, as it does in French (*cadre*).

The familiar language of "ready-made" clichés and phrases constructed collectively and historically over time (aside from any preliminary semiological planning) can thus be redefined not just as easy and banal, but as an *exact* "atmospheric" perception of a truth that goes lost in the thickness of things that have not yet been understood, and which appropriately situates the "everything moves too fast" within the space of big cities.

The aesthetic grasp of the political

Is the presence of centres of power in large capital cities related to this "hectic" urban rhythm? On a socio-psychological level, a heightened pace is also that of meeting a lover, of running to meet them, the social impatience of "seeing", of crowds that rush to theatrical premieres or to large squares where the powerful show themselves. The urgent desire to "know what is happening", to hear "the news", used to make city dwellers run – and, in old black and white films, people would open their newspapers as they walked on New York City's sidewalks. These are cultural images that circulate, and which can also be studied in quantitative terms. Will the difference in speed as a holistic sign that distinguishes the atmosphere of big cities from that of less dense, "slower" rural spaces disappear due to the current technological changes in how we exchange information? I consider this hypothesis to be plausible. During this first third of the twenty-first century, clicking on a screen suffices for this purpose, both in the big city and the countryside, where lifestyles have urbanized.

There is another potential acceleration that comes from "up above", the "Faster! Faster!" ordered, yelled by the warrant officer, the *kapo*, the second-in-command, the coach. It is an order or mandate for efficiency in the production of scores, and it goes hand in hand with a denial of the subjectivity of the social actor thus obliged *to hurry*. He has to sing louder, run faster, work more quickly, respond immediately, and so on. He is thereby cut off from his own inner self, from the warmth and the empty spaces of his private thoughts, from the permanent dialogue between him and himself. This mandate for speed is a hostile occupation of people's subjectivities; this sinister, diabolical, forced pace is a weapon for subduing people's minds. In certain places, the explicit extent of a power that commands the rhythms of the body, sounds, and the forms of the dominated subjects' mobility creates the conditions for an "unbreathable" atmosphere.

Regardless of any links between the nearby and centres of social domination and the fact of a collective acceleration in big cities that was not decided, the enactment of an imperceptible acceleration of the pace of pedestrians' steps presupposes a specific cognitive competency on the part of social actors. The neurosciences would be able to decipher this enigma of an individual and collective quasi-mimetic perception of a general but minor fact lacking in importance for the urban landscape that acts on the intensely personal way in which people take their next step at one pace instead of another. The imperceptible force of persuasion of "knowing how to walk in a city" is due not just to the mimetic imitation that has often been upheld in literature since Gabriel Tarde[13] but also – because it had to invent itself – to the effectiveness of a specific form of perception of a forceful, distinctive atmosphere that the *socialization* under way produces. Do all these striated diagonals of information, all these glances that meet and avoid each other, this entire mass of signs mixed together in the heart of a big city in such a peculiar manner, and this entire collective fabrication of an urban atmosphere perhaps constitute an initial form of an effective non-verbal acculturation and adaptation to the stage of the present that sketches out, in its own opaque and effective way, a possible canvas of *how to make here*? How do we anticipate what comes next? The unconscious freedom of the body comes into play when the self enters a given situation due to the instructive, immediate, blurry, but also complex and symphonic effectiveness (i.e., with all signs playing together) that the dominant atmosphere produces and emanates in a situation.

The minimal freedom of the subject

While we wait for the neurosciences to explain the different mechanisms at work in perception, phenomenological works[14] are helpful for our holistic and invisible subject, which remains hard to grasp, fleeting, but entirely distinctive and objective (like the aroma of a wine, as we mentioned). I take for granted that there is an effective mimetic competency to human perception. Immediate and holistic in case of a change in place, it can also remain suspended and dissolve in silence – look without seeing, express itself in the backs of signs of secondary things, and contribute to the perception of an atmosphere. Distraction is key here for the discreet presence of peripheral segments on the sidelines – just like, in the middle of the forest, the noise of a faraway large plane with its deep, grumbling harmonics, a barely audible powerful purr turning into vibrations, and troubling bass notes that imbue the reveries of the solitary walker with a hypothesis devoid of any interest, that of a large plane flying somewhere high overhead. The subject's entire subjective and physiological presence is captured and perhaps changed, unbeknownst to him, in a way that is impossible to define; he is fleetingly rooted in his contemporaneity, the purely acoustic signal of which reminds him that he is not walking in the eighteenth century.

At the other end of the spectrum, there are perceptions that are very focused on their target. Literature provides powerful, well-written observations of random perceptive fixations:

> Are you familiar with that state of mind where, alone with a cat busy licking her kitten on the rug in front of you, you watch the entire operation so intently that if the kitty cat were to miss even an ear, this would send you into a flight of rage?[15]

In a novel, an ethnologist devours the meaning of well-written reflections; he doesn't wonder about textual prowess. Mr Lockwood, our informant in *Wuthering Heights*, explains why the perception of a scene that does not concern him can move him to the point of irritation. A concentrated, slightly hypnotized perception may lead to a strange sense of excitation, in which the viewer is "captured", seized by something more than just watching a performance. He "puts something of himself in it", to quote the sociologist Jack Katz ("What is worthy of our attention is not just what we do, but the extent to which we put ourselves into it").[16] If there is something that contributes to the fabrication of a collective ambiance, it is the personal investment of actors in the scene that they are in the process of experiencing. The same festive, religious, political, or scientific sequence of events, the same crowds in a public square, in which everyone either gets bored and "is elsewhere", well protected within their own inner lives, or in which each and every person is rapt together and "gives themselves over fully" to what is happening, these events may resemble one another in terms of gestures and acts, songs, and postures, but they may be extremely different in terms of the atmosphere they generate.[17] It is not insignificant for an ethnologist that, within a given group that has assembled in the same place for a given collective act, the overwhelming majority of participants feel themselves to be intimately cut off from the world, unaffiliated, separated from their own contexts.

> Here I am in the world, wandering about alone in the middle of a crowd that means nothing to me, like a man who has long suffered from an accidental case of deafness, and whose avid eye stares at all these mute beings moving agitatedly before him. He sees everything, and everything is denied to him... he suffers the silence of all things as the world churns loudly around him.[18]

A temporary lack of affiliation, this "withdrawal", as Goffman called it, is not necessarily depressive or pathological, as in moments of affective abandon or in the classic sensation of estrangement. It may also be purely sociological, in the sense that the social actor, though present, remains withdrawn and silent within himself. Nothing draws him "outside" himself. He does not put anything of himself into his surroundings. He remains hidden at the bottom of his own inner world. He is "elsewhere", "as the world churns loudly around him", which does not prevent him from accomplishing what he must robotically do. The relationship between "atmosphere" and the form of the actors' presence or absence is thus tied to the question of ritualized obligations, permitted duties, and forms of acceptance or distance in relation to the constraints of the moment, which always depend on the existing structures of political domination. How much does one pretend? The physical response to this question contributes to the construction of the atmosphere in progress; everyone and each individual feels the quality of the presence of the others, be it artificial, "worldly", intense, vibrant, or full. Perceiving this, even fleetingly or blurrily, sometimes allows for a qualitative grasp that is vital to not dying of ridicule or suffering problems. An error in "atmospheric" perception, between a false sense and transgression, can be costly.

The entire intensity of what is happening comes into play. An atmosphere is not just signifying (in its own, non-verbal, and suspended way). It may or may not increase in intensity, overwhelm, and invade the inner space of participants, or it may fray into tattered shreds if everyone is elsewhere, distracted, or bored. The same waltz might not produce the same headiness; this doesn't change anything about the scene, instead everything about the atmosphere.

The minimal freedom of "not being there", within oneself, is a determinant fact that is only manifest through the most tenuous of signs. In a classroom, an inspiring teacher sees relaxed foreheads, bright eyes, and the class time passes very quickly, in an ambiance of positive tension; the students listen intently, and any atmosphere prevailing in that classroom during that hour is marked by that particular thickness, a brightening of the space that traverses the subjective intimacy of the students and the teacher's gaze, tinging the walls, the objects, and filling the entire air of the classroom. Or the teacher, too entrenched in his or her chair, no longer notices the boredom creeping over the youngsters' faces, or the oddly soft look of their distracted daydreaming, and time stands still for everyone. The intimate, un-self-conscious freedom that everyone has to inhabit one's own body and inner space in this or that manner concerns the specificity and the very subject of a collective ambiance.

Sensing, grasping, and translating into words

Describing the perception of an atmosphere is a very delicate undertaking, especially because the perceptive mechanisms involved can be extremely diverse in relation to an object whose signs may in turn be situated at utterly disparate, heterogeneous levels of reality. This holistic, perceptive configuration of an ambiance also entails a floating, peripheral, and barely conscious vision of an infinitesimally small detail that captures and focuses our attention (for reasons unknown) on a scene that is secondary and external to the main plot, and which has become a fascinating or even appalling spectacle. Also coming into play are the absences, the inductions tied to touch, negative, dark signals, contaminations of laughter or yawning, and the discomfort about the proper distances between bodies, which vary from one culture to the next.[19] I suppose that the several hundreds of millions of neurons in our belly that are by now well described,[20] as well as our neurosympathetic and parasympathetic systems, should also be considered in terms of the potential for blushing, going pale, experiencing nausea, and jumping and springing in case of a danger (the unconscious local circuit of fear). Lastly, the hypothetical a-perception of infinitesimally small vibrations and different electromagnetic levels – from the static electricity circles surrounding the body, perhaps inherited from perceptual systems that have fallen into disuse during our phylogenesis – should also be taken into account.

An alert consciousness and the evidence of its immediate perceptual horizon constitute the visible peak of this enormous, complex physiological and neurological continent, which enables an awareness of oneself in the world through the immediate synthesis of one's gaze.

It is impossible here to give in to the rage to conclude, to use Flaubert's expression, and to believe that it is possible to resolve the problem of human perceptive competence in relation to an atmosphere, this hazy, general non-object that glides across and cuts through any social situation in the process of unfolding.

So, how does politics act on these moving layers? The notion of control is useful to any ethnologist seeking to grasp an atmosphere that envelops a social situation in real time, its ambiance, which not only envelops and transverses it, imbuing and covering it, weighing on it and born from it, but which also remains in a state of flux and disorder, depending on how the situation unfolds. The critical moment is its translation into words when it occurs. Words resolve conflicts between atmospheres and often erase discordances that remain virtual and other possible interpretations that have remained in suspension.

The perceptual adventure of diving into an atmosphere as an internal adventure precedes the words. Once the words emerge from the opaque mass of things unsaid (but felt), they highlight certain, select forms in the soup of diffuse signs at the expense of other illegible traits, even a simple mark on a wall, which might also be meaningful. What the words clarify has an immediate, retroactive effect on what was felt in that situation. For example, once an oenologist utters certain words, all the tasters around him or her are absorbed by a kind of internal chiasmus, an immediate and intimate reflection that comes as close as possible to the internal adventure of tasting. They experience a recognition, one that is forceful, cognitive, sensual, and celebratory, of the flavour tasted and named, and still coated with a jumble of signs stuck to one's palate and cheeks. All of a sudden, with the word in one's mouth, the perceptual adventure is triggered and crowned by the right description. "But of course? That's what it is! *Flinty*!"

Glued to its present like its own shadow, its cape, and its halo, the ambient atmosphere is such an inextricably tangled mix of facts and signs that it froths until it dissolves at any given moment, *unless it is written down*. The damage, the distortions it can produce – like the sublime rescues it enables – often escape narration, yet they undoubtedly remain at a low frequency, acting on the interactions (or lack thereof) in progress.

The weight of hierarchies

The "political" sphere – understood here[21] generally as a form of an established collective organization that configures hierarchies of meaning, values, and positions – overlooks and runs through any social or historical situation about to happen. The minimal organization of power institutions vertically envelops all life sequences in progress in a given place ahead of time. For example, the hierarchies of social rank are forceful in a situation in progress as in a modelling structure ahead of time; they play their game of different weights, different intensities within the very heart of the virtual circle of possible interactions that remain in suspense.

In our social world, different concentrations of investment are made close to the throne, the altar, the street, around the body of a beautiful woman appearing in the doorway, and the guard in a school courtyard. A more or less democratic regime a priori frames and organizes the possible interactions between human beings present in a given situation through this immediate perceptual competence for each of one's own and others' positions, undoubtedly inherited from a childlike ontogenesis that is enriched through learning experiences. Not only does it invade the suspense of what has not yet happened with a plethora of expected constraints, but it also covers the most material framework of action with its own historically constructed semiological and aesthetic cape; in its choice of style in its palaces, its cemeteries, its public squares, and in these sites filled with emblems, it sets unforeseen atmospheric traps that are triggered before the utterance of a single word or the performance of any ritual or action. The inhabited social space contains a value of the centre, overloaded with social views and the aesthetic attempts of the powers that be, as well as a loss of the margins, which are abandoned. On the stage of the present, social separations are experienced in the street and on walls. Even in places that remain untouched by the political gaze, this abandonment produces its own ambiance that is as distinctive as the more emphatic one of preferred sites.

For an ethnologist of atmospheres, the presence of the political on the stage of the present at first concerns its distinctive aesthetic style. For example, it is odd that the more totalitarian the power, the more gigantic its choice of architectural scale.[22] Manipulating the effects of scale and form, appropriating luxury objects, lighting effects, controlling the soundscape with silence, or the blaring of trumpets constitutes an attempt by the powers that be to produce an atmosphere that invades the inner space of distracted passers-by, to surreptitiously enter into their intimate sphere through a visual sign, a choice of form, or scale that draws people's eyes, be it a staggeringly tall column, an immense flag, or an enormous square.

All these formal tactics for invading a space through an aesthetics of domination testify to this will to invade people's consciousness by mobilizing bodies and touching the slightest unconscious freedom of being oneself and inhabiting one's body, *on which the atmosphere first acts*. Restrictive regulations in the military, in athletic training fields, or even in the "urbane world" are "atmospheric" war machines for touching and disabling one's calm of belonging to oneself, deep within one's thoughts, in the empty space of one's own subjective intimacy. Actual equality in action, despite the different hierarchical positions of the social actors on stage, concerns, on the stage of the present, this minimal freedom of barely significant gesticulations – moving, smiling, laughing, sighing, scratching one's head, etc. – the collective effect of which contributes to the fabrication of an ambient atmosphere. A mandatory straitjacket that ranks the positions and beings from up on high abolishes this corporal freedom, one that fabricates a minimum of ease in one's sense of belonging to oneself. Changing places also represents a change in atmosphere. And due to the effectiveness and relevance of a comprehensive, almost immediate, non-verbal, and mimetic perception of signs, this change is perceivable to a passer-by. One can immediately breathe in a poor but lively neighbourhood in which kids from richer but more sinister neighbourhoods come for a night out. Ennui becomes fatal in an urbane salon, a political meeting, a classroom, and a mandatory dance, where the space grows between the visible surface of the stage and the inner appropriation of people's consciousness. This strange, barely visible absence of everyone within themselves reconfigures any atmosphere of the moment, to the point of producing an absolute, collective moroseness that is more social than psychological.

Atmosphere plays with the distance between oneself and the present world

During collective situations in the process of unfolding, unforeseen countercurrents and turbulences may occur:

> There is an atmosphere of ideas. In a courtroom, the ideas of the crowd weigh on judges and juries, and the opposite is also true.[23]

This phrase from Balzac introduces us to a particular situation, a court of assizes,[24] which persists and which is the site of powerful countercurrents in terms of "atmospheric" pressures. The forms and setting, materials and rhythms, and lights and memories evolve in this bath of signs that everyone and each person perceives and fabricates in a different way. The noises, sighs, the denied responses, the expectations, all the culture that predates this moment, the tacit knowledge[25] that acts by omission and is studied by anthropologists, the implicit and consensual moral norms, the figure of the defendant, the judge, or people startled by a noise all express themselves (literally, almost physically) in a symphonic manner during the hearing. At times, "ideas" (the ones Balzac talks about, which are true, collective thoughts) are shared by a majority of the "audience", and this is felt and produces "pressure" that is counter to the one coming from the institution and its representatives.

A sometimes-gripping conflict of atmospheres emerges during the hearings, which an after-the-fact narration will not account for; the sociological force of the "atmospheric" event will be erased. What the audience thinks, "the weight of everyone's and each person's ideas", and which has not yet coalesced into words, expresses itself as a situation, in the hollow space of its extreme present, in the form of tenuous, fleeting noises, signifying movements of faces and bodies, possible and/or constrained, depending on the degree of basic democracy at work as a framework for the proceeding. The infinitesimally small drops of the corners of mouths, the holding of a hand, the rubbing of a forehead, and collective sighs during key moments constitute the non-verbal language of a permanent collective exchange, and this represents an ambiance's form of action. One of the conditions for diverging atmospheres to arise and flourish is tied to this relative freedom (ease) of bodies insofar as they are a non-verbal system for a collective capacity to signify, because I don't think that there is any transmission of thought outside an actual, physical system of signs, even if it is undecidable and barely perceptible. The *socialization* in progress is caught in the powerful trap of a conflict of atmospheres, spaces of suspension that lead to the next thing, tensions without language, and an orientation of consolidated intensities framed by overarching institutional codes.

* * *

The production of an ambient atmosphere during the extreme present of what is in the process of happening can be defined as the major, collective invention of a collective system for exchanging comprehensive information in real time. It perhaps preceded language in its implementation of signifying gesticulations as socialization gradually became the first form of collective life, and undoubtedly the condition for everyone's survival. It enables the anticipation of possibilities that may not come about, but which remain suspended in the present. It reveals their range and thus contributes to the collective, cultural invention of situations. It becomes possible when a space of minimal freedom envelops human presences made of flesh and bone and a perceptual awareness. It is massacred by dictatorial political regimes or in specific totalitarian spaces that are sometimes created in the folds of various worlds (families, parties, villages, etc.). That is the moment when the specific "atmospheric" effort to *socialize* generated by a place finds itself hindered and blocked, as does its real-time conflict of turbulences. Within totalitarian spaces, regardless of their scale, the imposition of a specific aesthetic and semiological cape that is deaf, divisive, and often implacably perfect suffocates the slightest wisp of difference. That is when one is bored the most,[26] but demonstrating that is another project.

Notes

1 The field is that of our actual, familiar contemporaneity; the challenge resides in grasping a social situation in real time (ethnological observation) and pushing empathic mimetic thought to its limits. This "thick description" of a phenomenology that is fundamentally free in its thematic choices obviously takes the form of a mere hypothesis. There are many references. The unpublished courses given at the Collège de France by anthropologist Françoise Héritier provided me with an example of a methodology that was rigorous, expert, and wide-ranging

in its choice of references. The works of comprehensive sociology, from Max Weber and Georg Simmel to Isaac Joseph and Erving Goffman, are too many to cite. Several readings that made a huge impression on me have in my opinion generated the possibility of an "ethno-phenomenology", a discipline that does not yet exist. See Clifford Geertz, *The Interpretation of* Cultures (1973) (New York: Basic Books, 2000). See also the classic description of cockfighting in Bali in Clifford Geertz, Hildred Geertz, *Kinship in* Bali (1975) (Chicago: University of Chicago Press, 1978) and also by Clifford Geertz is *Local Knowledge: Further Essays in Interpretive Anthropology* (New York: Basic Books, 2000 (orig. published in 1983)); "Anti-Anti-Relativism", *American Anthropologist*, 86, 2 (1984): 263–278; *Works and Lives: The Anthropologist as Author* (Palo Alto: Stanford University Press, 1990); *Available Light: Anthropological Reflections on Philosophical Topics* (Princeton: Princeton University Press, 2000); and "An Inconstant Profession: The Anthropological Life in Interesting Times", *Annual Review of Anthropology*, 31 (2002): 1–19. On the comprehensive sociology of Isaac Joseph and Jack Katz, see "Ethnography in 3-D", an interview with Jack Katz conducted by Alexandra Bidet, Carole Gayet-Viaud, and Erwan Le Méner on 26 September 2011 and posted online on 21 May 2013 on laviedesidees.fr; Jack Katz, "On the Rhetoric and Politics of Ethnographic Methodology", *The Annals of the American Academy of Political and Social Science*, 595, 1 (2004): 280–308; *id.*, "Cooks Cooking Up Recipes: The Cash Value of Nouns, Verbs, and Grammar", *American Sociologist*, 43(1), (2012): 125–134. And of course, in France, Albert Piette, a pioneer already in his early works: *Le Mode mineur de la réalité. Paradoxes et photographies en anthropologie* ["The Minor Mode of Reality: Paradoxes and Photographs in Anthropology"] (Louvain: Peeters, 1992). All of these works are heuristic for scholars attempting to address the issue of atmosphere; see, for example, *Méditation pessoanienne. Sciences de l'existence et destin de l'Anthropologue* ["A Pessoanian Meditation: The Sciences of Existence and the Fate of the Anthropologist"] (Paris: Éditions Matériologiques, 2014).

2 Arletty's rejoinder in Marcel Carné's film *Hôtel du Nord* (1938, screenplay by Henri Jeanson), when Louis Jouvet (Mr Edmond) tells her that he can no longer breathe in her atmosphere.

3 See the general bibliography drawn up by Maxime Le Calvé during his 2015–2017 EHESS seminar. I will specify that I use the terms "ambiance" and "atmosphere" interchangeably; their theoretical distinction is justifiable elsewhere but not useful in this instance.

4 David Bowie, *All the Young Dudes* (1972).

5 David Bowie, *Rock'n'roll Suicide* (2004).

6 Concept is expressed in footnote 1 on page 252 of a text by Jean Starobinski, "Mémoire de Troie" ["Memory of Troy"], in *La Beauté du monde. La littérature et les arts* ["The Beauty of the World: Literature and the Arts"], Martin Rueff (ed.), (Paris: Gallimard, 2016).

7 Furio Jesi, *La Fête et la Machine mythologique*, translated into French by Fabien Vallos, and introduction by Giorgio Agamben (Paris: Éditions Mix, 2008), 77.

8 Furio Jesi, *Spartakus: The Symbology of Revolt*, translated by Alberto Toscano, with an introduction by Alberto Cavalletti, (Chicago: University of Chicago Press, 2014 (texts written in the 1960s)).

9 Marc Ferro, sentence spoken during a lecture at EHESS in 2016.

10 Geoffrey West, *Scale: The Universal Laws of Growth, Innovation, Sustainability, and the Pace of Life in Organisms, Cities, Economies, and Companies* (New York: Penguin Press, 2017) ("We need to understand how the dynamics of innovation, technological advances, urbanization, financial markets, social networks, and population dynamics are interconnected and how their evolving interrelationships

fuel growth and societal change—and, as manifestations of human endeavors, how they are all integrated into a holistic interacting systemic framework... and whether such a dynamically evolving system is ultimately sustainable").

11 For a lovely reflection on the contemporary urban situation that talks about this overall acceleration of a set of heterogeneous parameters within the urban space, see Olivier Mongin, *La Ville des flux. L'envers et l'endroit de la mondialisation urbaine* ["The City of Flows: The Opposite and the Site of Urban Globalisation"] (Paris: Fayard, 2012). See also the review of *Tous urbains* (https://www.cairninfo/revue-tous-urbains-2016-2.htm), which attempts to define the urban from the perspective of the humanities.

12 This expression appears constantly during the second half of the eighteenth century, and in *Le Tableau de Paris* (1782–1784) by Louis Sébastien Mercier. See also the historian Bernard Lepetit, *Les Villes dans la France moderne (1740–1840)* ["Cities in Modern France (1740–1840)"] (Paris: Albin Michel, 1988) and Bernard Lepetit, Denise Pumain (ed.), *Temporalités urbaines* ["Urban Temporalities"] (Paris: Anthropos, 1993).

13 Gabriel Tarde, *Les lois de l'imitation: étude sociologique* ["The Laws of Imitation: A Sociological Study"] (Paris: Félix Alcan, 1890).

14 An entire contemporary current of the neuro-cognitive sciences makes use of Maurice Merleau-Ponty's great work, *Phenomenology of Perception*, translated by Donald Landes (Abingdon-on-Thames: Routledge, 2013).

15 Emily Brontë, *Wuthering Heights* (Ware: Wordsworth Classics, 1997).

16 Jack Katz, "Ethnography in 3-D", interview.

17 In my research into both social boredom and conduct of drunkenness, this difference in "atmosphere" proved critical. See Véronique Nahoum-Grappe, *L'Ennui ordinaire* ["Ordinary Boredom"] (Paris: Austral, 1995) and *id.*, *Vertige de l'ivresse. Alcool et lien social* ["The Vertigo of Inebriation: Alcohol and Social Bonds"] (Paris: Descartes et Cie, 2010).

18 Étienne Pivert de Senancour, *Oberman*, (Paris: Charpentier, 1863 (1804)), cited in "La solitude", *Le Magazine littéraire*, 290 (1991): 36.

19 All the classic works on proxemics should be cited here; see, for example, Edward T. Hall, *The Silent Language*, (New York: Doubleday, 1959).

20 See "Le ventre, notre deuxième cerveau" ["The Belly: Our Second Brain"], Sciences et avenir, 784 (June 2012): 51: "Our enteric brain – our belly – produces 95% of our serotonin, a neurotransmitter that helps us manage our emotions". See also David Grundy, Michael Schemann, "Enteric Nervous System", Current Opinion in Gastroenterology, (2005). For more information, see http://www.lemonde.fr/culture/article/2014/01/31/le-ventre-notre-deuxième-cerveau_4354317_3246.html#9mUhZOJ3xrVzlpuO.99.

21 Ethnologists use concepts such as "tool", and not in relation to their theoretical definitions in political philosophy.

22 A classic topic for architects working with ambiance, and for the social sciences; for example, see number 42 of the review *Communications*, "Le gigantesque" ["The Gigantic"], directed by Bernard Paillard (1985).

23 Honoré de Balzac, *A Murky Business*, translated by Herbert J. Hunt (London: Penguin Classics, 1978).

24 An ethnologist had the opportunity of being a juror in a case before the Court of Assizes in January 2003. See Véronique Nahoum-Grappe's preface to the book by Sophie Képès, *Probe et libre. Un ecrivain juré d'assises* ["Properly and Freely: The Writer as Juror in the Court of Assizes"] (Paris: Buchet-Chastel, 2013). See also Aziz Jellab, Armelle Giglio-Jacquemot, "Les jurés populaires et les épreuves de la cour d'assises: entre légimité d'un regard profane et interpellation du povoir des juges" ["People's juries and the trials of the Court of Assizes: between the

legitimacy of a layperson's view and calling on the power of judges"], *L'Année Sociologique*, 62 (2012): 143–193.

25 See Christiane Chauviré's article "Ce qui va sans dire" ["That which remains tacit"], published online on 4 April 2016 in *Implications philosophiques, espaces de recherche et de de diffusion*.

26 Moravia's sentence in his novel *La Noia*, or The Empty Canvas, is "È sotto al fascismo che ci si annoia di più" ["It is under fascism that one is the most bored"]. This needs to be demonstrated with precise examples: lines in stores or government offices, interminable speeches and doublespeak, restrictions, mistrust, the end of horizontal connivance, and so on. The atmosphere of totalitarianism is a veritable topic unto itself.

5 Perceiving

Ambiances, places, and connections

François Laplantine

Walking, dancing, street art, singing, and music do not exist independently of a spatial and temporal context, and we will now occupy ourselves with this latter aspect of time. There are several words that express similar meanings, such as milieu, environment, ambiance, atmosphere, and landscape. I will not retain *milieu* and *environment*, because they represent geographical data that do not account for the process of transformation of the sensory (especially in aesthetic terms). *Landscape* is an exclusively visual notion. This leaves *ambiance* and *atmosphere*. The latter has a meteorological connotation that evokes something gaseous and vaporous, but the term interests me anyway because it provides us with fresh air and maintains a link with breathing.

Therefore, I will choose the notion of ambiance – like atmosphere, about as vague a notion as can be – which since the 1980s has been the subject of practical and theoretical developments in one specific domain: architecture, a particularly original form of human activity. It consists of a set of know-hows that have to take the *whole* of sensory modes into account: the visual, the acoustic, the thermal, the proprioceptive, and body movements – basically an awareness of everything that surrounds us (i.e., luminosity, aeration, and proximity). In fact, there is no such thing for architects as a minor sensibility, as the distance between designers and inhabitants must be abolished as much as possible.

Ambiance is a comprehensive notion, one that traverses disciplines and brings together fields of research that have traditionally remained separate: architecture, urban planning, physics, aesthetics, the humanities, and the social sciences. It represents a point of convergence between a milieu and a moment, i.e., between space and time. As regards the relationship to the space envisaged in terms of a place, a field, a site (for architects), or a stage as Erving Goffman terms it, we are dealing with a spatial mode that is local, even micro-local. And as regards the relationship to time, it occurs in the present, during an experience of presence. This is why a precise description of an ambiance recalls what Clifford Geertz means by the "thickening" of the present.

An ambiance is necessarily singular each time. It happens here and nowhere else, and it takes place now. This temporal aspect is important. Consider, for

DOI: 10.4324/9781032655307-10

example, the ambiance of Christmas in Manhattan, the quiet ambiance of a New Year's Day in Tokyo, and the supercharged ambiance of a Friday night in Rio. The ambiance of a city or a neighbourhood is not the same early in the morning, in the evening, or at night.

This singular encounter of a milieu and a moment occurs within a subjectivity and a perception, thus involving both memory and imagination. There is no ambiance without seeing and listening. And talking about ambiance means talking about the body.

By attempting to intertwine the moment, the milieu, perceptions/sensations, and subjectivity, we have by no means accounted for all the ingredients that go into an ambiance. We must specify that the subject that perceives, or rather that receives, an ambiance is collective. An ambiance cannot be individual. It is always the common ambiance of a scene where we happen to be, which predicates a sharing of the sensory. It provokes a set of sensations that are experienced together, not alone; there is not necessarily a uniformity, but there is a temporality, which is the unifying factor. In sum, at this early stage, I believe that we can specify three notions: a shared place, a shared time, and a connection that creates a sense of connection.

Having said this, I believe that the dominant trend of Western rationality hardly helps account for this eminently complex phenomenon; it might even be an obstacle, for the following two reasons.

1 The Western or westernizing thought in which we have been educated privileges decomposition and analysis. But an ambiance is not an equation of sounds + images + forms + colours + smells + a certain thermal intensity, instead a set of indivisible relationships that themselves form a relationship (or resonance) with the whole of the subject's sensibility.

   We should thus imbue the aforementioned notions of present, presence, and local with a sense of belonging to a whole. An ambiance is the whole of what is produced in a space-time continuum, and it is also experienced by a sensibility that itself is also a whole. But this whole is not integral. Said another way, an ambiance cannot be said to have an integrity that calls for a constant, all-at-once, and all-comprising integralism. The logic of designating/determining/defining a permanent "object" that consists of extracting, abstracting, and extracting oneself from a relationship is inadequate in this instance. Because ambiance comes before language and resists verbalization, we have what Max Weber called a "remainder". This remainder is what resists being said. So, what we feel is a discrepancy between experiencing and saying something (as the words fail us), a gap between the certainties of perception and the approximations of expression.

   An ambiance is indivisible and immediate. It is at first grasped as a totality or a configuration, and here gestalt theory can be of some help. Ultimately, though, it remains of limited assistance, as the experience in question is not just of a totality but also of an intensity of sensory variations.

2 The second reason for our difficulties in describing what we call ambiance is much more determinant and concerns the particularity of Indo-European languages. In this linguistic family, the grammatical construction of a sentence begins with a subject, which always holds a position of anteriority, exteriority, and superiority.

It is thoroughly impossible to truly explain ambiance in a European-subjective-centred way, that is, based on a "I-me" that predates a context and can abstract itself from it. To see this more clearly, let us take a detour through Japanese, a language in which, for example, one does not say "I'm cold". Why? Because in the "I'm cold", the *I* precedes the cold and is some way above the cold. This does not hold true for *samui dessu*, which we would translate as "I'm cold", but which does not have the same meaning. There is no *I* in *samui dessu*, only cold. The cold imbues the situation, but this is not at "my" initiative. Similarly, we can't say "I'm soaked" in Japanese, a somewhat bizarre pseudo-passive for the Japanese, who understand this as "I was rained on". And one final example. Descartes' "I think, therefore I am" is thoroughly untranslatable in Japanese (or Chinese). This statement sounds like a fool's declaration. The expression is so pretentious that it comes across as a gag.

Two of the most often-used terms in Japan are *arimassu* (for objects) and *imassu* (for living beings). *Arimassu, imassu*: it happens, it occurs, it arises, and it comes to be. These words express an impersonal, a-substantive dimension of our experience that is very close to what Emmanuel Lévinas means by the French *il y a*, or "there is". From this standpoint, it is impossible to say "I am going to Tokyo" in Japanese. In *Tokyo ni ikamassu*, which literally translates as "To go to Tokyo", the subject is Tokyo, not me. Similarly, when we say "I love saké" in Japanese, *o saké ga ski dessu*, the subject is the sake.

There is something not thought out, or insufficiently thought out in our common way, but which is also wise in our envisioning our relationship to everything that surrounds us, namely the inseparability of the subject and the activity (which we may call intentionality or will). The social sciences were formed in the knowledge of creating society. They engaged in the study of social forms of acting, and certainly not non-acting or being subjected. This is beyond discussion or questioning (or insufficiently questioned), it being said (or implied) once and for all, and it won't come up again. With a few exceptions such as psychoanalysis, it is always "me" who takes the initiative – me, me, me, and again me – and what we call "milieu", "atmosphere", or "ambiance" is simply a motif to accompany one's actions, a backdrop of sorts.

The Japanese perception of subject and milieu, and Japanese itself as a language, allows us to sketch out a theory of ambiances and to understand that they exist independently from an initiative and an appropriation by an

"I-me" (or "we, the others"). They do not refer to a subject, even though the subject obviously participates in them. It is not abolished, but it can no longer be considered as prior, preeminent, and determinant. Said another way, "I" do not at all have the initiative of an ambiance. I am taken by an ambiance that may touch me, upset me, or discomfort me. What is primary is the interaction in which I might well be the actor, but not acting in the linguistic sense of a process initiated by a subject who directs a verb towards an object, i.e., towards an objective.

In my opinion, Japanese accounts much better for this experience than European languages. The grammatical subject is most often not explicit, as modest as a turtle withdrawing its head into its shell. As to the social subject, it tends to be self-effacing, adaptive (to the ambient milieu), and to excuse itself (*sumimassen*), which is only an advantage in Japan, because the corollary to this conformity is a certain conformism.

What we can perceive in the Japanese sentence and also in how people conduct themselves in Japan is that what is determinant and determined is in fact the place and the milieu, which we will call the ambient milieu. The sentence is structured according to something that for us is merely a context. Said another way, the subject and the milieu have a greater ability to form a resonance and even a closeness with one another. One sees this in speaking, reading, and writing. They are placed side by side, one next to the other, but never in a relationship of subordination as in European languages and literatures.

Therefore, we should not reverse, but instead break the one-way flow of our consciousness from subject to object. We can all have this experience very concretely by using the most concise literary form in the world: the haiku, which is only seventeen syllables long. Basho is one of the most famous haiku authors. He wrote: *Matso no koto wa matsu ni narahé* ("What a pine tree is, learn it from the pine tree"). Or: *Také no koto wa také ni narahé* ("What a bamboo is, learn it from the bamboo").

There is something resembling passivity in this that may look like a resignation; however, Japanese does not have an exact equivalent of our passive voice. The closest thing are the forms *réru* and *raréru*, which do not have the (negative) connotation of subjection, instead one of acquiescing to what comes from the outside. To be more precise, this quasi-passive voice in Japanese associates three significations (which are perceptions, first and foremost): spontaneity, potentiality, and respect.

This linguistic or rather lingual digression lets us take a few steps ahead, past our initial propositions: an ambiance is a *place* where, *at a certain moment*, a unique encounter happens between the individual and the collective, the subjective and the objective, and nature and culture. However, this formulation remains insufficient due to its somewhat asymmetrical nature. An ambiance does not occur in space and time. *It generates a unique space-time continuum.* We understand it and feel it first when the dancing spills out into the street and transforms it. I am thinking of the G. Bistaki dance company's performance of

The Baina Trampà Fritz Fallen at Atelier 231 in Sotteville-lès-Rouen. This is a choreographic art of permanent metamorphosis that in turn asks us to move physically; the experience represents a total rupture with the static, traditional conception of hall and stage. To express this experience, we must therefore renounce several verbal constructs. We do not "attend" an ambiance. We do not "have" an ambiance. We do not "go" to an ambiance. An ambiance comes to us that is indivisible, unspeakable, and, yet, undeniable.

What we must reconsider then is the distinction between subject and object and the very notions of observer and observed. An object is either the objective of the subject, or the act of obstructing or objecting to the latter, which opposes it. And in this sense, ambiance eludes observation. It is not an object. In a European conception of ambiance or milieu, the subject and the object tend towards a clear distinction that can lead to an opposition. In a Japanese conception, they tend towards indistinction and fusion.

The two cases are not free of inconveniences for the construction of a theory of knowledge. Fusion may anaesthetize thought, which is then replaced by meditation. But the absolute distinction specifically entails breaking the ambiance and abstracting oneself from it. We can no longer feel an ambiance if we adopt an attitude of clear distinction between subject and object and between the individual and the collective. This is because an ambiance is not an object. It is that which, from the outside, momentarily and locally affects and transforms a subject. It cannot be grasped within an objective and objectal (in the psychoanalytical sense) relationship because it presupposes a relationship of inclusion.

Let us resume. Ambiance includes the subject. It is enveloping. It envelops the subject and the object, blurring the separation between them without totally eliminating it. From a Western standpoint, this is tantamount to hallucination or delirium. It suspends separation.

In my opinion, there is no word in French or English that accounts for such an experience. On the contrary, I do believe that the Japanese language is much more at ease in describing the intransitive and non-vectorial nature of a subject that does not direct itself towards an ambiance because the ambiance arises from the exterior. Japanese allows one to not separate the place, the milieu, and the connection.

Thinking in Japanese, i.e., in a non-anthropomorphic, non-subjective-centric, and non-logocentric manner comes closer to what we call ambiance. *Ba* and *basho* (place), *fôdo* (milieu), and *bamen* (ambiance) are notions that are not exclusively topographic, instead choreographic and choral in an etymological sense. They do not disassociate the notions of time and space, or, more precisely, the notions of circumstance (*tsugo*) and situation. *Bamen* is hard to translate into French or English because it unites what we tend to separate: a situation and a circumstance in a unique space-time continuum.

Other terms close to it do not have an exact equivalent in French or English either. Thus, *keikan*, which refers to both a landscape and a painting of a landscape (be it Japanese or Chinese), unites two characters, the first of which signifies the scene (*kei*) and the second one's gaze (*kan*). The Chinese pictogram *fengjing* (landscape), which becomes *fûkei* in Japanese, might be even more interesting because it associates wind (*feng*) with a view (*jing*).

Let us recap our progress to date and subject it to questioning. As we saw, the notion of ambiance is hard to analyse – or describe – within the context of a Western duality between subject and object. It instead conforms much more to a Japanese conception-perception, in which the subject is never abstracted from a place, a milieu, and a connection.

An ambiance is a relationship (to a milieu) that calls for reflection and not a sterile reversal of the Western paradigm. This is why we should not renounce the development of a theory of ambiances. Such a theory must redefine the notions of passivity, intransitivity, and non-vectoriality, but it must do so with other criteria than those of individual autonomy and the absolute distinction between subject and object, but without advocating for the Buddhist lack of a distinction between the me and the non-me.

As we have said, ambiance is a blurry notion. Its anthropological elaboration leads not to a dissolution, rather to a polemicization of the relationship between subject and object. In fact, we are not in a relationship of thingness with an ambiance, instead one of a transformation of subjectivity. To analyse an ambiance, we must first feel it, which means suspending the oppositional relationship between subject and object, interiority and exteriority. There is no "in front of", no "behind", or "next to" in such an experience. An ambiance "touches us", but it is impossible to identify the source of its emanation (except during concerts and performances of various kinds). Nor can we envisage it in terms of functionality or usefulness. It serves no purpose, except of course in the case of ambiances that are provoked, organized, and normalized for commercial ends. Our era, which is not at all that of the night birds of yore, now sees a prevalence of hucksters and "ambiancers" (!) at all-night parties and trips to London or Amsterdam. They know-how to flirt with our desires – and to neutralize and orient them towards a gregarious form of consumerism.

This perspective – which could be qualified as an aesthetics of the sensory in an etymological sense – is a call both to suspend functionality and to polemicize expressiveness. But this attitude should not lead us to dismiss intelligence, especially discursive intelligence; instead, it posits the need for extreme precision.

Any ambiance has an intangible aspect, but this intangibility is not at all formless. At the same time, it is the opposite of uniformity. Each ambiance has its own coherence, singularity, and specificity, and it creates a relationship of differentiation. This is why it is never hard for us to distinguish one ambiance from the other. Each one has limits in terms of both space and time. Thus, the ambiance of the city of Tokyo is nothing like that of Beijing. And within Tokyo, the neighbourhoods of Shinjuku, Shibuya, Harajuku, and Ebisu all generate very different ambiances from one another.

Some of the world's largest cities lend themselves particularly well to this differentiation, which we all notice, especially in terms of music. Among the cities where I have stayed, I am thinking in particular of Salvador de Bahia and its *axé* music, Havana and its salsa, Buenos Aires and the tango, and New Orleans and jazz. But we should pay attention especially to the micro-variations between different neighbourhoods and different eras. In Paris, to the Latin Quarter in the years when I was a student at the Sorbonne, before it had undergone its current commercial destruction, and going back in time, to Saint-Germain-des-Prés (at the time of Sartre and Simone de Beauvoir), Montparnasse (when American writers would meet at La Coupole), Ménilmontant (at the time of Bruant), and Montmartre (at the time of Utrillo).

We cannot fail to mention the major contribution of literature here – for Paris, Victor Hugo, Eugène Sue, Émile Zola, and others; for Rio, Lima Barreto and his precise descriptions of the ambiance of the Lapa neighbourhood and of Rua do Ouvidor; for Beijing, Lao She and daily life in the Hutongs in the 1930s. In Faulkner, the humid atmosphere of Louisiana and Mississippi. In Mo Yan, the canton of Gaomi. For Yonville in Normandy, Flaubert and his flamboyant description of the agricultural fair in *Madame Bovary*. And for Bordeaux, François Mauriac. We should also note the contribution made by cinema: Claude Chabrol (the middle-class provincial lady) and Jia Zhangke (the petty bourgeoisie of Fenyang), among so many others.

Not all ambiances are spectacles (consider Guy Debord), and not all of them are as big, festive, outstanding, and showy as the expression "What an ambiance!" suggests – far from it. Some of them may be gloomy (the United States as painted by Hopper, or Sweden as filmed by Kaurismaki) or discreet (Simenon's provinces or Modiano's Paris). We should also account for what Raymond Depardon calls the "weak times". This expression does not summon grandiose words that reiterate this figure. Without ignoring those ambiances described as "exceptional" that entail a rupture and provoke a surprise, we should also explore those to which we hardly pay any attention, such as the familiarity of daily ambiances and their ordinary uses, or street life, for example.

I am not advocating for swapping an anthropological approach for poetry, much less for leaving language behind altogether; instead, I believe we must refine our ethnographic description, which must be as precise as possible. Such a description is necessarily historical, partial, and provisional. It serves not to coincide with the reality of an ambiance, but simply to get us closer to it. Researchers might not want to espouse such an approach, but they may change their relationship to it. They could adopt multiple approaches in how they use language as well as sound and image.

We should not give in to the laziness of untranslatability (the unnameable, the unspeakable, the ineffable, mystery, and what all else). We must polemicize our modes of enunciation. As we have seen, an ambiance contains a *where* and a *when*. *Where* is in a certain place and *when* is at a certain point in time. An ambiance necessarily occurs (and becomes and recurs) in a *where* (in this place, in this neighbourhood) and at a *when* (something is happening in this place). In fact, it is nothing more than the perception of this here and now that, unlike an object, has no stability and no invariability. Any ambiance is relative to a movement, a variation that evolves through modulations (that are most often infinitesimal). Any ambiance has a beginning, a growth, phases of transformation, decline, and then of disappearance. But what disappears (the ambiance of the Latin Quarter) is likely to appear in another place and in another form.

However, the conjunction of a *where* and a *when* does not suffice to describe an experience that remains dissatisfied with definition-type statements and nouns. In fact, nouns are perfectly suited to name substances or essences, i.e., relations of thingness or ideality, but they are of limited interest when it comes to describing the phenomenon at hand. Qualifications more in the spirit of connotation than denotation are more likely to provide clarification; hence, the tendency of overusing adjectives, of adding terms like "festive", "explosive", "relaxed", "spirited", "cheerful", "crazy", or "discreet", "muted", or "subdued" to the word "ambiance".

Adjectives specify, but they tend to enclose the singularity and also the variability of a given ambiance within stereotypes. This raises once more the delicate issue that has characterized our thinking about ambiances from the outset, but which we will now formulate in a different way. Because we cannot resign ourselves to silence without claiming to be saturated with knowledge, what attitude should we adopt in relation to the undeniable, implicit nature of a given ambiance? How do we polemicize this undeniable and render this implicit explicit? It is a perilous but by no means hopeless undertaking. We can engage in it carefully, but by renouncing the syntax of the "I-me", which is not just inappropriate and illusory but also indecent. This represents a call to invent parataxic forms.

The limits of what ordinary language cannot say (and what traditional, scholarly language can't either) also impel us to explore *what can be shown* (Wittgenstein). In being confronted with an ambiance and transformed by it, we experience the limits of language, or at least of certain linguistic denotations, connotations, and constructs, but not of gestures, sounds, and images. An ambiance is not merely given; it may also be performatively transformed, both in J.L. Austin's linguistic sense and in terms of the performing arts.

It may be hard to describe an ambiance using common language that is analysed (i.e., deconstructed) using scholarly language, but it can be narrated, sung, danced, filmed, and staged. But what is thereby recreated instead of being represented is yet another ambiance.

6 Phenomenology and ontology of ambiances. Some clarifications

Bruce Bégout

Introduction

When we are seriously interested in ambiances,[1] which are particular phenomena that are perhaps difficult to define, one aspect may nevertheless seem obvious to us: it is the strong contrast that exists between the immediately obvious character of the experience that we have of it and the difficulties that we feel to name it. For, in fact, there are few people, even among scholars, who dispute the existence of this type of phenomena. On the other hand, there are many who interpret them in a very different way. It is as if there were a *gap* between the commonly accepted atmospheric impression and its very varied descriptions. In a way, we understand without difficulty what an ambiance is when it manifests itself to us, but as soon as we try to define it, we find ourselves in a quandary.

In truth, if we take a closer look at the scholarly studies that have been done over the past thirty years on phenomena such as ambiances and atmospheres, we notice that this gap is not as wide as we usually think. In fact, most of the thinkers who have taken a close interest in these types of experiences – through different approaches in the human sciences (sociology, anthropology, psychiatry, aesthetic theory, etc.) – manage to describe in an almost identical way what an atmosphere is and how it manifests itself. Schmitz, Böhme, Anderson, Hauskeller, Griffero, and Thibaud, to name but a few of the contemporary theorists of atmospheres, all give very similar presentations. In reading their works, we notice that it is a sensorial, diffuse, and impalpable phenomenon for them, which envelops the individuals in a subtle way and makes them feel the particular affective tone of the situation in which they find themselves. However, we must admit that, although their descriptions of atmospheres align with one another, the explanations that they each provide of this phenomenon do vary. Some of them consider atmosphere as an effect of the environment on our sensibility (introjection), others – on the contrary – as the result of a transfer of some of our moods on that environment (projection), and some finally consider that atmosphere is in fact a simultaneity of these two aspects, thus combining in an ambiguous phenomenon objective elements and subjective reactions (mixture). We

DOI: 10.4324/9781032655307-11

would thus be dealing with a double contrast: *first*, between the mode of feeling and that of saying and, *second*, between the mode of saying and that of interpreting.

In both cases, what would be at stake is the phenomenological gap between the common character of an experience (the phenomenal dimension) and the diversity of the explanations concerning it (the type of *logos* called upon). Moreover, it is not certain that the apparently neutral way of describing these phenomena does not already include notions that are already part of an explanatory scheme.

How is it possible then that thinkers coming from different theoretical horizons describe atmosphere in an almost identical way but, when delving further in their analyses and studying it in a more profound way, describe it from very different explanatory angles? To answer this problem, we will adopt a phenomenological approach. It seems to us indeed – for reasons that we will have the occasion to specify later on – that phenomenology is the most capable method for grasping the specific mode of being of these familiar and yet strange phenomena that are ambiances. The fact, for example, that phenomenology does not dissociate in general the being and the appearing of a thing, but that it draws the former from the latter, to the point of confusing them, proves to be a decisive advantage for a philosophy of ambiances or atmospheres. Even though the field of atmosphere studies is vast and yet expanding, it faces multiple problems, the most important being this astonishing disparity, not between the observations, but between the explanations provided. Of course, we do not pretend to eliminate this diversity by proposing a singularly valid theory, and it is likely that our own theoretical way of grasping the being of an ambiance will itself become the object of discussions and objections made from competing theories, but we are nevertheless convinced that this phenomenological approach will greatly help to reduce the gap between description and understanding.

Neither lived experience nor thing

To approach ambiances in a phenomenological way, it is important – even before ruling on their essence – to specify their particular mode of givenness. Ambiance is nothing else than its manifestation in the here and now. What we can know of it, we can only know on the basis of the particular way in which it appears. It belongs to a type of being that exists only as it manifests itself. Now, if we rely on what Husserl has established in his works, there are basically only two great regions of appearing: consciousness and reality. Everything that appears is thus distinguished according to two distinct types of being: the immanent experience and the transcendent thing. The non-spatial and non-adumbrated givenness, namely the immediate, immanent, and adequate givenness, belongs to the being of the lived state of consciousness, whereas the adumbrated givenness and the integration in an indeterminate horizon and a coherent course of perception belong to the being of the transcendent things. The phenomenal field is thus divided between (a) the region of experiences, which itself can be divided according to subcategories (e.g., intentional and non-intentional experiences), experiences of acts of apprehension, and merely "presentative" (*darstellende*) hyletic experiences and (b) the region of material things, which, in the same way, admits subdivisions according to the type of thing that appears (natural, cultural, artistic thing, etc.).

Where does ambiance fit into this scheme? Does it have the mode of being of lived experience, and in particular of affectivity? Or, on the contrary, due to its connectedness to spatial situations, does it belong to the realm of material things? First of all, it seems evident, for reasons that we will discuss and on which several thinkers insisted, that the phenomenon of ambiance is not reducible to a psychic state. Nothing in its very appearance, its spatiality, its temporality, its relief, and its character, refers directly to an experiencing subject. That an ambiance is felt by a subject does not mean that it is ultimately only an immanent and psychic content of this subject. On the contrary, what is immediately manifested in this phenomenon is its subjective character, if by this we understand the fact that this character is not directly referring to a subject. The atmosphere I feel is not my atmosphere, it is the atmosphere that appears around me and that others can feel as well. While I automatically attribute to myself an organic affective sensation or an intentional feeling, I feel the atmosphere as immediately characterizing the surrounding world. If, therefore, an organic experience or a feeling refer immediately to the one who feels them, so that the subjective character of these experiences constitutes a manifest quality, in the sense that I cannot like something without knowing that I like it and without knowing that this immediate relation to myself manifests itself in the feeling, this does not hold for the affective tone of the ambiance. The gloomy atmosphere of a dark and rainy alley on a winter evening does not express an affective modification of the subjectivity nor does it refer to it – rather, it testifies directly to an expressive quality of the situation itself. What is gloomy for the subject is the situation itself, not its feeling of the situation.

Now, for Husserl, lived experiences are always immanent contents of consciousness, namely real parts of its *continuum*, which refer to a centre of experience called the subject. There is obviously nothing like this in the case of ambiances or atmospheres. Phenomenologically speaking, it is thus erroneous to relate ambiances as immediate and holistic phenomena to internal contents of consciousness. Furthermore, the subject of experience immediately makes the distinction between an internal affective experience, intentional (feeling) or not (organic sensation), and an atmosphere. When it experiences an ambiance in situ, it feels something floating around it that, while enveloping it, decentres it, crosses it, and exceeds it, something that seems to belong rather to the ambient environment in which it is inserted than to its personal sensitivity. This is not simply a question concerning the localization of such an affect, within me or outside of me, since an intentional feeling is also not localizable in my body or in the world, but a question concerning the affective quality itself. The phenomenal properties of ambiances, notably their spatial and temporal dimensions, manifest themselves in an autonomous way, without referring to an internal centre of experience that would be the self or consciousness.

Unlike any concrete experience, intentional or not, the phenomenon of ambiance does not belong to the real immanent sphere[2] and, as such, it does not refer in its mode of being to a subject-pole whose experience it would be. For a lived experience is not only a real content of consciousness, it is also a phenomenon that manifests, in its very appearance, its dependence on the self. This dependence is clearly expressed, for example, in its mode of temporalization, which is absolutely identical to that of the consciousness of which it forms a concrete part. However, on this point, the temporal character of ambiances has nothing in common with that of immanent experiences flowing in the stream of internal consciousness according to the laws of impression, retention, and protention. There again its temporal givenness is singularized from that of the states of consciousness. Thus, this phenomenon also lacks the adequate, even apodictic givenness of an internal experience, which cannot fail to be grasped in its totality by the subject who lives it at the moment when it lives it. While "any immanent perception necessarily guarantees the existence of its object",[3] and that therefore conflict, simulacrum, and otherness are excluded for "the absolute sphere" of lived experience, the seizure of an atmosphere does not concern this unperceived lived experience that was there before the glance of reflection turned towards it and did not perceive it as an absolute *datum*. In this respect, if the phenomenological reduction consists, as Husserl still indicates in §37 of his *Phänomenologische Psychologie* from 1925, in turning the gaze back towards the really immanent components of consciousness, which are hyletic data and intentional acts,[4] then this cannot be applied to ambiances that, because of their non-immanent and asubjective character, are not susceptible to be reduced to such phenomenological data.

Does this mean then that atmospheres have the mode of being of things? Not really either. In their case, it is indeed impossible to attribute to them the mode of spatial and material givenness, notably of the givenness by adumbrations (*Abschattungen*) and within a progressive and regulated horizon of unfolding. That atmospheres possess spatial qualities, especially envelopment and amplitude, does not mean that this spatiality amounts to the spatiality of things.

Ambiances do not present themselves as things, i.e., bodily substances limited by more or less precise contours and possessing internal consistency. Their style of appearing is completely different. If a thing can appear to us only from one point of view, from one side,[5] and if this aspectual appearance implies other such appearances, according to the perspectival relation between them and the subject that perceives them, then atmospheres absolutely do not possess this unilateral mode of givenness. One cannot walk around an ambiance to discover new hidden facets. It would make no sense to try and do so because an ambiance does not appear to us as a thing. It is not a matter of the combination of intuitively felt data (the perceived facet) with a set of co-data only perceived in a unity of a thing. If the perception of a thing is characterized by a mixture of perceived and glimpsed, of data and co-data in a process that associates an intuition with an idea (that of the teleological unity of the thing), then the phenomenon of ambiance presents itself in a total and tonal givenness. In a sense, an ambiance or atmosphere always gives itself as a non-decomposable whole, and if it can change in time – and it is indeed in this manner that it usually gives itself in a transient and fleeting character – it does not do so according to the rules of the thing's gradual modifications. Also, an ambiance cannot disappoint an expectation in a way that a perception can, since it delivers itself entirely in its phenomenal presence and does not contain any process of confirmation or invalidation of its givenness. Whether they are pleasant or unpleasant, atmospheres thus manifest themselves as non-figurative and non-substantial spatial totalities. They do not have facets and, for this reason, they do not unfold according to a regulated course of gradual appearances as do spatial things. Since they do not have a figure, but an air, a look, that is to say a specific tonality, they rather manifest the affective and expressive background of the situation than any quality of the various substances that compose this situation. They are phenomena that express in an affective mode the always variable ground of experience – not its perceivable forms.

Ambiances and third being

Should we then consider that ambiances are mixed phenomena, borrowing their qualities from immanent experiences, on the one hand, and from transcendent things, on the other? Or should we rather admit, by respecting the phenomenological notion of the equivalence between mode of appearing and type of being, that they designate a new phenomenal region, irreducible to either consciousness or objectual reality?

If ambiances do not possess the mode of manifestation of a lived experience nor that of a spatial thing, would it not be surprising, to say the least, that they result from the mixture of the two? Moreover, how could something be the combination of two absolutely contradictory types of being? What exactly do we mean when we say that atmospheres would be a phenomenon that is at the same time subjective and objective, half lived experience, half thing? What is certain is that this theory of a mixture, presenting atmospheres as ambiguous phenomena pertaining at the same time to lived experiences and to things, must, on pain of remaining enigmatic, answer two fundamental questions:

First, how could a phenomenon derive from the effective synthesis of two phenomena with which it has no relation? Second, how can these two opposite types of phenomena – lived experience and transcendent thing – be merged into a third type of phenomenon, even though they have nothing in common? It seems to us, therefore, that for phenomenological and eidetic reasons that have to do with modes of givenness, ambiances are neither quasi-lived experiences nor quasi-things nor – and even less likely – phenomena that would be at the same time lived experiences and things. If an atmosphere seems to express something that belongs to the one and to the other, it is perhaps because it manifests something that, preceding them, belongs neither to the one nor to the other. It is thus advisable to bring to light *a third way of being* that does not belong to the first two and which is neither a combination of them. Many researchers who are interested in atmospheres and their phenomenological and ontological specificity underline the necessity to admit this third type of being.[6] Conscious of acting in a spirit contrary to that of the ontological sobriety of the Ockhamian type, they support a *widening* of the field of being. In a certain way, they subscribe to what Griffero terms a necessary "ontological inflationism".[7] If ambiances indeed exist, if we believe to feel them every day, and if the known types of being (mental and physical) do not suit them, then it is necessary to recognize an original mode of being of atmospheres, on the unique and undeniable basis of their mode of appearing. There are thus ambient phenomena that belong to the *medium* of experience and which refer neither to mental states nor to transcendent things. If, as Sartre writes, "one knows that there is not, on the one hand, one for oneself and, on the other hand, the world, like two closed wholes of whom it would be necessary to seek how they communicate",[8] one knows it all the better when one experiences an environment. It is necessary to add to that the phenomenological pact, which methodologically insists on the necessity to draw from the phenomenal data all that we can learn about their mode of appearance as well as their appearing contents, motivates us to take seriously this mode of givenness of ambiances that is neither psychic nor physical.

There seem to be two possible ways to proceed here. (a) Either we contest, in the name of pure experience, the idea that the respective categories of lived experience and thing refer to something existing. From here on, the phenomenon of atmosphere refers to this unique dimension of the being which is that of the ground of experience itself, before it is separated into the regions of consciousness and world. On this account, we end up with a phenomenal monism, which seeks to reveal underneath traditional theorizations an original and pure experience not yet met by dualistic conceptions and which involves thinkers as different as Richard Avenarius, William James, or Renaud Barbaras. (b) Or, on the basis of differentiated regional eidetics, we accept the joint existences of a type of lived being and a type of objectual being, which are no longer *flatus vocis* but true expressions of reality. Then, however, we must also recognize a new original type of being, that of ambiances and atmospheres, which is radically different from the two other ones and which is not their synthesis.

First of all, we can underline that the thinkers who are interested in the phenomena of ambiances and atmospheres postulate the necessity of allowing this *third type of being* and of thus recognizing a mode of existence that is neither that of psychic states nor that of physical entities. Atmospheres certainly refer to a type of phenomena that slips between psychic phenomenality and objectual phenomenality.[9] It is typical of experiences of atmospheres that they are experiences of something that stands *between.* On the one hand, *between* the things themselves as the non-objectual background of perception and, on the other hand, *between* these things distinct from their background and ourselves. But there again everything depends on the way we conceive this dimension of the *Between,* of the interstitial, and of intermediate spaces. It is not enough to say that atmospheres belong to these phenomena that slip between the things and us, it is necessary to specify the nature of "this new ontological category".[10] In our opinion, we can, on the one hand, seize the *Between* as that which connects the separated entities in as much as it unifies them into a common dimension. Here, the *Between* assumes the role of an intermediary that melts together elements into a superior unity. But we can, on the other hand, and in a more profound way, grasp this *Between* not as that which would belong to the distinct elements while linking them but as that which precedes them. In the first case, the *Between* stems from a synthesis that, mixing the separated (lived states and things), makes them communicative by a common dimension. The third is the result of the mixture of the first two types of being. One exceeds the opposition of subject and object by creating a kind of hybrid being: the "subjobjet". In the second case, the *Between* does not unify the parts distant from one another by the dualism, *rather it erases them to reveal the common ontological ground that precedes them and which is not them.* The third does not result from a meeting of the two existing elements, it appears as a dimension common to these elements, referring neither to the one nor to the other. While in the first case, the *Between* plays the role of a *fusion,* in the second case it does not merge anything, but drives the distinct entities into a homogeneous and original experience. The common that precedes any thesis is opposed to the common born of a synthesis.[11] If, consequently, lived states and things can communicate and give place to experiences that seem to blur their borders, it is not because lived states of mind possess in them something of things and things something of lived states, but because lived states and things possess within themselves something common that is neither lived experience nor external thing.

It goes without saying that, for us, in the mersive and anti-jective logic that we deploy, the experience of ambiance reveals the common background of the situation that does not result from a fusion of the elements present in this situation in a shared atmosphere. However, would this not amount to positing the existence of an ambiance that de facto precedes the psychic and objectual elements that seem to compose it, and to conceive in a pre-subjective way an ambient experience of situations? How can one affirm the presence of an atmosphere *in itself,* of an atmosphere as an intrinsic quality of the background of the situation, without a subject who could experience it at all in a fusional or non-fusional mode?

As we have said, the phenomenology of ambiances implies both a *de-psychologization* and a *de-objectification* of their understanding. This means, very simply put, that the analysis of atmospheres cannot immediately refer to the presence of living subjectivities, on the one hand, and objective situations, on the other, in order to grasp them. It makes headway towards a third type of being that relativizes both the psychic and the physical. If it is not difficult to admit that ambiances form non-objectual phenomena, as we have shown above, insofar as they are not substantial, limited, and durable entities, it seems much more arduous to detach them totally from a subjective experience. What is the meaning of concrete ambiances, with their characteristic tone, which would occur outside the presence of subjects who we suppose to feel these ambiances? Can we really affirm that atmospheres form phenomena independent of all subjects, and manifest this pre-subjective background of all experience? Isn't every atmosphere, happy or sad, temporary or lasting, someone's atmosphere? Doesn't it require the presence of a moved person to occur and appear? What is the meaning of a joy or a sadness that is not correlated to a subject who experiences them?

To answer these questions, it is first necessary to show that not all phenomena are necessarily subjective and that multiple things are continuously manifested in the world without living subjects being in proximity to them, who would be capable of perceiving them and of bringing them into being. There is thus an intrinsic phenomenal dimension to the world and to the things that do not wait for its reception in and by a subject to appear. The phenomenal field does not occur only by our intervention, rather it is pre-existing with regards to it. This is the reason why we can argue that phenomenality is a property of things and non-things, and that it is not reduced to its subjective translation as phenomenon for us and in us. In other words, before the *phantasma* there is *phainomenon*, before the subjective representation there is something that appears in and of itself. Thus, instructed by the asubjective phenomenology of Patočka, Arendt, and Barbaras, it does not seem aberrant to say that there can be an appearance of an atmosphere without a person for whom there would be such an atmosphere. Ambiance, whatever they are, qualifies as the expressive and tonal quality of a situation in which no human witness is present. This can be the case in a landscape, a moment of the world that possesses an atmosphere, i.e., expresses in an atmospheric and physiognomic way singular tonal qualities, which belong only to them and which no present or absent subject can create itself. The systematic emphasis of the subjective presence, and of its role as phenomenal converter, hides the proper phenomenality of ambiances and the fact that their affective and expressive qualities do not depend in any way on the presence of receiving subjects. In the experience of an ambiance or atmosphere, the self is not pushed to reflect and take position. This does not necessarily mean, however, that it coincides with the environment and loses itself entirely in it. It rather means that what is felt does not *ipso facto* refer to the one who feels it, but essentially to what is felt, namely the particular atmosphere of the situation.

Ambiance in itself, ambiance for us?

But should we then distinguish between this original *ambiance in itself*, this immediate and autonomous expressiveness of situations, on the one hand, and the derived *ambiance for us*, the one we feel and which embraces us, on the other? If we proceed in this way, for example as Gernot Böhme often does,[12] distinguishing between the "atmospheric element", i.e., the objective atmosphere not yet felt and the "atmosphere" proper – where this atmospheric element is really felt by an individual – are we not exposing ourselves to the introduction of a new *duality*, consisting no longer of experiences and things, but now of ambiance in itself and ambiance for us? One could very well imagine, from this perspective, that the primary ambiances would be only ambiances in power, in the sense that the situation would possess a certain expressive and tonal potential character, which would be actualized only by the subjects who feel them and make them reach the real phenomenality. Thus, there would be factors in the world objective that would predispose towards the appearance of ambiances (light, sound, odours, landscape, etc.) and only when these factors act on present subjectivities, the emergence then of real ambiances would occur. However, this way of proceeding does not really convince us for the following reasons: mainly, because, from the phenomenological point of view, *there is no derivation leading from one to the other.* Nowhere do we make the effective encounter with this distinction and with its two elements. We feel an atmosphere, we do not feel the transition from an atmosphere in itself to an atmosphere for us. We are thus led to say that the ambiance for us is nothing else than the ambiance in itself: first, because we have no way of showing in what way they differ, given that we are unable to distinguish them as types of experience to then compare them with one another, and second, and in a more phenomenological fashion, because the way in which the first one is given is not different from the way the second one is. It is indeed the fidelity to things themselves that leads us to posit that the ambiance we feel is precisely the ambiance in itself, because – usually – this is the way we feel it. We feel in a subjective way the absolutely asubjective character of the atmosphere, whereas we live it as an atmosphere that belongs to the situation and not to us feeling this situation.

In an ambiance, neither of the two supposed terms of the relation (lived experience and thing) appear, nor do they – and even less this relation without or before the related terms – appear to us. The atmosphere does not indeed present itself as a relation, but as a total, enveloping, and penetrating presence of the surrounding world. Not only do things and situations exist without us, but they already exist without us with their own qualities. In these conditions, the ambiance does not derive from the relation that would be established between a worldly situation and us; rather, it is already present in an irrelative way in this situation. If we can and must, in a certain way, think of the primary ambiance as characterizing situations in the world without us, we must also recognize that the so-called secondary ambiances that we experience concretely at every moment of our lives present themselves as these primary ones. From then on, without entering further into the debate of the distinction between pre-human original ambiances and human-derived ambiances and of the way of thinking the passage from one to the other as emanation, reproduction, internal duplication, etc., it is enough for us to note that every experience of ambiances poses its own ambiance with its own tone and relief as a phenomenal and tonal property of the situation itself outside of us (the melancholic character of the twilight, e.g., is *an intrinsic quality* of this moment, not a sensorial modification of our sensitivity). What appears – the atmosphere qualified by a particular air – reveals itself as irreducible to what we feel of it, precisely because we cannot feel it otherwise than how we feel it, namely as a characteristic of what surrounds us.

Thus, we can affirm that the anteposition of primary ambiances is justified phenomenologically by the particular way in which secondary ambiances are themselves experienced, namely as primary ambiances. In this case, the subjective experience of ambiances does not distinguish between the original and the derivative, the in-itself and the for-us, and it delivers the whole ambient content of experience as a global, impersonal, and anonymous character. Here, the subject of the atmospheres feels them as characters of the background of the situation, even when these characters are oppressive. If the subject reacts to a negative atmosphere (sad, hostile, distressing, etc.) with an attitude of resistance (opposition, flight, intervention), it is not the author of this atmosphere – and it is aware of this. The atmosphere does not become atmosphere when I feel it and react to it, it becomes atmosphere when it is felt; and when it is felt, it is felt as an atmosphere that always precedes my feeling and refers to a tonal expression of what surrounds me. This is the very way in which our sense of smell grasps it. It understands the ambiance, in its particular air linked to a particular situation, as expressing a homogeneous background of experience that precedes any division between the thing-beings separated from this background, on the one hand, and these thing-beings and ourselves, on the other. The ambiance is one and the same, whether there is nobody to feel it or a crowd transfixed by it. When we ask ourselves what an ambiance existing before subjects capable of feeling it could look like, we can always say to ourselves that it is useless, and even sometimes implausible, to pose a difference between a primary and a secondary ambiance from the moment when what they manifest *in concreto* is identical from the phenomenological point of view.

For it is the same phenomenality with or without us, and if the ambiances linked to the collective experience possess a specific character, this does not mean that the group that feels a certain atmosphere necessarily relates it to itself. It is more carried by it than carrying it. It is not an ontological thesis positing the being before the appearing and affirming the reality in itself of the atmospheres, even when nobody would be there to experience them, but a phenomenological thesis stipulating that the experience of the atmospheres makes us at once experience them as expressive and independent characters of situations.

However, what we learn from the experience of ambiances is that we feel that its characters belong to the situation itself and not to those who feel it. It immediately gives an account of an autonomous expressiveness of situations, namely of the cuts of worldly time and space. By expressivity, we should understand the fact that these variable situations of life, considered as units of experience, possess physiognomic and tonal qualities that belong to them in their own right and which are not the more or less conscious result of psychic projections on the part of the individuals who would add affective or aesthetic meanings to them. The introduction of subjectivity is of no help here. The original expressiveness of situations is the opposite of a metaphor, of a transfer of meaning from one sphere of reality to another, namely from mind to matter. In the case of atmospheres, their particular tones and variable intensities manifest themselves directly and autonomously in what appears. They are immediately felt, so that the subjectivation of the phenomenon, if one understands by this its relation to a subject, does not subjectivize it in the sense that it would confer a phenomenally subjective dimension to it. The opposite is in fact the case: the subjective experience as *condition* is not subjective as *dimension*. If a subject is indeed necessary to feel an atmosphere (who would deny it?), this does not imply that the content of what is felt is independent of the one who feels it. In the phenomenon of ambiances, there is a fundamental disengagement of the condition and the dimension (the phenomenal content), so that what is given in feeling is given as independent of the one to whom it is given. In a sense, the experience of ambiances confirms the realistic character of the appearance for us, which is basically nothing less than an appearance in itself. The atmosphere does not disappear behind the consciousness that we have of it, nor does it reveal itself there. For this reason, the experience we have of the ambiance and the ambiance outside the experience we have of it should not be considered as different. Otherwise, it would be necessary once again, according to the scheme of the passage from power to act, to think that the ambiance really manifests itself only when the subject feeling it intervenes. But when this subject of experience feels it, it does not feel as if it were bringing it into being in itself and by itself. It receives it as if it were already there, as if it were an atmosphere that does not apply to the subject in particular and not even to anyone. The tonal presence of the world without us is precisely what we experience and feel in the ambiance.

As we can see, ambiances do not designate the way in which we feel the variable situations of our experience, but the expressive and ambient qualities of these situations themselves. As such, even the way we feel them individually confirms this phenomenological fact, as we feel them as natural, floating affective presences around us – presences that do not depend at all on the fact that we feel them. Moreover, we realize the fact that us actually feeling them adds nothing to what they are and to the way they appear. Thus, from the phenomenological point of view, it does not matter much to distinguish the *in itself* and the *for us* of ambiances, since any experience of ambiances confuses these two dimensions and confuses them to the benefit of the first one, that of the *in itself*. From there, it follows that the pre-dualistic character of ambiances is clearly vindicated in their experience and that any genetic and causal explanation of their appearance misses this state of affairs by referring, in an abstract way, to the unit of experience thus present as preliminary elements that do not appear in it and which, therefore, in the phenomenological perspective, where being and appearing coincide, do not exist.

Notes

1 We prefer to use the term ambiance rather than atmosphere, for two reasons: (1) the term atmosphere is metaphorical, since it designates first of all the layer of air that surrounds the earth and only then, by extension of meaning, the affective tonality of what surrounds us as humans and (2) thinkers who deal with atmospheres often tend to conceive them as a dimension that is first of all sensitive and perceptive (night, twilight, wind, etc.) and then to add to it in a second move an affective colouring. Thus, for them, the atmospheres constitute first and foremost sensorial phenomena, which do not possess a figure and have a non-localizable spatiality, and only then produce affective resonances in us. However, according to our perspective, the ambiances, seized as effectively atmospheric phenomena, possess a pathic presence from the start. By doing so, they do not belong to the register of perception, but to that of affectivity, and it is indeed this affective tonality that is felt in an ambiance, not the purely perceptual qualities that would belong to the domains of light, sound, odours, etc. Certainly, all sensorial environments without objects, such as light, heat, sound, smell, or air, possess an ambient mark; but it is thus from this mark, from this affective imprint, that it is necessary to start studying them – not from an enumeration that is affectively neutralized of their perceptual properties. On this question, cf. Bruce Bégout, *Le concept d'ambiance. Essai d'éco-phénoménologie*. Paris 2020, 17 f. However, we must note that atmosphere and ambiance are very close terms and that, sometimes, to avoid repetition, we use them as synonyms.

2 Edmund Husserl, *Ideen zu einer reinen Phänomenologie und phänomenologischen Philosophie. Erstes Buch: Allgemeine Einführung in die reine Phänomenologie. 1. Halbband: Text der 1.–3. Auflage*. Husserliana III/1, ed. by Karl Schuhmann. The Hague, 1976, §36, 65:

> By the name lived experience in the broadest sense of the term, we mean everything that is found in the flow of lived experience; thus, not only intentional experiences, actual and potential cogitations, taken in all their concrete content, but also everything that can be found of any real moments in this flow and its concrete parts.

3 Edmund Husserl, *Ideen zu einer reinen Phänomenologie und phänomenologischen Philosophie. Erstes Buch: Allgemeine Einführung in die reine Phänomenologie. 1. Halbband: Text der 1.–3. Auflage*. Husserliana III/1, ed. by Karl Schuhmann. The Hague, 1976, §47, 85.

4 Edmund Husserl, *Phänomenologische Psychologie: Vorlesungen Sommersemester 1925*. Husserliana IX, ed. by Walter Biemel. The Hague, 1968, 188.

5 Cf. Hua III/1, §44, 80 f.

6 Cf. Tonino Griffero, *Quasi-Things. The Paradigm of Atmospheres*. Albany 2017, XIII: "tertiary qualities or sentimental (and therefore atmospheric) ones, which permeate the space in which they are perceived".

7 Cf. Tonino Griffero, *Quasi-Things. The Paradigm of Atmospheres.* Albany 2017, XIII, 53.

8 Jean-Paul Sartre: *L'être et le néant. Essai d'ontologie phénoménologique.* Paris 1943, 345.

9 Griffero: *Quasi-Things*, XIV: "So in most cases, in our everyday life, atmospheres exist 'between' the object (or rather, the environmental *qualia*) and the subject (or rather, the felt body)". Same idea, Tadashi Ogawa: *Phenomenology of Wind and Atmosphere.* Milan 2021, 16: "Contemporary phenomenology is increasingly turning its attention towards what is 'between' human beings and things; to what is between subject and object".

10 Griffero: *Quasi-Things*, XV.

11 Inspired by Schmitz and Griffero, the work of Tadashi Ogawa seeks to bring to light, under the name of Ki (spirit, air, wind, blow, etc.), this medial presence of the atmosphere that would precede the ontological cutting and coupling between the subjects, on the one hand, and the objects, on the other, would flow like a vital energy between them; cf. Ogawa: *Phenomenology of Wind and Atmosphere*, 21: "my thesis is the following: the totality of atmosphere, manifesting itself within the human existence before anything else as world-horizon [...] is the pre-predicative, pre-logic dimension of what is 'between' human and world". The same idea appears on p. 21: "atmosphere determines human beings before our identification with a subject and our independence from objects".

12 Gernot Böhme, *Aisthétique. Pour une esthétique de l'expérience sensible.* Dijon 2020, 85: "Let us recall that we establish here a difference between the atmosphere, as a characteristic modality of the co-presence, and the atmospheric element which forms a certain class of natural phenomena".

7 Milieu, ambiance, and atmosphere

New perspectives on experience

Yves Michaud

1

As Stendhal stepped out of Santa Croce in Florence on 22 January 1817, he felt faint because the beauty he had experienced. It had overpowered him:

> As I stepped out of Santa Croce, I could feel my heart beating, what in Berlin they call a case of nerves. I was exhausted, and as I walked, I was afraid I might fall down.
>
> I sat down on one of the benches in the Piazza Santa Croce. I was delighted to reread Foscolo's verses that I had kept in my wallet. I didn't see their faults at all. I needed to hear the voice of a friend who shared my feelings.[1]

What matters is not the experience of a perfume; the perfume is the experience. It's not the experience of visiting a city that matters; visiting the city is the experience. Later on, when someone thinks about it, when they look back, they will talk about the perfume's qualities and the city's peculiarities. What matters during the actual experience is not the experience's reference, rather its tenor, its emotional tonality, to which a name will be assigned, be it "Picasso" or "Venice".

We have an almost too perfect example of this in the "Ferran Adrià Experience" at the 2007 edition of Dokumenta in Kassel.

At the time, Ferran Adrià was the most creative chef in the world, who had broken every culinary and gastronomic convention imaginable, a kind of hyper-chef. At his restaurant El Bulli outside Rosas, in Catalonia, he had transformed cuisine into fireworks of flavours, consistencies, aromas, vapours, and images, because each one of his creations was photographed before being catalogued – which diners would experience during a meal lasting several hours and consisting of some 60 micro-tapas, some of them solid, others gelatinous, and yet others gaseous, dissipating in just a few seconds. One did not so much eat at this restaurant as live the experience of taste. This lasted until 2011, when Adrià decided to close the restaurant.

DOI: 10.4324/9781032655307-12

In 2007, Adrià was chosen along with other contemporary artists – who suddenly found themselves shipped off from the "modern" avant-garde they were so proud to represent to the most academic, hackneyed kind of art – to represent Spain at Documenta in Kassel, Germany, an exhibition held once every five years during the summer.

Obviously he didn't take his ovens with him

Two visitors to Documenta would periodically be selected at random, based on their admission ticket, and invited, all expenses paid, to Ferran Adrià's restaurant to experience his festival of tapas. The experience proposed was a space-time bubble to travel by plane and taxi round trip from Kassel in the region of Hessen in Germany to Rosas in Catalonia, some 1,500 km away, a trip that culminated in an experience of rare flavours, colours, and aromas as mad as those of the aesthete Des Esseintes in Huysmans' *Against the Grain*.

2

"Experience" is not meant here in the cognitive sense.

We must specify this, because philosophical tradition has long linked experience with testing, experience with knowledge, and an experimental system with the validation of a hypothesis.

This view of experience as experiment corresponded to its Ancient Greek etymology, which associates experience and the test to which one submits something to establish its nature; hence, the scientific sense of the test characteristic of modern science that we say began with Galileo. This notion, with all its successes, significantly influenced the idea of the scientific method and the vocabulary used to describe it. These effects extended as far as Bergson, who only used the word "experience" in the cognitive sense, even though he was a philosopher of lived experience and of time, and despite the fact that at the same time, almost all German philosophers spoke of *Erfahrung*, of lived experience.

This notion of experience has been erased. We could even say that it's dead–and not just due to the progress of aestheticization.

The respective roles of hypothesis and experience/experiment have changed considerably in science, which is now produced industrially by a massified scientific community. Hypotheses are no longer operating frameworks defined by one person and their team and discussed at length within small groups of researchers. They function as collective projects that are often tied to large facilities and research centres that are open to many teams. Experiences/experiments in the sense of tests have nothing crucial or decisive about them; they are submerged amidst an excess of protocols to be evaluated and heaps of data to be used. They have been replaced by procedures for selecting and evaluating observations of all kinds.[2]

At the same time, another notion of experience has come to prevail, no longer that of a test, and instead what is felt, of a subjective experience, a way of living and feeling a situation. It isn't fabricated as a mechanism for verifying things, and it isn't used to test hypotheses. It's something that one has and that one lives, *and if there is anything that it makes known, it is itself.*

This sense that has become current – "live out an incredible experience!" – lies at the heart of our daily life, in its most authentic moments and the most artificial ones. An experience is a more or less articulated and conscious whole – which at the same time is more or less unarticulated and unconscious – of sensations, ideas, emotions, feelings, and sentiments accompanied by memories and desires.

The cliché prototype of this experience is obviously Proust's madeleine, with its world of sensations and memories:

> Soon thereafter, overwhelmed by the dreariness of the day and the prospect of a sad tomorrow, I mechanically lifted a spoonful of tea to my lips, on which I had left a bit of a madeleine to soften. But at the very instant at which the sip mixed with the cake crumbs touched my palate, I began to quiver in the awareness of the extraordinary thing happening within me. I had been invaded by a delicious pleasure, without any sense of why. Soon it rendered the vicissitudes of life indifferent, its disasters inoffensive, its brevity an illusion, in the same way as love, by filling me with a precious essence. Or rather this essence was not in me, it was me. I no longer felt mediocre, contingent, and mortal.[3]

Philosophers have analysed such bundles of sensations and ideas using a different set of conceptual tools.

Cartesian philosophy distinguishes between the perceptions of senses and of ideas by trying to classify our ideas according to their objectivity. John Locke extended the undertaking by distinguishing between ideas of primary qualities, ideas of secondary qualities, and ideas of power. Hume distinguished between impressions and ideas, and Kant, between intuitions and concepts. These all represented different ways of deconstructing experiences by attempting to weigh and distinguish between the cognitive and emotional quotients, the clarity and obscurity, and the distinctness or confusion of their elements.

Until at least the end of the nineteenth century and in classical philosophy,[4] these facts of perception were related to a subject, be this called *cogito*, "I think", me, self, transcendental subject, original unit of apperception, or transcendental ego, a subject who knows he/she has these perceptions, who knows that he/she has them and not someone else, who neither wants nor is afraid of losing him/herself in them.

On the other hand, our "experiences-totalities" correspond to modes of existence of subjects who are different from those of the receiving subject or the centre of representations. The relationship of belonging or of a face-to-face between the subject and the object leads to a fusion: "this essence was not in me, it was me," as Proust tellingly wrote.

3

In *Art in a Gaseous State*, I discussed these kinds of experiences in the field of contemporary art.

One can try to describe them literally, to imagine the correspondences and analogies, for example the poetic ones, but the essence remains fleeting. A good analogy for this ensuing diminishment would be the difference between being in a nightclub in full swing, with a crowd of people pressing against one another, perspiring, and blending into a single, moving mass, amidst an interplay of lights, video projections, and lasers, techno music beating against people's chests, and then visiting the same place in the early morning, when the cleaning crews come in, with its leprous walls, cables running everywhere, empty electronic consoles, and go-go dancer podiums with their scratched paint.[5]

I used the term "gaseous" to talk about these new experiences, in the same way that Zygmunt Bauman talks about the "liquid state" of modernity.[6] It's a question of describing situations where things matter less than processes, objects less than ambiances, qualities less than gradients of quality, facts less than their effects and residual echoes, substances less than flows, and entities less than relations.

This notion necessarily reaches past Art with a capital A for two obvious reasons.

First of all, there are more and more artistic fields in which these experiences play an essential role, especially music and architecture.

Music, whether experimental, commercial techno or disco, classic or electric-acoustic, ambient or as an individualized streaming, envelops and penetrates the body without representing or telling.

Architecture is a building to be looked at (from where, though?), a site to be visited, lived in, inhabited, or used. It is an element within the space of a city or a landscape to be traversed, whose ambiance it helps define as much as it is in turn defined by it. The experience of an architecture, even a villa by Alberti, is never of an object, rather of a movement whose duration is marked by changes in perspective, blanks, moments of attention and inattention, physical sensations of motion and of variations in light, smells, and sounds.

On the other hand, as we saw with aestheticization and the intensification it exacerbates, there are more and more domains in which we live experiences marked by a particular emotional tonality that appeals to several senses and to a synaesthesia, to several faculties (memory, imagination, and sensation), and which have a significant sensuous and sensory value. These lie at the heart of consumption (including the most essential items), of the economy of intangibility, pleasure, tourism, political life, and social interactions. At times still natural, they are most often fabricated, controlled, and technically guided. They are experienced as "everythings" in a new milieu of life to which a new sensitivity adapts.

4

Whether at a concert, on a promenade, a path around a monument or through a historical neighbourhood, while walking through an empty museum,[7] or a crowded nineteenth-century Salon, we are dealing with atmosphere. Our attention might suddenly focus and dwell, fascinated, on a detail, a tree, a painting, an architectural fragment, signs, symbols, and significations that point to the "old" uni-sensorial or cognitive modes of perception, but what comes first, chronologically and conceptually, is the atmosphere, the ambiance, the tonality of the surroundings, the milieu, the *Umwelt*.

In cities, these atmospheres are produced by the buildings, streets, cafes, stores, public transportation, and traffic, noises and sounds, smells (good, bad, and unique), people walking by, pressing against one another, gathering.

In landscapes, there is vegetation, animals, birdsong, gusts of air, humidity or dryness, reliefs, perspectives, and backgrounds. Everything blends into a "tonality" that is as undefinable as its recognizable.

At tourist sites, there are other tourists, souvenir sellers, musicians, pickpockets, guides, local smells, and cafes – and yes, even monuments to visit.

In places of consumption, there are lights, music, other consumers, bystanders, sales staff, people to welcome you, smells to draw in regular clients, and advertising.

In railway stations and airports, there is a blending of light and sound-based announcements, the noises of machines and means of transport, the colours of the uniforms of the flight attendants and air traffic controllers, stalls in front of the stores, the silhouettes of travellers who are anxious, in a rush, or already on vacation, families with children who play or cry out, the smell of food, the coolness of air conditioning, or an overheated environment.

When you decide to go see Giotto's frescoes in the Scrovegni Chapel in Padua, you first have to reserve a spot, and then, on the given day, cross the gardens of the old Roman Arena to arrive on time, joining a group of some 30 people together with whom you will be admitted. You then enter into a pressurized airlock that serves as an antechamber to your visit, and then, finally, you are in the chapel with its frescoes, which are protected from the atmosphere of the city of Padua, from the humidity of that climate and its bacteria, all through this artificial suppression.

There is little chance that, under such conditions, you won't be at once collected within yourself, impressed, and distracted. You have decided to have this experience, and you form part of a privileged group that is admitted to this chapel that is by now consecrated solely to the religion of art. All the circumstances have prepared you. You will rediscover the "normal" world once the time of the visit has ended. Perhaps you wish that you could extend the experience by holding on to some of that sense of collection and reflection as you walk down the city's streets to the Caffè Pedrocchi, where you will order a coffee and a pastry and think of Stendhal or Byron.

One might object that this atmosphere has nothing to do with a "true" aesthetic perception, namely a careful, collected examination of Giotto's frescoes.

Such an objection reduces the aesthetic experience to a perceptive core that is hardly ever attained, and which is in any case never maintained for long. It cuts it off from its conditions of possibility – those of a pathway to and through the experience, followed by one's exiting outside the experience.

Most of all, it reduces it to a moment of contemplation and deciphering, similarly to the folly of Theodor W. Adorno saying that one can only appreciate music if one follows its execution by reading the score. He was right! One can see Giotto's frescoes so much better when they have been reproduced in the pages of a coffee table book published by Taschen! Similarly, sunsets are so much more beautiful on the postcards of yore or in an Instagram post!

In fact, ambiances and atmospheres are the reality of the experience, and they demand that we shift our interpretive paradigm.[8]

The difficulty is that notions of experience, ambiance, atmospheres, and milieus are first and foremost hazy and vague; they are manners of speaking. As to the gaseous, it's no surprise that it's hard to grasp.

To give such expressions their due, we have to try to define them against the conceptual background that justifies their relevance.

5

Let us begin with the more general term that has been used for the longest: "milieu".

Leo Spitzer provided an impressive history of the notions of milieu and ambiance. He showed how we have gone from a concept of the physical that accounts for the transmission of actions (Newton's ambient "aethereal medium"[9]) to one concerning the relationship between a living being and its environment.[10] Spitzer quotes an entry from Lalande's dictionary of philosophy that summarizes this evolution:

> From the language of physicists, this word passed to the language of biologists, under the influence of Geoffroy Saint-Hilaire, one of whose principal ideas was applying the concepts used by physics and chemistry to the study of living things. He routinely referred to the "ambient milieu" in this sense. It then entered the language of the moral sciences through two independent paths. Auguste Comte, who had borrowed it from the naturalist Blainville, used it frequently. If one looks at the *Course of Positive Philosophy*, Lesson XL, p. 13 *et seq.* in particular, "Milieu" is initially printed in italics, and it is only used following a preliminary explanation of the idea it represents. On the other hand, Taine, who did more than anyone else to popularize this term, borrowed it from the introduction to Balzac's *Comédie Humaine* of 1841, which compares society to nature and the individual varieties of

humans to zoological species that depend on their "milieu". Balzac had himself taken it directly from Geoffroy Saint-Hilaire.[11]

In passing from the domain of physics to those of biology and then sociology, the term "milieu" became dynamic in its meaning. Instead of being a medium of transmission for actions that itself does not exert any action (Newton's ether), it became a field of forces and conditions on which living beings depend and to which they react. It came to signify the ensemble of external circumstances on which an organism's existence depends.

Thus, the milieu determines, acts, and constrains. This is the (caricatural) use that Hippolyte Taine makes of it when he adopts Auguste Comte's positions, reducing the three factors that – in his opinion – govern history and civilization to race, milieu, and the moment, which is not a time factor, rather an impulsion acquired through inertia, a momentum, and the quantity of a dynamic's motion.

This is not the occasion to engage in a detailed discussion of whether or not such a milieu is determinist. The interest in doing so would at this point be merely historical, because the notion has lost its determinist nature, which was still strong enough that a philosopher of biology and of living things such as Georges Canguilhem expressed strong reservations about this concept in the 1950s because he wanted to save the possibility of action and creation for living things.[12]

It was precisely in opposition to this determinism that Jakob von Uexküll proposed the notion of *Umwelt*, which is often identified with that of milieu,[13] and which renewed interest in it.[14]

Aside from the polemical context in which this notion was formulated (against the monist materialism of Ratzel and Haeckel), it takes into account the reciprocity of action between living things and their milieu.

Every animal species constructs a world unique to it by selecting certain stimulations received from the outside world. Its sensory organs are not mere receptors. By selecting stimulations, they isolate a certain excerpt (the *Ausschnitt*) from the outside world. An *Umwelt* is not universally shared; it is the correlation of the species' sensory organization.

Human beings also have their own, human *Umwelt*, which Uexküll calls their "home" (*Heimat*), with an explicit criticism of the destruction of this habitat at the hands of modern urban living, which denatures the human animal.

In the first article in which he addresses these questions in 1907, Uexküll termed milieu as "that excerpt of the outside world that is different for each animal and which is unique to it".[15] Starting in 1909, he replaced milieu with *Umwelt* and then *Merkwelt* (the characteristic world), which is connected to the *Wirkungswelt*, the world of action developed by the living being.

The stimuli of the *Umwelt* to which an animal reacts may be olfactory, luminous, tactile, vibratory, hygrometric, temperature-related, or chemical. They can be analysed in terms of the physical or the chemical. A living being responds to those stimuli that are decisive to its life.

An example provided by Uexküll has become a classic, that of the *Umwelt* of a tick, whose life cycle is governed by just three stimuli: an olfactory stimulus concerning the perception of butyric acid secreted by the glands of the mammals on which it lands, a tactile stimulus that helps the tick head towards a portion of skin free of hair where it can attach itself, and, finally, a signal coming from the animal's blood and temperature, which triggers the tick's own egg laying.

Aside from the fact that our experiential world is less simple than that of the tick, there is a clear interest in using the notion of *Umwelt* to think about ambiances, atmospheres, and milieus.

Human milieus consist of stimuli whose sensory relevance and degrees of intensity condition our experiences. The difference from the tick lies in the fact that these stimuli are not merely physical or chemical. They also include symbols and representations, even if positivists might always ascribe a symbol's effects to mathematical, physical, or chemical facts. After all, the sensuality emanating from a pair of Louboutin luxury high-heel shoes, symbols of a sexy elegance, wealth, and fashion, can be reduced to a red stimulus of a shoe bottom, to an equation that describes the high-heel's vertiginous instep, and to the modelling of the legs resting on them.

The human *Umwelt* is thus rich in signs and stimuli, from the symbols attached to objects to the stimuli in their environment, with at times (as for the tick) a reduction solely to physical, environmental stimuli. Sometimes the experience is individual, and sometimes, it is shared, not necessarily in unison with others, instead with different degrees of engagement, depending on whether one is attending a symphonic or electronic music concert, a movie screening, an opera performance, a football match, experiencing the tension of an election evening, or participating in a funeral.

For those who live these experiences, the multi-sensorial and often intersubjective emotional tonalities may not be easy to describe. This is not true, however, as we possess a very rich repertoire for doing so, one we use without difficulty, and in which we see many of the aesthetic categories we addressed in Chapter 2. A reception may be cold, tepid, enthusiastic, warm, or glacial. One can enter into a trance, ecstasy, be left indifferent, or made to feel aggressive, experience an adrenaline rush, and have to calm down and "de-escalate".

In terms of the mechanisms for aestheticization, the dimensions of the stimuli are identified, measured, and controlled by the producers of the human *Umwelt*, by the "aestheticizers", be they artists, musicians, sound engineers, lighting engineers, specialists in climate control, videographers, producers of virtual images and immersive environments, stage designers, orchestra conductors, political PR advisors, or producers of all kinds. All these "aesthetic workers" work on the degrees of stimuli of a milieu or *Umwelt* by combining them with one another.

6

The notion of ambiance does not disqualify nor supersede that of milieu; it instead adds a dimension of affect, feeling, and envelopment.

In the aforementioned article by Leo Spitzer,[16] written in 1942, which, for reasons of political circumstance that are readily understandable, offers a veiled criticism of the determinism of "milieu", ambiance is "a word that evokes a climate or spiritual atmosphere emanating from a milieu or even a thing," whose connotation of embrace, of envelopment is rather positive. We talk about the lively ambiance of a get-together, but about the awful climate of other kinds of meetings. As to milieu, it can host all values; proper milieus contribute to education, while others lead to delinquency.

The noun ambiance began to be used in the 1890s: in 1885, in a short story by Villiers de L'Isle-Adam titled "Sublime Love",[17] in 1891 in André Gide's "Treaty of Narcissus",[18] and in the *Journal* of the Brothers Goncourt.[19]

The adjective ambient was already current, generally associated with milieu, but symbolist writers frequently used the suffix -ance to suggest duration and persistence; ambiance was the ambient becoming a duration, extending and persisting, an enveloping and surrounding environment.[20] Ambiances played an even more important role for existential psychoanalysis (*Daseinanalyse*) in the 1930s.[21] Eugène Minkowski, Ludwig Binswanger, Hubertus Tellenbach, and Erwin Strauss questioned behaviourist psychology and its coupling of stimulus and response, adding an ontological-existential dimension to behaviours to interpret mental illness and modes of being in the world in which the psychological and the physical merged.

Minkowski thus talked about the syntony that "vibrates in unison with ambiance" and "the schizophrenia that separates us from it". He defined ambiance as follows:

> Ambiance should not be compared to what, in talking about perceptions, we call the "outside world", nor its purported elements into which it is consequently broken down. It must be considered as a vast, living whole, with all of its primitive dynamism, in which only humans, with the assistance of the analytical processes at their disposal, manage to discern living beings, objects, and even physiological excitants. In all its details, an ambiance is not a thing that is made. It makes itself through contact with humans, in the same way that humans make themselves. Ambiance is primitively a moving ocean. It is the future.[22]

In *The Case of Suzanne Urban*,[23] Ludwig Binswanger discusses a woman who is unable to keep the world at a distance and is thereby dispossessed of herself.

Instead of sticking to a space of objects and stimuli within which we act, these authors believe that it is essential that we consider another space, the affective one of our feelings and sensations, of the ambiances that surrounds us, and the sentiments that run through us and which to us represent the physiognomy of the world.

This way of thinking does not exclude the sensory data of ambiances, which have, in addition to their informational value, an existential scope. Hearing is the retention of the world; smell, its expansion; touch, a grasping or a fear of the world. Similarly, our expressions about our relationship to space explain our ways of being in the world: falling, going up, embracing, distancing oneself, withdrawing, or shrivelling up.

Erwin Strauss dwelled even more explicitly on the aesthetic aspect of our presence in the world.[24]

For him "feeling" means our original way of relating to the world, thus insisting on the sensory dimension of our lived experience. This feeling is not a precise, mono-sensorial sensation, rather an immediate, comprehensive means of communication with the world. It is through feeling that a living being forms a whole together with the world. Anticipating views that would later be expressed by Henri Maldiney and Michel Henry,[25] Straus identified feeling as the pathic moment preceding any and all objectivation (and as the condition for any objectivation) of perception that has a cognitive dimension.[26] Empathy is the broader concept that encompasses both the act of separating and reuniting, fleeing or following, fear or attraction, and which therefore also includes like and dislike.

7

Milieu and *Umwelt* consist of physical and chemical stimuli of a measurable intensity that condition our behaviours. By making so much room for psychological tonalities, the notion of ambiance rebalances the approach, however with the risk of assigning too much importance to lived experience and to existential situations.

The notion of atmosphere seems to allow for an intermediate conception. We must nevertheless be sensitive to the risks of linguistic bias and not precipitously "reify" atmospheres as if they were things. French, with its symbolist and aesthete heritage, renders ambiances more familiar, while "atmospheres" in German and English provide a framework better suited to the *Gestaltung* ("configuration" or "formation") of architects and designers.

The word atmosphere comes from meteorology and designates the gaseous layer roughly thirty kilometres thick that envelops the earth. According to Littré's French dictionary, the word has been used figuratively since the nineteenth century, when people began to speak of the atmosphere of a landscape, a room, or of a family reunion.

The notion was systematically developed in the late 1980s by German philosopher Gernot Böhme, in dialogue with the treatment provided by Hermann Schmitz in his *New Phenomenology* published in 1980, which comprised the ten volumes of his *The System of Philosophy* published between 1964 and 1980.[27] It was then commented on, developed, and at times rendered more complex starting in the 2010s by architectural theorists (such as Jean-Paul Thibaud and Christian Borch[28]), Italian philosophers of aesthetics (Tonino Griffero in particular[29]), while Bruce Bégout in France opted for the concept of ambiance and the ontology that this entails.[30]

The fact is that Böhme is not a phenomenologist. He does not base himself on lived awareness, instead on an ecology that takes humans into account in terms of ecological, natural, material, and aesthetic factors.

In his 1989 book *For an Ecological Aesthetics of Nature*,[31] he found that an environment produces impressions on our feelings, thus reiterating the Heideggerian concept of *Befindlichkeit*, the sensory disposition characterizing the human situation and reflected in our moods, feelings, and affectivity.[32] This results in "atmospheres" that may be considered both as feelings and as sources producing feelings.

Böhme's main assertion is that atmospheres are objective, subjective, and intersubjective, and that they are produced by, yes, the natural environment, but even more so by humans acting on their milieu. It is this aspect of production, doing, and technique that renders this concept so interesting.

In terms of this production, the operating paradigms were, in the past, the art of the garden and that of the stage.[33]

In my mind, the poietic, practical, instrumental, and technical aspects of atmospheres render the notion more relevant for addressing a hyper-aestheticized world than that of an ambiance closer to natural and poetic experiences. It is also more relevant than that of the milieu, which retains traces of its mechanistic origins and of its reference to merely physical stimuli. Advertising images, gestures, and words also form part of atmospheres.

These are the result of engineering processes that act on a variety of technically controlled factors, even though they are felt and experienced as psychological and physical ambiances. They have an objectivity and a subjectivity, and they express the co-presence of subject and objects, an inseparable co-existence.

What was first expressed in gardens and theatres and operas is now omnipresent in our consumption, advertising, sports, culture, architecture and design, art, politics, and social life. We live atmospheres and we produce them.

8

If we do not totally reject the classic, objectifying approach, we must at least relativize it. Our perception is composite. It is made up of atmospheres, ambiances, things, people, appearances, disappearances, and processes. It consists of the ever-shifting mixing of all these things. In its details, Proust's passage about the madeleine exemplifies this mixing together of perceptions that range from the precise to the atmospheric.

The great interest of the notions of *Umwelt*, ambiance, and atmosphere lies in their ability to explain a vast number of experiences from both the past and the present, in the field of art as well as in daily life, in terms of their experience and their production: the art of gardens and walking, the art of landscape and the experience of moving through them, musical production and immersive listening, architectural construction and modes of inhabiting, the promotion of tourism and the experience of a holiday, and the design of experiences and their consumption.

The experience must be related both to the techniques and the producers who generate them: landscape architecture and landscape architects, composition and composers, musical performance and musicians and orchestra conductors, stage productions and stage designers, architectural projects and architects, product and interior design and designers, the construction of tourist packages and tourist trade operators, the design and construction of public spaces and urban planners, and the staging of political communications and the communicators and spin doctors.

This is a matter not of states, rather of passages, transformations, and processes that depend on technical progress and on ideas for using and perfecting them.[34]

One can only understand and explain the number of "natural", sensory, poetic, and aesthetic and aesthetic experiences fully by invoking these notions.

This is even more so the case for the experiences that have been designed, produced, and implemented in complex technical environments, like most of the ones in which we live. The perception of a bit of wax (Descartes), a ball rolling down the smooth groove of an inclined plane covered with leather (Galileo), or of two billiard balls striking one another (Hume) is a somewhat limited aspect of our perceptive conduct.

John Locke, who, to my knowledge, never once talked about art or beauty, distinguished between ideas of primary qualities, secondary qualities, and power. He thus unwittingly opened the door for a new domain of experience to a new kind of knowledge. Ideas of primary qualities reveal certain aspects of the nature of things (form, solidity, and movement), while those of secondary qualities (colours, smells, and tastes) affect us without our understanding their correspondence to things. As to the powers, as the term indicates, these are tendencies to exercise actions and effects on us and on other things. Therefore, there are ideas that we truly have, which we know we have, but for which we do not know what their counterpart is in things.

In bringing together his thoughts on Locke and on the clarity and distinction of Descartes' ideas, Leibniz took one step further in identifying this domain. In his opinion, we do in certain cases have clear knowledge (we know that we have this) and distinct but confused knowledge (we don't know what its elements are). This is the case of our knowledge "of colours, smells, tastes, and the experiences that painters and artists give us".[35] We recognize the thing without being able to say what the differences and properties are. By means of these clear and confused ideas, the mind enters into states that are not logical, rather aesthetic and sensory. This was precisely the field of "lower gnoseology" for Baumgarten. We have an experience, but we can't clearly say what our knowledge of it is, nor how it is different from others. Hence, the meaning of the "je-ne-sais-quoi" that became so current in the second half of the seventeenth century.

In many ways, ambiances and atmospheres are our very own "je-ne-sais-quois", except that we are more and more capable of producing and manipulating them. We are virtuosos when it comes to producing such "je-ne-sais-quois".

The passage to this new type of approach has several implications.

The first is an expansion of aesthetic theory that will bypass the domain of Great Art, which will become just one aspect of this theory.

We must then see to what extent we can elaborate these notions of atmosphere and ambiance. Until now, we have merely located them. This elaboration is not easy, theoretically or culturally speaking, as we continue to reason within the classical paradigm established by Descartes, Locke, and Leibniz, which identifies sensations and representations in relation to a subject. How can we escape this, and with which concepts and using which language?

Lastly, the recognition of the affective and pre-reflective aspect of atmospheres requires us, sooner or later, to consider the issue of the type of pathic or aesthetic subject experiencing them. That will specifically be the subject of the third and final part of this book.

Notes

1 Stendhal, *Rome, Naples and Florence*, trans. Richard N. Coe (London: Alma Classics, 2019).

2 *Le Renouvellement de l'observation dans les sciences* ["The Renewal of Observation in the Sciences"], Yves Michaud (ed.) (Paris: éditions Odile Jacob, Université de tous les savoirs collection, 2003). See my introduction in particular.
See also the recent lecture by Anouk Barberousse before the French Philosophy Society on 16 November 2019, "What role does data play in scientific reasoning?", the gist of which – with which I am in full agreement – was summarized as follows:

> Philosophical theories of scientific reasoning traditionally emphasize the relationship of confirmation between data and hypotheses. In so doing, they consciously compare data to utterances so that they can opine on their logical or probabilistic relationships to hypotheses. Such a comparison does not do justice either to the diversity of data considered today in the sciences, or to the ways in which they are processed by computers.

The goal of this essay is to consider in depth the nature of contemporary scientific data and study the transformations they cause to scientific reasoning. What can we call "reasoning" when the bodies of data are not just digital, but also gigantic? When are the data accompanied by measures of uncertainty? When do they come from digital simulations and not from physical interactions with a detection or measurement instrument? The current avalanche of data seems to render traditional forms of scientific reasoning obsolete, and yet, it is precisely on the basis of data that we establish the facts that we can try to explain using hypotheses and theories. The road from data to hypothesis is longer than classical approaches to scientific reasoning presume. We must assess their credibility and the uncertainties associated with them. At times we have to complete them with the assistance of models of the phenomena being studied. Models are introduced not only into the reasonings that go from the data to the hypotheses, but also into those that help render the scientific data acceptable. Scientific reasoning is constructed today on the basis of a dialogue between data and models, with the help of computers; however, this does not mean that the intellectual control exercized by the human mind has become superfluous."

3 Marcel Proust, *Swann's Way: In Search of Lost Time, Volume* 1, transl. Lydia Davis (London: Penguin Classics, 2004).

4 We must leave out of this critique the pre-Romantic and Romantic texts on states of consciousness and sensibility, whether by Rousseau, pre-Romantic and Romantic poets, or Maine de Biran. They testify to a current that represents

a counter-history of subjectivity in opposition to a history of philosophy that focuses on the concerns of objective knowledge. See Sections 3–5 of Chapter 10 of this book.

5 This difficulty explains why much of what we now call art criticism is a kind of para-text, a quirky commentary that assumes that it is impossible to describe and analyse what one sees, or rather, what one feels – when it's not just an advertising blurb devoid of any meaning.

It also explains why contemporary artists no longer assign any importance to critical texts, other than as promotional material, without any regard for their intellectual or literary calibre. Some go as far as prohibit commentary and criticism, because only an experience is possible.

Frank Stella was a precursor of this when he said of his paintings of minimalist black stripes in 1964: "When pressed, people always say that there is something more than just painting in the work. My painting is based on the fact that only what one can see is there. What you see is what you see". His phrase still holds true if we adjust it slightly: "The experience you have is the experience you have". Even more radical was the initiative undertaken at the Guggenheim Museum in New York in January 2010 by English artist Tino Sehgal, who exhibited a work without an object, without an introductory text, without a catalogue, without an opening, with a prohibition of photographing (although one should read Holland Cotter's article in the *New York Times* of 31 January 2010, because, despite everything, we cannot prohibit journalists from doing their job). The "work" consisted solely of visiting the museum emptied of its works, occupied by two dancers who would occasionally embrace each other during a performance with the unexpected encounter of paid interviewers who would strike up a conversation with visitors. The work was the experience as a whole, with its full and empty volumes, visitor expectations, which were disappointed yes and no, their surprises, curiosity, or boredom, their pleasure in participating in this experience, or their scorn for having been guinea pigs or feeling duped.

6 Zygmunt Bauman, *Liquid Life* (Cambridge: Polity Press, 2005).

7 I refer to the hilarious instance of bravura in the visit to the Louvre by the guests at the wedding of Gervaise Macquart and Coupeau in Chapter III of Emile Zola, *L'Assommoir,* trans. Brian Nelson (Oxford: Oxford World Classics, 2021, first published in French in 1877).

8 This view is also shared by Bruce Bégout in his book *Le Concept d'ambiance,* with whose analyses I agree and view as essential – without appropriating his attempt at a neo-phenomenology culminating in an ontology of ambiances. I find his choice of a concept of ambiance indicative of his "fundamentalist" option, whereas the privilege I assign to "atmosphere" refers to the technical and cultural conditions that format this new sensibility. As we will see, this is the divergence we find between Schmitz (neo-phenomenology) and Böhme (techno-aesthetics), and I accordingly situate myself on Böhme's side.

9 For example, Isaac Newton:

> I here use the word attraction in general for any endeavour, of what kind soever, made by bodies to approach to each other; whether that endeavour arise from the action of the bodies themselves, as tending mutually to or agitating each other by spirits emitted; or whether it arises from the action of the aether or of the air, or of any medium whatsoever, whether corporeal or incorporeal, anyhow impelling bodies placed therein towards each other.

Book I, Proposition 69, Theorem 29,
Scholium of Isaac Newton,
Principia mathematica Philosophiae naturalis,
translated by Andrew Motte (London: Benjamin Motte, 1746)

10 Leo Spitzer, "Milieu and Ambiance: An Essay in Historical Semantics", in *Philosophy and Phenomenological Research,* Vol. 3, No. 2, (December 1942): 169–218.

11 *Vocabulaire de la Philosophie* ["Dictionary of Philosophy"] by Lalande, entry for "Milieu", cited by Leo Spitzer, "Milieu and Ambiance: An Essay in Historical Semantics", in *Philosophy and Phenomenological Research*, Vol. 3, No. 2, (December 1942), p. 174.

12 I refer to Jean-François Braunstein's article "Canguilhem avant Canguilhem" ["Canguilhem before Canguilhem"], in *Revue d'histoire des sciences*, Vol. 53, No. 1 (January-March 2000): 9–26, on Georges Canguilhem's reservations on the determinism inherent to the notion of milieu. Canguilhem's text "Le vivant et son milieu" ["Living Things and their Milieu"] was presented at the Collège philosophique in Paris in 1946 and published in Georges Canguilhem, *La Connaissance de la vie* ["A Knowledge of Life"] (Paris: Vrin, 1952).

13 In the above article, Georges Canguilhem often mentions "milieu" as an equivalent to *Umwelt.* See Wolf Feuerhahn, "Du milieu à l'Umwelt: enjeux d'un changement terminologique" ["From Milieu to Umwelt: the Challenges of a Change in Terminology"], in *Revue philosophique de la France et de l'étranger*, "L'homme et la nature" ["Man and Nature"], Vol. 134, No. 4 (October-December 2009): 420.

14 See Wolf Feuerhahn, "A Specter is Haunting Germany – the French Specter of "Milieu": On the Nomadicity and Nationality of Cultural Vocabularies", *Contributions to the History of Concepts*, Vol. 9, No. 2 (Winter 2014): 33–50.

15 Jakob von Uexküll, "Die Umrisse einer kommenden Weltanschauung" ["The Outlining of a New Worldview"] in *Die Neue Rundschau*, Vol. 18 (1907): 649, cited by Wolf Feuerhahn in "Du milieu à l'Umwelt: enjeux d'un changement terminologique".

16 Leo Spitzer, "Milieu and Ambiance: an Essay in Historical Semantics", 188.

17 Villiers de L'Isle-Adam, "L'Amour sublime" ["Sublime Love"] (1889):

> Évariste – as we have implied – was one of those people who comes to understand things without ever getting worked up. His energetic but calm demeanour gave him the constant and serial ability to *label* any fact without isolating it from its ambiance – and as a result, to control it and even use it, if possible – to an extent that was proper, of course.

Auguste Villiers de L'Isle-Adam, *Oeuvres complètes*, Volume II, Pierre-Georges Castex and Allan Raitt with Jean-Marie Bellefroid, ed. (Paris: Gallimard, Bibliothèque de la Pléiade collection, 1986), 717.

Villiers' piece describes an adulterous affair that was nevertheless platonic, and which stayed that way, despite the traps repeatedly set by the husband. He used metaphors of the skies, vapours, clouds, and ethereal spheres.

18 André Gide, *Le Traité du Narcisse* (1891):

> Narcissus stopped at the edge of the river of time, that fatal, illusory waterway where the years flow and pass by. Simple edges, like a rough frame, enshrined the water, like a one-way sheet of ice, with nothing to see behind it, where the emptiness of ennui unfurled itself. A sad, sluggish canal, almost a mirror lying horizontally. And nothing distinguished this drab water from its monochrome ambiance, other than the sound of its flowing by.

André Gide, *Romans et récits*,
Volume I, Pierre Masson,
ed. (Paris: Gallimard,
Bibliothèque de la Pléiade collection, 2009), 170

19 Edmond and Jules Goncourt, *Journal*, 20 March 1891:

> This evening, at the *Dîner des Spartiates*, people argued that Westerners had a more complete and detached sense of individuality that existed with greater relief in relation to nature, less consumed by the ambiance of milieus, an individuality which thus held more of a will of one's own than Easterners, whose individuality is lost, melted, drowned in the Great Big Everything, in its exuberance for plant-ness and animal-ness, which renders them prey to Nirvanaism, this cowardly, smiling spineless will which seems resigned to the nothingness that is humanity in these exotic lands.

20 See Leo Spitzer, 189:

> For -ance (derived from a substantivization of the neuter plural of the participle)... has ever suggested the perpetuation of a state of being, even though this state may be the result of an act itself transient; *la souvenance* is an eternal residue. Revived by the Romanticists, who delighted in seeing the ternal even in the passing, this suffix was kept alive by the Symbolists and the Impressionists who followed them. Thus Goncourt was enabled to create a word which would represent not only an essence, but an essence given form, a quality with an existence, a subsistency, a consistency of its own.

21 Jean-Paul Thibaud, "Petite archéologie de la notion d'ambiance" ["A Brief Archaeology of the Notion of Ambiance"], in *Communications*, "Les bruits de la ville" ["The Noises of a City"], No. 890 (2012): 155–174. Bruce Bégout, *Le Concept d'Ambiance,* 89–120, on Minkowksi, Straus, and Bollnow.

22 Eugène Minkowski, "Constitution et conflit" ["Constitution and Conflict"] in *Écrits cliniques*, Bernard Granger, ed. (Ramonville-Saint-Agne: Erès, 2002), 67–79, cited by Jean-Paul Thibaud, 162–163.

23 Ludwig Binswanger, "Der Fall Suzanne Urban: Studien zum Schizophrenieproblem [IV Studie]", translated into French as *Le cas Suzanne Urban: Étude sur la schizophrénie* [Sn#19] ["The Case of Suzanne Urban: A Study of Schizophrenia"] trans. Jacqueline Verdeaux (Paris: Desclée de Brouwer, 1957) (new edition – Paris: Gérard Monfort, 2002).

24 Erwin Straus, *Du sens des sens. Contribution à l'étude des fondements de la psychologie* ["The Sense of the Senses: A Contribution to the Study of the Foundations of Psychology"] (1935), translated into French by Georges Thinès and Jean-Pierre Legrand (Grenoble: Jérôme Millon, 2000).

25 Henri Maldiney provided many examples of this pathic moment of feeling. For example, when one enters a cathedral, one's sensibility, the ability to feel is first ordered and situated according to a sense of attraction and fear, simply due to the interplay of colours and forms in the stained-glass windows. Before engaging in any iconographic interpretation of the stained-glass windows, our sensibilities our affected and ordered according to what we like or don't like. Henri Maldiney, "Le dévoilement de la dimension esthétique dans la phénoménologie d'Erwin Straus" ["The Revealing of the Aesthetic Dimension in Erwin Straus' Phenomenology"], in *Regard. Parole. Espace*, Lausanne, L'Âge d'Homme, coll. Amers (1994): 137–138. For Michel Henry, see *Incarnation: a Philosophy of the Flesh*, trans. Karl Hefty (Evanston, IL: Northwestern University Press, 2015), especially Chapter 15 "The original self-revelation of life as the foundation of the phenomenological method", and the conclusion of *Philosophy and Phenomenology of the Body*, trans. G.J. Etzkorn (Amsterdam: Kluwer Academic Publishers, 1975).

26 Straus understands human life on the basis of two radically different modes of being: the world of feeling and the world of perception, which are distinguished from one another the way the space of a landscape is distinct from that of its

geography. The space of the landscape is felt pathically, according to frameworks of like and dislike. The space of the geography is the space of knowledge and of language; it is an overall marker of the world in relation to geometric coordinates.

27 Hermann Schmitz, *System der Philosophie* ["The System of Philosophy"] (Bonn: Bouvier, 1964–1980): I. *Die* Gegenwart ["The Present"], 1964; II 1. *Der Leib* ["The Body"], 1965; II 2. *Der Leib im Spiegel der Kunst* ["The Body Mirrored in Art"], 1966; III. *Der Raum* ["Space"], 1. *Der leibliche Raum* ["Bodily Space"], 1967; III. *Der Raum* 2, *Der Gefühlsraum* ["Emotional Space"], 1969; III. *Der Raum 3, Der Rechtsraum, Praktische Philosophie* ["The Legal Space: A Practical Philosophy"], 1973; III. *Der Raum, 4. Das Göttliche and der Raum* ["The Divinity and Space"], 1977; III. *Der Raum*, 5, *Die Wahrnehmung* ["Perception"], 1978; IV. Die Person ["The Person"], 1980; *Die Aufhebung der Gegenwart* ["The Abolition of the Present"], 1980. See. Republication of the 10 volumes (Freiburg-im-Breisgau: Alber, 2019) and as *Neue Phänomenologie* ["The New Phenomenology"] (Bonn: Bouvier, 1980).

28 Jean-Paul Thibaud, "Petite archéologie de la notion d'ambiance"; *En quête d'ambiances. Éprouver la ville en passant* ["In Search of Ambiance: Experiencing the City in Passing"] (Geneva: Métis-Presses, 2015); *Ambiances en débat* ["The Debate over Ambiances"], Pascal Amphoux, Jean-Paul Thibaud, and Grégoire Chelkoff, ed. (Bernin: À la croisée, 2004); and *Architectural Atmospheres: On the Experience and Politics of Architecture*, Christian Borch, ed., (Zurich: Birkhauser Verlag, 2014).

29 Tonino Griffero, *Quasi-Things: The Paradigm of Atmospheres* (Albany: SUNY Press, 2017); *Atmospheres: The Aesthetics of Emotional Spaces* (Abingdon-on-Thames: Routledge, 2016); and *Il pensiero dei sensi. Atmosfere ed estetica patica* ["The Thought of the Senses: Atmsopheres and Pathic Aesthetics"] (Milan: Guerini Scientifica, 2016).

30 Bruce Bégout, Le Concept d'ambiance.

31 Gernot Böhme, *Für eine ökologische Naturästhetik* ["Towards an Ecological Aesthetics of Nature"] (Frankfurt-am-Main: Suhrkamp, 1989).

32 Martin Heidegger, Being and Time (1927), trans. John Macquarrie and Edward Robinson (New York: Harper Perennial Classics, 2008).

33 Gernot Böhme, *Atmsophäre. Essays zur Neuen Ästhetik* ["Atmospheres: Essays on a New Aesthetics"] (Frankfurt-am-Main: Suhrkamp, 1995) I, "Atmospheres", Section 6, "The Making of Atmospheres", 34 *et seq*.

34 If we consider the case of music, past and present, its formal and technical evolution has expanded musical materials and the experiences made from them. Not only has there been an evolution in ranges and intervals, but also the invention of many new instruments, the use of pure sounds and noises, the production of synthesized sounds using first magnetic tape and then synthesizers and beat-boxes, and the use of sampling and deformations. Work on the spatial environment, already present in the composition and arrangement of musical formations, has become an integral part of the work of music. To this we must add the manipulation of the light environment and the projection of images.

Art installations use all these techniques and are becoming increasingly immersive, like operas, be this in the work of Matthew Barney or Laure Prouvost.

At the same time, concerts have become artistic performances, like the concerts given by Pierre Henry in the 2010s or those by Kraftwerk during their 2017 reunion tour. The acoustic space becomes atmospheric, with an ambiance.

The phenomenon is so widespread that spaces in general have all come to have sound signatures. These "signatures" were for a long time barely noticed, if at all, except when people reiterated "the noise of Paris". Sound specialists now record soundscapes to sell them: the acoustic atmosphere of a city, a landscape, a forest, or a jungle, with its birds and animals. A landscape also consists of smells, lights, murmurs, and cries.

We thus come to the ubiquitousness of music today, be this in the form of ambient music in places on moves through or what one takes along to colour one's experiences – or to cut oneself off from some of their aspects (the presence of others, extraneous noise, or aggressive stares).

If instead of building aesthetic theory on the basis of literature and the visual arts, we had constructed it using music and sound, it would have been entirely different. This hypothesis is obviously pure fantasy, because the theory could not avoid ignoring the artistic productions of its time, or overturn their hierarchy, which depended on the technical, social, and even political conditions of the time. On the other hand, it is imperative that we begin to develop an aesthetics of ambiances based on these new experiences and their technical conditions.

35 Gottfried Wilhelm Leibniz, *New Essays on Human Understanding*, trans. Peter Remnant and Jonathan Bennett (Cambridge: Cambridge University Press, 1996), Book II, Chapter 29, and especially the *Meditationes de Cognitione, Veritate et ideis* (1684), in *Opuscules philosophiques choisis* ["Select Philosophical Opuscules"], Paul Schrecker, ed. (Paris: Vrin, 1969), 10.

When Leibniz was expounding his ideas (these works date from the 1680–1690s), in France Père Bouhours was developing in his *Entretiens d'Ariste et Eugène* ["Conversations between Ariste and Eugène"] (Père Dominique Bouhours, *Les Entretiens d'Ariste et d'Eugène,* Paris, Chez Sébastien Mabre-Cramoisy, 1671 and 1687) the idea of a "je-ne-sais-quoi" at the heart of our sensory and emotional experience. When we perceive something beautiful, we feel a je-ne-sais-quoi. In other words, we clearly perceive a feeling the cause of which we are unable to grasp. The "je-ne-sais-quoi" to which the fifth conversation is devoted is "something that only exists because we can't say what it is" (p. 257). It is interesting to note that Ariste and Eugène, though agreeing on the importance of the je-ne-sais-quoi, instead differ on its presence in art: Eugène believes that all beauties in artworks are marked (thus, there is a distinct knowledge of these beauties), whereas Ariste, unwittingly a Leibnizian, believes that there is also a "je-ne-sais-quoi" in artworks. Leibniz explicitly uses this idea of a "je-ne-sais-quoi" to discuss the experience of clear and confused ideas.

SCHIN
SCHIN
SCHIN

Part 3

Ambiances, dialogues, and resonances

8 Exercises in ambiances

A journey across definitions for a multi-perspectivist anthropology of atmospheres

Olivier Gaudin and Maxime Le Calvé

From boredom to feverish passion, suffocation to euphoria, from fear to indignation and to sudden appetite: we move through ambiances, and ambiances move through us. Certain atmospheres make us feel as if we are compromising, while others make us feel welcome and ask us to stay. Their sensory presence might also ask us to act, to laugh, or to run away. Ambiances vary and take many forms, from the tenacious to the diffuse. We cultivate some while eliminating others. We are witnessing global contagions that tend to render our sensibilities uniform. And yet, changes in perspective may arise, surprising us and rattling our dispositions.

For almost a century, social and environmental psychology, the philosophy of perception – especially its phenomenological and pragmatist strains – and the social sciences, ethnography in particular, have helped to make ambiances and atmospheres into topics of study. The first works on the ecology of sensorial perception[1]and inhabited space[2] built upon the foundational work of John Dewey and George Herbert Mead in social psychology, the shared conditions of experience, and learning environments. Researchers in the domains of architecture, urban planning,[3] and geography[4] have also addressed the notion of ambiance. At the same time, with the development of cinema, the adventure of architectural modernism, and the advent of graphic and industrial design, new tensions have come to permeate the aesthetics of everyday life. Thus, thinking about ambiances now also extends to the study of literature[5] and the theory of aesthetics,[6] as well as the sociology of public spaces,[7] the anthropology of the ordinary,[8] urban history,[9] and even the history of sensibilities and emotions.[10]

With the notions of atmosphere and ambiance, ethnographers, sociologists, and historians have sought to highlight the perceptible effectiveness of situations on individuals, both those observed and observing. In anthropological research, the person conducting the investigations experiences a multitude of perspectives and attachments and is affected by various atmospheres, even if their existence remains uncertain. The journey "across ambiances" that we trace within this chapter begins with a brief historical analysis that demonstrates the abundance and ambivalence with which this notion has been used. It then considers ambiance and atmosphere as major themes

DOI: 10.4324/9781032655307-14

that in turn travel through our individual and collective experiences. This dual movement reveals "exercises in ambiance" in the most empirical of contexts.[11] The expertise of artisans who create atmospheres – be they ritual specialists, experts, architects, or artists – can elicit misleading manipulations as well as productive shifts in our perceptive habits. We believe that taking these usages and their ambivalences seriously bring us to a new position, as it opens the way to new (multi)perspectives in social sciences.

Ambiances and atmospheres in art and in life: a historical journey

"Ambiance": French seems to have exclusivity over the current usage of this neologism, while the term "atmosphere" is widespread in most European languages. When and why did we begin to distinguish between these two notions? As in other domains, the work of artists and scholars predates that of philosophers and the refining of the terms by several decades. In French, the noun "ambiance" stems from the proximity of scientific and literary language. We find one of its first appearances in *L'amour sublime*, a posthumous novella by Auguste de Villiers de L'Isle-Adam published in 1893. The word also appears in the Goncourt's *Journal* (1891) and Mallarmé's *Divagations* (1897). These initial usages were as a quasi-synonym to the term "atmosphere", widely used by the symbolists. Baudelaire, who used it on several occasions in his *Salons*,[12] discovered its expressive force when he was translating Poe, whose works are laced with recurring detailed descriptions of various atmospheres, each endowed with specific "influences". A word that usually qualified the properties of the air we breathe came to designate, in resonance with the natural science of the time, an active ingredient of our sensorial milieu. Poe's originality lay in personifying atmospheres, in accentuating their unique properties, and in dramatizing their effects on those who move through them.

Even though "atmosphere" found its way into art criticism and eventually the vocabulary of poets who reformulated it as "ambiance", it was primarily a scientific concept.[13] Mathematical calculations of variations in air pressure attest to the fact that atmosphere is something material, tangible, and quantifiable. Research into the mechanics of fluids and the movements of air converged with the concerns of naturalists and especially physicians, particularly in terms of our circulation. This interest in the concept of atmosphere suggested the possibility of articulating an anatomical knowledge of the internal workings of organisms and their relation to the outside world, that is, their immediate surroundings (*ambiens*) as well as unlimited cosmological space. Pascal's famous "two infinities" are also two *invisibles* to the naked eye that a scientific description of atmosphere is instead able to grasp, at least in part.

Starting in the mid-eighteenth century, the word "atmosphere" began to appear everywhere. We find its first aesthetic uses in Diderot's *Salons* (written between 1759 and 1781). Landscape paintings rendered visible different, singular atmospheres. In their cultivation of art criticism as an authentic literary genre, materialist philosophers showed that a descriptive use of the term "atmosphere" had begun to spread beyond the academic domain.

Thus, the appearance of the word "ambiance" is tied to the spread of this concept in the eighteenth and nineteenth centuries. By detaching itself from the scientific connotation associated with the atmosphere and referring to the term milieu and its English equivalent, "environment",[14] ambiance introduced the possibility of a more specifically artistic designation. For Mallarmé,[15] as for Villiers de L'Isle-Adam, a literary appropriation sought to preserve the formal force or "influence" associated with atmosphere while introducing the notion of a recomposition of the emotional, sensorial, and moral qualities of a lived or envisioned situation. Ambiance was simply an atmosphere shaped or fabricated by a painter, poet, or architect. But it could be perceived by viewers in a supposedly non-reflective way, regardless of their degree of artistic literacy; it could be sensed immediately and appreciated for its aesthetic qualities, to the point of making a deep impression.

From the presence of atmospheres to exercises in ambiances

The possibility of distinguishing between ambiance and atmosphere in French should in no way contribute to a sense of opposition between them. It should instead lead us to consider the resources offered by translation and a collective reflection on exchanges between languages and cultural traditions.[16] A first definition of ambiance could be as a *kind* of atmosphere generated by interactions that affect the perceptible forms of a situation with its existing atmospheric processes, but which also alters and directs the flow of perceptions, emotions, and interactions. As a descriptive resource, ambiance encloses within a specific time and place the prospect of a broader study of the atmosphere – while energizing and complexifying it, because it embraces a whole range of emotional processes and gestures perceived by a situated observer. Ambiance is characterized by its experiential aspect, by what one experiences synesthetically, as highlighted in the "new phenomenology" of Hermann Schmitz.[17] Based on the difference between these two terms in French, we can draw a distinction between ambiance and atmosphere by specifying the former in relation to the latter according to the following two criteria:

– its limited dimension in terms of space and time: it is hard to separate an ambiance from its specific spatial and temporal context, whereas we can analyse the qualities and properties of an atmosphere more abstractly, as a fluctuating whole that presides over a stretch of time and space that exceeds the observer's immediate range of perception;

– its formal character: the appearance and persistence of an ambiance is the result of a set of interactions between the people present, the perceptible qualities of a place, and the different emotional circumstances of a situation. However, there can be recurrent, stable ambiances. This property distinguishes ambiance from atmosphere, milieu, and landscape – three notions that incorporate living things and natural phenomena such as climate, weather, topography, and physical geography. This aspect allows us to envisage that ambiances can be controlled or monitored for purposes ranging from the social and the political (gatherings, protests, riots, as well as civic, neighbourhood, and political participation meetings) to the religious (ceremonies, masses), the artistic (performances and live spectacles), and the commercial (amusement and theme parks, commercial centres, retail stores, and sporting events). The expert practices of ordinary workers in the realm of the aesthetic prove that they can be studied from a naturalist perspective. Such an approach, which would, for example, employ the methodological tools of description and quantitative measurement of atmospheres, would nevertheless remain contingent on the complexity of ambiance-related processes, as we are never totally immune from the "shattering of an ambiance". The existence of such processes depends on a dense atmospheric substratum inhabited by living bodies that are moving and becoming. These active forms of life are endowed with organs of perception and a capacity for resistance. They are sentient, thinking beings that technological protocols, the calculated workings of machines, and the application of probabilistic statistics will never be able to capture or subjugate fully, however perfect they may seem.

None of these criteria can settle for a dualist divide between the subjective and the objective, representation and experience, or nature and culture. Ambiances are the result of living activities, both present and future, which are both lived experiences and states of interaction and communication. They coexist, recur, or reform at intervals that may or may not be regular, one potentially masking the persistence of the other. To make things more complicated, the actors and sentient forms in movement who produce ambiances never cease to change and be affected – just like the landscapes we move through, though at rhythms of varying intensity. For example, the progression between the different ambiances in a garden is based on their simultaneous coexistence. We could even consider gardeners as an archetype of workers who curate an ambiance, as they are in touch with living materials whose ongoing transformations throughout the seasons they must take into account. Plants are living things that create themselves, and by orienting their own growth, they invent forms.[18] The formation of a garden is the result of living processes that create work for the gardener. The pleasure and joy of losing oneself in a garden as one moves from one spot to another are no exception to such a situation: between immersion and contemplation, a visit to such a place is capable, like the experience of a forest, of confronting us with the plants' "point of view", that is, of the living matter that makes up the "fluid" world of which we too are a part.[19] The notion of ambiance may shed some light on this kind of aesthetic experience that is grounded in a sense of immersion and encounter. Visiting a garden procures a specific kind of happiness when we search for motifs of uncertainty and surprising effects; it stimulates our imaginations just as the arrangement of these spaces can throw off our perceptive habits of identifying, categorizing, and summarizing, all the while making us aware of the fluidity of our world as something that is both tangible and alive. Fabricating these ambiances is a subtle art that is not limited to the expertise of engineering, horticulture, or architecture. It instead requires that we listen to and work with the processes of change, of which we are – just as they are – ultimately constituted.

The concrete nature of ambiances and their capacity to affect those who move through them become all the more obvious when such phenomena involve diversions and disjunctions of varying degrees of predictability or sudden changes in rhythm.[20] Situations may change because of a change in ambiance: a street protest can turn into a party, and a scene of collective jubilation can prove the start of a riot. At times, the sensory conditions and perceptible means of interaction deteriorate from one moment to the next, as a result of which the ambiance "skids out of control". More generally, we might say that an atmosphere becomes an ambiance of a certain type when it is established by a dominant political or economic actor, who uses it to obtain a sense of "authority" capable of affecting us and influencing our sensations and emotions, if not also our opinions.[21] The radical transformation of nocturnal urban ambiances in many European cities during the last two centuries yields various examples of this phenomenon.[22] But the trend is not limited to the night hours. In Berlin, the reconstruction of the Potsdamer Platz, the redevelopment of the area around the Zoologischer Garten railway station, and the erection in the city centre of a pastiche structure that mimics the architecture of the Hohenzollern Castle on the very site of the former palace of the German Democratic Republic, these cases provide particularly intense examples.[23]

The fabrication of ambiances may have limiting, repressive intentions, but such an effort is never guaranteed. Ambiances frequently resist control; when the movement is reversed, lived situations become active responses to the conditions that certain people try to impose. Now, eclectic expertise may seed and nurture trouble, but the fabric of an ambiance remains a process of sensorial configuration that is diffuse and in motion; therefore, it does not grant our intentions a firm purchase.[24] It morphs, in either a torrential or capillary fashion, lending a varying texture, colouring, and intensity to our experiences. To understand this complexity and these ambivalences better, emerging frameworks in anthropology can offer a trove of resources.

That which exerts itself on us: feeling, inquiring, and moving through

In this final section, we argue that atmosphere and ambiance can serve as precise operative concepts, despite their significant instability. They help articulate the notion of presence using ambiguous mechanisms and situations in which living beings form resonances with one another to compose the emotion of a moment or place. These concepts displace aesthetics from its role of a simple purveyor of the vocabulary of art history and art criticism, offering it instead other spheres of competence and forsaking the notions of the beautiful or of the sublime, or even of the artwork and the art institution. As Gernot Böhme noted, this discipline could then regain its former status as the science of the senses. By joining forces with other intellectual approaches to social issues, it is able to describe phenomena neglected by the main currents in academia.[25]

More than a parameter of our perceptive milieu and of spaces that have been constructed or arranged, ambiance is mainly a situated process where living beings interact. When we study its emergence in the coexistence, activities, and interactions of social beings, we focus on the environmental anchoring and lived experience of atmospheric phenomena. Such processes take different forms, depending on the configurations encountered by ethnographers – ranging from the exotic to the ordinary, and often somewhere in between – discovered by historians in sources and archives, or studied and commented on by scholars of aesthetics. If we take John Dewey's idea that a "diffuse quality" is what gives any situation its unity, ambiance may trigger those ordinary processes of inquiry that everyone engages in when faced with a challenge.[26] This parameter of action and interaction often remains poorly explicit for participants and hard to perceive for outsiders – all the more so for an ethnographer coming from elsewhere. How can one observe or grasp it?

Investigations of ambiance approach most sensory experiences as lived worlds and perspectives that actually exist. A simple diversity of experiences is not set against a single, objective reality that is itself insert and immune to the senses. The anthropologist Eduardo Viveiros de Castro provided a cornerstone case study of the possibilities opened by the notion of multiple perspectives. He takes Amazonians at their word when they state that the form of the sensory world, and of bodies in particular, depends on each being's rung on the ladder of predation; thus, the jaguar sees blood as manioc beer, and a vulture perceives worms as grilled fish.[27] The multitude of perspectives and lived experiences radically inflates the number of sensorial realities. We can see this multitude in Véronique Nahoum-Grappe's description of ordinary boredom: the same ambiance of a Sunday afternoon that seems so successful and restful to parents provokes a deep existential malaise in their adolescent son.[28] The dual reality of that Sunday afternoon, like the different aspects of the same blood and the same worms in the Amazon, does not ask us to subscribe to a subjectivist relativism that would simply state that things appear differently to everyone, without any further consequence to the supposedly "objective" materiality of situations. We should instead adopt an ontological multi-perspectivism that is more than just a simple matter of vantage points to describe simultaneous and at times dissonant experiences.[29] It is an adolescent's nature to be bored around his parents, just as it is in a jaguar's to rejoice in the spilling of blood, and in a vulture's to delight in the living agents of decomposition.

In conclusion, we begin to speak about ambiance when ambivalence arises, when something enters into play in the selection of a perspective for the time of observation or action ("ambiance" and "ambivalence" share the Latin prefix *ambi-*, which means "around" or "on either side"). This is undoubtedly why the consolidation of domains of shared experiences – which Mead called the "organization of perspectives"[30] and which Sloterdijk would call "climatic spheres"[31] – is explicitly or implicitly programmed into most human activities and institutions. We must still take into account the multitude of these ways of constructing and maintaining the world, leaving any totalizing or reductionist ambitions by the wayside. More than anything, the divergent sensing of perceived atmospheres and the situatedness of ambiances as phenomena, in and of themselves, attest to the multiplicity of living perspectives and speak at length of the methodological challenges that we face in attempting to produce accurate sensorial descriptions.

Notes

1 From Kurt Koffka and the theoreticians of *Gestalttheorie* to Kurt Lewin, James J. Gibson, and Edward T. Hall.
2 During these same years, Edmund Husserl, Martin Heidegger, and then Maurice Merleau-Ponty thought innovatively about our inhabited space and the situated temporality of the flows of consciousness, going so far as to open a dialogue with animal ethology and experimental psychology.
3 In particular, the laboratory at CNRS CRESSON (established by Jean-François Augoyard); for a critical approach, see Olivier Chadoin, "La notion d'ambiance. Contribution à l'examen d'une invention intellectuelle postmoderne dans le monde de la recherche architecturale et urbaine" ["The Notion of Ambiance: A Contribution to the Examination of a Postmodern Intellectual Invention in the World of Architectural and Urban Planning Research"], *Annales de la recherche urbaine*, 106 (2010): 153–159.
4 Rainer Kazig, Damien Masson, "L'ambiance comme concept de la géographie culturelle francophone" ["Ambiance as a Concept in Francophone Cultural Geography"], *Géographie et cultures*, 93–94 (2015): 215–232.
5 Hans Ulrich Gumbrecht, *Atmosphere, Mood, Stimmung. On a Hidden Potential of Literature*, trans. Erik Butler (Palo Alto: Stanford University Press, 2012).
6 Tonino Griffero, *Atmosferologia. Estetica degli spazi emozionali* ["Atmospherology: The Aesthetics of Emotional Spaces"] (Rome/Bari: Laterza, 2010); Anna-Katharina Gisbertz (ed.), *Stimmung. Zur Wiederkehr einer ästhetischen Kategorie* ["Atmosphere: The Return of an Aesthetic Category"] (Paderborn: Wilhelm Fink, 2011).
7 For example, Virginie Milliot, "Pluralistic Ambiance and Urban Socialization", *Ambiances* (on-line) (2013), http://journals.openedition.org/ambiances/223.
8 Éric Chauvier, *Contre Télérama* ["Against Télérama"] (Paris: Allia, 2010); *id.*, *Les Nouvelles Métropoles du désir* ["The New Metropolises of Desire"] (Paris: Allia, 2016).
9 Olivier Balaÿ, "L'ambiance comme objet historique" ["Ambiance as a Historical Object"], in Robert Beck, Ulrike Krampl, Emmanuelle Retaillaud-Bajac (ed.), *Les Cinq Sens de la ville du Moyen Âge à nos jours* ["The Five Senses of the City, from the Middle Ages to the Present Day"] (Tours: Presses universitaires François-Rabelais, 2013), 29–51.

10 See especially Jan Plamper, *The History of Emotions: An Introduction* (Oxford: Oxford University Press, 2015); Alain Corbin, Jean-Jacques Courtine, Georges Vigarello (ed.), *Histoire des émotions* ["A History of Emotions"] (Paris: Seuil, 2016–2017).
11 This was the spirit of the contributions to a volume of the review *Communications*, published in 2018 under the title "Exercices d'ambiances" ["Exercises in Ambiances"] (https://doi.org/10.3917/commu.102.0005).
12 "The marvellous envelops us, showering us like an atmosphere, but we don't see it" (Charles Baudelaire, *Salon de 1846*, XVIII, "De l'héroïsme de la vie moderne" ["The heroism of modern life"], in *Critique d'art* (Paris: Gallimard, 1976) 156; see also 285.
13 This neologism based on two Greek words (*atmos* for "vapour" and *sphaira* for "sphere") appeared in the seventeenth century. In the wake of the experiments conducted on atmospheric pressure by the physicists Evangelista Torricelli, Giovanni Borelli, Robert Boyle, Blaise Pascal, and Daniel Bernoulli, the article devoted to the term in Diderot's *Encyclopedia* (1751) shows that air is an environment of substance and weight, "full" of masses of water and what would soon come to be called "gasses" ("vapor", "spirits", and "exhalations").
14 See Leo Spitzer, *Milieu and ambiance: An essay in historical semantics* (Baltimore, MD: Johns Hopkins University Press,1942).
15 Stéphane Mallarmé, *Divagations*, trans. Barbara Johnson (Cambridge, MA: Harvard University Press, 2009).
16 This avenue of thought is under way, thanks to the work of the Ambiances international network, established in 2013 and as testified to by this very edition.
17 Hermann Schmitz, *Der Leib, der Raum und die Gefühle* ["The Body, Space, and Emotions"] (Bielefeld/Locarno: Sirius, 1998) and *Atmosphären* ["Atmospheres"] (Freiburg im Bresgau: Karl Alber, 2014).
18 See Emanuele Coccia, *The Life of Plants: A Metaphysics of Mixture*, trans. Dylan J. Montanari (Cambridge, UK: Polity, 2018).
19 Emanuele Coccia, *The Life of Plants: A Metaphysics of Mixture*, trans. Dylan J. Montanari (Cambridge, UK: Polity, 2018), p. 46.
20 See Olivier Labussière, "Éléments pour une symptomatologie des ambiances urbaines. L'exemple de Venise, à la lumière de Ruskin et de Proust" ["Elements for a symptomatology of urban ambiances: the example of Venice through the eyes of Ruskin and Proust"], *Articulo – Journal of Urban Research* (on-line), Special issue 2 (2009), http://journals.openedition.org/articulo/1153.
21 Tonino Griffero, "Who's Afraid of Atmospheres (And of Their Authority)?", Lebenswelt, 4, (2014) 193–213.
22 Schivelbusch, Wolfgang. *Disenchanted Night: The Industrialization of Light in the Nineteenth Century* (Berkeley, CA: University of California Press, 1995); Luc Gwiazdzinski, *La Nuit, dernière frontière de la ville* "Night: The City's Final Frontier"] (La Tour-d'Aigues: L'Aube, 2005).
23 See in particular Christof Forderer, "Images vides ou présence pleine? À propos du paysage urbain du Berlin actuel entre mémoire et imaginaire" ["Empty images or full presence? On Berlin's current urban landscape, between memory and imagination"], *Allemagne d'aujourd'hui*, 211 (2015): 81–94 and Marie Hocquet, "Les effets d'exclusion du geste destructeur: le cas du Palais de la République à Berlin" ["The excluding effects of the destructive gesture: the case of the Palace of the Republic"], *ethnographiques.org* (on-line), 24 (2012), http://www.ethnographiques.org/2012/Hocquet.
24 See especially Matthieu Duperrex and Mikaëla Le Meur (Ed.), "Semer le trouble" ["Sowing trouble"], *Techniques & Culture*, No. 74 (2020).

25 Silvia Mancini, "Les civilisations comme 'absolu esthétique': l'approche morphologique de la Mittel-Europa" ["Civilizations as an 'aesthetic absolute': the morphological approach of Mittel Europa"], *Diogène*, 186 (1999): 83–109.

26 Jean-Paul Thibaud, "De la qualité diffuse aux ambiances situées" ["From diffuse quality to situated ambiances"], in Bruno Karsenti, Louis Quéré (ed.), *La Croyance et l'Enquête. Aux sources du pragmatism* ["Belief and Inquiry" at the Roots of Pragmatism"] (Paris: Éditions de l'EHESS, "Raisons pratiques", 2004), 227–253.

27 Eduardo Viveiros de Castro, "Perspectivisme et multinaturalisme en Amérique indigene" ["Perspectivism and Multinaturalism in Indigenous America"], *Journal des anthropologues*, 138–139 (2015): 174–175.

28 Véronique Nahoum-Grappe, *L'Ennui ordinaire. Essai de phénoménologie sociale* ["Ordinary Bordeom: An Essay on Social Phenomenology"] (Paris: Austral, 1995).

29 Eduardo Kohn, *How Forests Think: Towards an Anthropology Beyond the Human* (Berkeley, CA: University of California Press, 2013).

30 George Herbert Mead, *Mind, Self, and Society* (Chicago: University of Chicago Press, 2015) (Enlarged edition, originally published in 1934).

31 Peter Sloterdijk, *Bubbles (Spheres I)*, trans. Wieland Hoban (Cambridge, MA: MIT Press, 2011).

9 Returning to atmosphere

Derek McCormack

In December 2021, I travelled from London to Oxford by train. It was the first time I had visited London for almost two years. It was also, as it happens, the first time I had taken any public transport during the same period, with the exception of a ferry. And it was also the first time I had visited a theatre. None of these things could be considered remarkable. But one of the many consequences of the more severe periods of the pandemic was a withdrawal from many taken-for-granted atmospheres or atmospheric situations, and most obviously from those involving the presence of groups or crowds of people. This withdrawal effected people very differently, of course. For many it was forced, either because of government intervention or as a strategy of self-preservation in the absence of effective protections. For others, such withdrawal was and continues to be unavailable because it is unaffordable. And for some, including maybe myself, this withdrawal perhaps lasted too long. It became a habit and persisted as an entrenched reorientation to the everyday, characterized by, amongst other things, a combination of inertia and residual fear.

The pandemic was – and continues to be – a reminder of how the capacity to withdraw from atmospheres, and to modify exposure to their effects and affects, is distributed differentially (Philippopoulos-Mihalopoulos, 2016). It is also a reminder that prolonged withdrawal has the effect of defamiliarizing taken-for-granted atmospheres and atmospheric experiences. Returning to those situations after a prolonged absence when the time came was therefore tempered by all kinds of affects. There was joy, reassurance, and comfort, to be sure. But there was also anxiety, trepidation, and fear. The return to atmospheres from which one had become estranged could feel like a plunge back into worlds of unfamiliar density and intensity. In those circumstances, atmospheres could thicken, press, and envelop with greater proximity and palpability than before. They could become more obviously foregrounded and feel more intrusive. Their composition, and especially the participation of the breath of others in that composition, could become an unavoidable focus of concern.

These were all part of my experience on the train from Oxford to London in December 2021. A journey tensed by the mere presence of other passengers, interrupted by the sound of coughs, influenced by the minor discomfort, and reassurance of masks. And my experience of the theatre, and all the

DOI: 10.4324/9781032655307-15

work that went into the production and staging of its atmospherics, was also coloured by these feelings. Again, nothing especially remarkable beyond a reminder that returning to atmosphere is more than a conceptual encounter; that it involves the ongoing process of being and becoming immersed and exposed and of different capacities to resist this process. But between train carriage and theatre there was another encounter that seemed to open up different and more surprising possibilities for returning to atmospheres. This one took place in the turbine hall of the Tate Modern, where a work by the Korean-born, New York-based artist Anicka Yi was installed. *In Love with the World* featured floating creatures moving in the air above visitors. Called "aerobes" by Yi, and equipped with discrete propellors, these creatures had the capacity to respond to the presence of visitors and each other through a complex algorithmic system. In this chapter, an encounter with Yi's atmospheric experiment in the Tate becomes, in the context of a return to atmosphere in the wake of the pandemic, an occasion to concentrate attention on the relations between concepts of atmosphere, ambiance, and the ambient. The focus here is less on teasing out the different and entangled traditions and lineages of these concepts, not least because this task has been undertaken elsewhere by (e.g., Thibaud, 2015; Böhme and Thibaud, 2016; Griffero, 2017; Riedel, 2019). Instead, my point of departure for thinking about these concepts is their difference in *tone*. The key claim here is relatively straightforward: the term *ambiance* (or ambience) in English has a more benign tone than *atmosphere*, which results in the former having a more limited affective range. Informed by this, in this chapter, I consider how the broadly phenomenological orientations of both concepts are being extended, noting how *atmosphere* reaches further in drawing attention to the affective materiality of the non-human world. This is both amplified and complicated by considering atmosphere in relation to ongoing efforts across the social sciences and humanities to make sense of elemental media (Peters, 2015; Starosielski, 2019; Jue, 2020). This chapter concludes by drawing some of these themes together through returning briefly to Yi's installation as a speculative thinking space.

Conceptual tone

Thinking between atmosphere and ambiance does not necessarily mean trafficking between incommensurable concepts but involves working in a space of potentially generative tension and difference. Jean-Paul Thibaud (2015; 2011) has argued that despite variations in empirical emphasis and theoretical influence, atmosphere and ambiance share an orientation towards the affective and material qualities of the surroundings in which bodies of different kinds find themselves. At their most basic, both concepts indicate the sensed quality and situational intensity of the mood or feeling of these surroundings. As a point of departure, this claim is a helpful one. It also suggests that atmosphere and ambiance are reasonably interchangeable as concepts.

And that atmosphere, as a concept more prominent in an anglophone context, is more or less translatable into ambiance as a concept employed more frequently in a francophone context. There are certainly overlaps in emphasis between these concepts, as Thibaud and others (Adey et al, 2013) have shown. There are also limits.

The similarities and differences between these concepts were central themes of the 2018 event in Cerisy from which an earlier version of this chapter emerged. Unsurprisingly, much of the event was conducted in French. For some of the sessions, the organizers generously provided an interpreter to assist participants, including me, whose French is not what it should be. In those sessions without an interpreter, I found myself trying to follow the discussion as best I could. When I lost the thread of a sentence or the line of an argument, I filled in the gaps by attuning to the mood or atmosphere of the session or event. I made assumptions about content and meaning on the basis of rhythm, gesture, and intensity. In the sessions with an interpreter, I still attuned to those phenomena while making sense of meaning via the inevitable delay and difference of the translation. Nothing especially remarkable about this. But the experience of both kinds of session (with and without interpreter) served to generate moments in which I sometimes listened or attuned as much and sometimes more to the sound of a concept than to its meaning.

The sound of a concept is related to its tone, that is, to its overall feeling or mood, or the "cluster of more or less vague affective impressions" that accompany it (Anderson, 2016: 744). Admittedly, tone – like atmosphere and ambiance – is a vague concept, referring to the quality of a situation that is indeterminate, "diffuse and immanent" (MacDowell, 2012: 14). Like other concepts for making sense of affective life, a combination of vagueness and indeterminacy is one of the reasons tone tended to be relatively ignored until the more recent renewal of sustained interest across many disciplines in affect, feeling, mood, emotion, etc. For Ngai (2005: 42), however, tone is useful for thinking about the affective dimensions of cultural experience, providing a concept for grasping "compressed assessments of complex 'situations'". Ngai's comments echo the use of tone in Raymond Williams' writing, and especially his analysis of structures of feeling (see Williams, 1977; 1981; See also Highmore, 2016). In Williams' work, tone describes one of the ways in which these structures show up as emerging patterns of feeling in everyday life; structures of feeling are characterized by Williams as patterns of "impulses, restraints, tones" (1977: 159). Both Ngai and Williams therefore use tone in the context of the effort to make sense of the wider sensibility of historical junctures. Tone, writes Ngai, is a cultural artefact's "global or organizing affects, its general disposition or orientation towards its audience and the world" (2005: 28).

This understanding of tone helps inform understandings of the affective dimensions of more traditional cultural artefacts including literary texts, speeches, films, and images (e.g., Anderson, 2016). It can and has also been used to inform analyses of newer kinds of cultural objects, including digital devices and interfaces. Here, tone becomes part of a wider vocabulary for thinking about how affective spacetimes emerge through the design and engineering of the qualities and capacities of technical objects. As Ash et al. (2018: 176) note in the context of an analysis of digital interfaces, "tone can be defined as the effect created by a combination of vibrations and how these combinations are designed to shape intuitive and usually un-reflected upon forms of response and action from users of an interface". This emphasis on the tone of experience as something pre-reflective is also a feature of Alfred North Whitehead's (1978) analysis of experience in terms of feeling prior to and excessive of cognition. For Whitehead, every experience is characterized by a tonal feeling preceding cognition.

As cultural artefacts, concepts can also be grasped in terms of tone. This means that a concept itself has an overall mood or feeling that is part of how it addresses those who think about, with, and against it. The tone of a concept is an important aspect of that concept to which those who encounter and think with it respond in ways that are sometimes excessive of cognition (as Whitehead suggests). This is why the question of how a concept sounds matters. Crucially, because every encounter with a concept is a question of feeling as much as thought, tone is not necessarily something to be stripped away from concepts in order to reveal a kernel of meaning. Foregrounding tone therefore emphasizes differences between concepts in ways that are supplementary to concerns solely with their content or the effort to disentangle their respective histories and theoretical inheritances. Conceptual tone draws attention to how concepts sound and feel in use, and to the feeling of being addressed by them. This emphasis on use is obviously important. The tone of concepts is not fixed but has to be performed through use – the feeling of a concept is *intoned* by repeated invocation and deployment. If all concepts have a tone or feel, the tone of some concepts is however more complex, or multivalent, than others. And where the tone of concepts is more overdetermined, their range of affective possibilities is also more limited as a result. Crucially, and as with other cultural artefacts, tone is central to how concepts resonate (or not) with wider structures of feeling and conjunctural sensibilities.

This emphasis on tone might tell us something about the differences between ambiance and atmosphere in their everyday usage in English. In this respect, the tone of ambiance has a number of characteristics. First, it is relatively benign, suggestive even of something pleasant. Certain usages of atmosphere share this, of course. If I think of the event at Cerisy in September 2018, both terms might be used interchangeably to describe it. But it is harder to imagine applying ambiance to situations with a more negative valence or unpleasant feeling. The journey in the train carriage to London, for instance, does not sound right when described in terms of an ambiance. In English, ambiance just sounds too nice to convey the feelings associated with being enveloped in that carriage and being so close to other passengers.

This is hardly surprising. As Malcolm McCullough notes, ambiance has "less than helpful connotations, for it sounds like a luxury. It suggests a cultivated atmosphere that you might pay for in a restaurant" (2013: 17). If ambiance is relatively benign, it also feels relatively low in intensity. Ambiance is suggestive of a settled *medium* that lingers in the background with an unintrusive gentleness. Depending on the context of its use atmosphere also shares something with ambiance in this regard. But its range is wider. It can feel intense, intrusive, and uncomfortably palpable enough to disrupt or radically transform a situation. Ambiance also feels relatively calm or at least less turbulent than atmosphere. Atmosphere may suggest something febrile, for instance, but it is difficult to think of ambiance in the same way. At a stretch, it is possible to imagine a weird, unsettling, or disquieting ambiance. But the idea of a violent ambience just does not work in the same way as the idea of "violent atmosphere" (Stanley, 2021) does. Its tone means it sounds and feels wrong. In short, in an anglophone context at least, the tone of ambiance arguably means the concept has a rather more circumscribed affective range than atmosphere, even if there are situations when their range overlaps. This matters because translation is always a matter of meaning *and* tone, always about how concepts sound when they address us through their use.

Orientations

Differences in tone colour shape how concepts orient thinking. One of these orientations concerns the respective opportunities that atmosphere and ambiance afford for stretching the focus of perception and feeling beyond the human subject and the phenomenological traditions of thinking that inform this focus. Atmosphere and ambiance both supplement a concern with the perception by the subject of discrete objects. It emphasizes the qualities of the subject's surrounds that are vaguer, more diffuse, barely palpable, etc. (see Peterson, 2021). Nevertheless, atmosphere and ambiance can both be understood as qualities of surrounds towards or away from which subjects are just as "oriented" as they are towards or away from more discrete features such as trees, tables, or mountains (Ahmed, 2006). However, atmosphere and ambiance are not merely things towards or away from which subjects are oriented. They are also part of the conditions or fields in which perception and orientation take place. In relation to ambiance, Thibaud (2011: 212) writes that the "phenomenological approach to perception leads us to define ambiance as a sensory background that specifies the conditions under which phenomena emerge and appear". In turn, this means that atmosphere and ambiance are usually taken to emerge and reside somewhere between subject and object (Anderson, 2009), taking form as what Tonino Griffero (2017: 14) calls "quasi-things".

A phenomenology of atmosphere and ambiance emphasizes the active role of the perceiving subject, even if the exact framing or conceptualization of this role varies. The influence of this way of thinking remains important.

But in some work, a point of bifurcation between both concepts emerges in terms of their relation to the human subject. Ambiance is arguably more tethered than atmosphere in this respect. As Roquet (2016: 4) suggests, "ambience emphasizes the mediating role of human sense perception in a person's relationship to the surrounding world". This approach certainly serves to denaturalize the surrounding world by emphasizing the role of human sense perception. Conversely, it also potentially relegates the role of non-human objects in constituting those surrounding worlds. The less determinate character of atmosphere arguably keeps the non-human in the foreground to a greater extent, however. Efforts to elaborate a post-phenomenological account of atmosphere (Ash and Simpson, 2016) exemplify this. Whether the prefix "post" in post-phenomenology indicates "after", or "beyond", or some kind of break from phenomenology is moot and is outside the remit of this discussion. It is enough to note here that one of the aims of this work is to develop a "a distributed account of human experience by expanding what we mean by the human and re-evaluating the role non-human objects play in the construction of experience" (Ash et al, 2018: 169; for a critique of some aspects of this work see Kinkaid, 2021). That said, the point is not necessarily to become post-subject or even anti-subject in theoretical, empirical, or ethico-political orientation. The key aim instead is to qualify any attempt to take the human subject as a stable point of departure for thinking about affective spacetimes (in this, of course, it is by no means the only perspective) or to think of those spacetimes primarily in terms of how they are perceived by human subjects.

The upshot of this for thinking about atmospheres and ambiance is at least threefold. First, it encourages attention to how atmospheres are generated via and between different kinds of objects and entities in ways that can but do not need to cross a threshold of subjectivity in order to be sensed (see Anderson and Ash, 2015). Second, it is a reminder, if one is needed, that non-human objects and devices are central to how affective conditions are sensed. This sensing can be facilitated, for instance, by a range of devices powered by computational processes, algorithms, and codes (Ash, 2013) that track patterns of feeling across groups or populations (Amoore, 2020). In turn, this kind of data generates a range of solicitations on myriad platforms that amplify tendencies in those conditions. Having long been the focus of various artistic experiments, especially through collaborations with science and engineering, today these processes provide the conditions for various forms of responsive, ambient, atmospheric, or climatic media across different domains of life as part of a political ecology of attention (e.g., Roquet, 2016; Furuhata, 2022). Third, and in turn, this means that subjectivity itself might need to be understood in atmospheric or ambient terms. Put another way, the question here is not so much how atmosphere or ambience provides a field within which subjects sense their surrounds, or how atmospheres emerge between subjects and objects. Instead, it is about how and in what ways subjectivities are already atmospheric through their distribution across objects, bodies, media, platforms, and a range of non-human processes. To borrow vocabulary from the meteorological context from which the concept of atmosphere emerged, subjectivities can be understood in terms of situated processes of precipitation, condensation, and crystallization in relation to memory, habit, affect, etc.

If post-phenomenological work provides one important impetus to extend the conceptual and empirical orbit of atmosphere, a second is provided by efforts to challenge the anthropocentric emphasis of much of the literature on atmospheres and ambiance through a focus on animals. Again, the claim here is not that human perception, sensing, and feeling is unimportant, but that the task of understanding the diverse worlds of non-human animals also requires grappling with their capacities to sense affective spacetimes. Lorimer, Hodgetts, and Barua (2019: 28), notably, have argued therefore for more "awareness of the capacities of animals to sense the atmospheric in much more radical ways than existing work on humans has imagined". This goes beyond a theoretical claim. It also poses methodological challenges, while drawing attention to "how the engineering of animals' atmospheres [...] highlights the entangled fates of marginal human and animal populations in the biopolitics of the Anthropocene" (2019: 28). Part of this argument is speculative, as it must inevitably be. But it also draws upon research in ethology about the sensory habits and powers of perception of non-human animals that demonstrate their capacities to attune to atmospheres and their modifications. While the argument offered by Lorimer, Hodgetts, and Barua is not necessarily framed in this way, their exploration of animals' atmospheres shares something with parallel and ongoing efforts to develop an expanded domain of media studies. This involves a return to an older and more environmental sense of media defined as the influential surrounding in relation to which bodies, techniques, and technologies emerge and are articulated (Peters, 2015).

Elements

Arguments for this expanded definition of media studies are often framed in terms of the elements (Peters, 2015; Starosielski, 2019). This is itself a complex and differentiated category ranging across diverse materials and milieus. In relation to elemental media, atmosphere and ambiance share something obvious: both have an affinity for the milieu of air and the aerial, and both are entangled in traditions of thinking of surrounds as airy. Ambiance is especially suggestive of something olfactory – a benign fragrance, scent, or smell. The aerial associations of atmosphere are perhaps even more obvious (Adey, 2013). The term draws together the sense of an affective spacetime and an envelope of gases surrounding a body. There is, in other words, a certain kind of milieu-specificity (Jue, 2020) to how both concepts contribute to understandings of affective spacetimes in atmospheric terms. Any understanding of either term, no matter how post-phenomenological or more-than-human in orientation, remains circumscribed to some degree by the properties and qualities of the elemental milieu of air and the distinctive capacities of the bodies immersed in that milieu.

This might not seem so significant if the focus remains on anthropocentric accounts of atmosphere and ambiance for which air is the primary elemental milieu in which *human* bodies are immersed. However, it is more obvious if those animals whose primary milieu is water rather than air are considered. Does it make sense, for instance, to think of the affective and sensory spacetimes of those animals primarily in terms of atmosphere? Probably not. As Peters (2015) suggests, something more milieu- or medium-specific would be required to understand the particular ways in which cetaceans incorporate their environment through their sensory capacities. The example of cetaceans is also a reminder, however, of some of the limits of thinking in terms of milieu- or medium-specificity. Milieus or mediums are more mixed in terms of their composition than the term specificity sometimes suggests, and many forms of life move between them, including cetaceans and other mammals.

Attending to the miscibility of elemental media is particularly important as part of how atmospheres are understood in relation to a cluster of other concepts like climate and weather. Indeed, any attempt to understand atmosphere (and ambiance) probably needs to account for these relations. Thinking atmosphere in relation to climate and weather emphasizes the elemental materiality of atmospheres as affective spacetimes in ways that are more than metaphorical. Atmosphere, weather, and climate are not just analogies for affective spacetimes; they are concepts for grasping the relational materiality of elemental milieus or environmental surrounds in ways that acknowledge processes of circulation, storage, and transfer (of heat, for instance) across those milieus (see Neimanis, 2020). Highlighting this also encourages attention to the processuality of atmosphere. While atmosphere is often deployed in a way that suggests a condition of immersion in which primacy is placed on immediacy, it is a dynamic condition in which multiple temporalities and spatialities fold together. As Stacey Alaimo (2010; 2016) has argued, bodies (human and non-human) are implicated in these conditions transcorporeally, and in ways that involve different degrees of exposure. Others have understood this condition of exposure in terms of weathering (Neimanis and Walker, 2014; Sharpe, 2016; Neimanis and Hamilton, 2018). Weathering highlights how surrounds and bodies mutually compose one another in ways that are variously nourishing or toxic. Weathering, in this context, is also the condition of enduring different forms of environmental, political, and racial violence. If the relation between climate and weather is understood in this way, then atmospheres can be grasped in terms of how their material composition is shaped by those relations and how, in turn, that composition can be sensed (Shapiro, 2015). Understood in terms of elemental media, atmospheres are not just objects of sensing or orientation but conditions in whose composition bodies are implicated and to which they are exposed.

Thinking atmospheres in these elemental terms requires a couple of further qualifications, however. The first concerns the relatively limited scope of the anglophone conceptual vocabulary that frames this discussion. It is one thing to consider differences in how ambiance might sound between French and English. One of the challenges of thinking in terms of elemental media, however, is the question of how to translate across different ontologies and epistemologies as part of a process of controlled equivocation (see de La Cadena, 2015). To do so without appropriating concepts or reductively deploying them out of context is no small task. In addition, broad terms like atmosphere simply cannot convey the complexity and nuance of the myriad "elemental forms" and their vernacular vocabularies that populate diverse worlds (Fujikane, 2021). A second qualification concerns the persistent tendency, exemplified admittedly in the present chapter, to think of atmosphere and ambiance as things or entities. Both can be understood this way, but it is important to supplement this with a reminder of the value of the atmospheric and the ambient as qualifying yet indeterminate adjectives. Here, paradoxically, the milieu-specificity of the atmosphere circumscribes the term in ways that are unhelpful. The idea of the ambient, in contrast, stretches further, going beyond the aerial associations of the atmospheric while retaining an emphasis on elemental media. For instance, there may be ambient heat in different elemental milieus including water, earth, and air. The tone of ambient is also a little more open-ended than that of ambiance, with the former less immediately suggestive of something pleasant or benign. And, finally, the ambient is less entangled than ambiance to the sensory capacities of the human subject. As Morton (2007) argues, the ambient is a background described in relation to a particular entity, and not exclusively the human subject.[1] It is also a background whose existence does not depend on it being sensed in human experience – in this respect it contrasts with the definitions of ambiance noted above. Morton argues therefore for an ambient poetics that articulates an aesthetics in which diverse entities resound (2015). Experimenting with this aesthetics might involve exploring the capacities of non-human materials and devices to sense, register, and present changes in the ambient conditions of their milieus. This aesthetics does not however preclude the possibility that the ambient might cross a threshold in a way that means it is sensed or felt in human bodies as something atmospheric. It points instead towards speculative experiments tensed between the ambient conditions of elemental milieus and atmospheric experiences distributed across different bodies (see Engelmann, 2015; 2021).

Elemental media ecologies

Perhaps Yi's installation in the turbine hall of the Tate Modern offers one such experiment. The air overhead is populated by two types of aerobes, their shapes based on "ocean life forms and mushrooms" (see Figure 9.1).[2] The first type, "xenojellies", resemble translucent jellyfish-like balloons

Figure 9.1 *In Love with the World.* Anicka Yi. Tate Modern, London. December 2021.

Source: Photo by Author.

with gently moving tentacular limbs. The other type, *planulae*, is yellowed, opaque, and blob-like bladders. The text accompanying the exhibition suggests the aerobes offer a "vision of a new ecosystem", providing an invitation to think about what it would "feel like to share the world with machines that could live in the wild and evolve on their own". Yi's work stages scenes of a world in which non-human entities move in relation to each other and the conditions of the elemental milieu in which they move. As a staged work, there is, of course, some obvious input from humans (including the ongoing maintenance and tending to the aerobes that take place in a corral in the corner of the turbine hall – see Figure 9.2) to make this world technically possible. However, even if visitors are told that the aerobes can and respond to the presence of humans, particularly their heat, the creatures move as if indifferent to the presence of humans. Part of the allure of the aerobes is their withdrawal from human attention.

Yi's work offers a zoomorphic vision of a post-phenomenological elemental world (see Engelmann and McCormack, 2021). By incorporating entities that appear particularly oceanic, *In Love with the World* conjures an elemental world that is neither milieu- nor medium-specific. Nevertheless, the primary elemental milieu of the work is air. In that sense, it is a distinctly atmospheric work in a simultaneously affective and gaseous sense. The aerobes are relationally implicated in the gaseous atmosphere of the turbine hall, existing as bodies of enveloped helium of variably buoyancy that are

Figure 9.2 Technical zone as part of *In Love with the World*. Anicka Yi. Tate Modern, London. December 2021.

Source: Photo by Author.

responsive to the conditions in which they are immersed. The diffuse affective atmosphere of the space is a product of the dynamic composition of the work and the scattered positions of visitors in a voluminous space of transitions involving arrivals, departures, and waiting. This dynamism means that *In Love with the World* does not provide a single point of orientation or attention for visitors. The movement of the aerobes mingles with, modifies, and modulates the atmosphere in the hall without overdetermining it. As a result, while it can be grasped as a post-phenomenological atmospheric work, *In Love with the World* is also a field of atmospheric experience, albeit a field that is diffuse and relatively low key – almost, indeed, like an ambiance.

In Love with the World also operates in a way that is more ambient, modifying the air in the turbine hall to a degree that just about reaches a threshold of palpability. Yi released into the turbine hall a series of smells characteristic of different times and spaces. In the process of infusing the space with various scents, the ambient conditions of this volume become multi-spatial and multi-temporal. This is not just because the scents change on a weekly basis, but also because into this volume they fold radically different contexts – the Black Death, 20th-century coal use, and vegetation from the cretaceous period. These scents are not especially intrusive, however. At most, they are present as faint hints that don't really register in an obvious way. Modifying the ambient olfactory qualities of spaces is something Yi has explored in many of her other works, often through the microbial generation of distinctive smells linked with specific bodies or sites (see Lee, 2019). Yi does this in part to problematize what she considers to be the masculinist visual bias of odourless contemporary exhibition spaces. This problematization is sometimes difficult to discern, however; the scents in the spaces within which Yi works operate as subtle chemical influences rather than as intrusive odours. Through the taken-for-granted process of respiration and breathing, these scents become implicated in the bodies of visitors even if most visitors are not necessarily aware of this implication, generating connections across bodies, times, and spaces. Like her other works, *In Love with the World* operates in a way that does not need to cross a threshold of cognitive awareness in order to exert some influence, however subtle, on the conditions of atmospheric experience. This hints at a disquieting tone to the work. As Lee observes, "by incorporating scent signals in her installations, Yi might be construed as setting up stealth events of nasal ingestion" (2019: 694) that imbricate breath, body, and the chemical in the trans-corporeality (Alaimo, 2010) of ambient conditions. The tone of the work shifts from disquieting to potentially sinister if the aerobes are considered in relation to speculative depictions of autonomous machines. These depictions are by no means new, but they are often dystopian. Understood as UAVs, it is certainly possible to imagine Yi's aerobes as deceptively benign drone-like surveillance devices in the sky above a docile population.

Coda

Ultimately, however, the tone of the work is not overdetermined by these possibilities. It remains open, relatively light, and low in density. The ambient tone of the work is certainly influenced by the subtle modification – through scent and heat – of the conditions in which the bodies of visitors and aerobes move. But it is also shaped by the capacities and qualities associated with the aerobes. To some extent, these bodies appear whimsical – a quirky collection of balloon-like creatures. As atmospheric things (McCormack, 2018), a balloon can appear benign and inconsequential, especially if it merely drifts with the movement of the air in which it is immersed. Yi's aerobes retain some of these associations but complicate them somewhat by virtue of the fact that they don't merely drift. They are autonomous and dirigible. This does not turn them into surveillance drones, however. Jellyfish, after all, don't just drift. Some of them, at least, can exert a degree of influence on their direction of travel.

Because of this, it might seem easy to read too much into *In Love with the World*. And it is important to remember its limits – it is a staged, speculative thinking-space. Politically, the scope of *In Love with the World* is also relatively muted, ambient even, at least when compared with lots of recent writing on the politics of atmosphere and the atmospheric. It certainly does have a political dimension – like other examples of Yi's work that emphasize the olfactory, her installation at the Tate Modern can be grasped as a political comment on the importance of expanding the sensing spaces of contemporary aesthetic experience. For my purposes here, however, and in the context of a return to atmospheres from which I and others had withdrawn, an encounter with *In Love with the World* offers something more modest: a dynamic relational work that generates a speculative space in which to move between the concepts of atmosphere, ambiance, and the ambient. It foregrounds differences between these concepts without necessarily suggesting that these differences exist because these concepts refer to radically different things or processes. Like the movement of Yi's aerobes, thinking between atmosphere, ambiance, and the ambient is a dynamic process of feeling within a space of conceptual orientations. It offers a way of thinking about and with these concepts in relation to an expanded sense of elemental media. More precisely, *In Love with the World* is a zoomorphic vision of an elemental media ecology. This ecology is composed of entities with capacities to sense modifications in their milieus while also modifying those milieus in ways that are generative of a field of differential atmospheric experience.

For me, however, an encounter with Yi's work also offered something else. In the context of a fraught return to atmospheres, a return that had been postponed and delayed, it offered the luxury of shared spaciousness. In the volume of the turbine hall, it offered the relief of distance, the prospect of well-defined separation between bodies in the context of a fear of the proximate and worries about aerial transmissions between bodies. For me at least, *In Love with the World* offered a low-density atmospheric experiment in the midst of which to continue the process of reorientation associated with the return to everyday atmospheres. Part of this reorientation is conceptual. Yi's installation is also a reminder, however, that much of this reorientation involves fine recalibrations of relations between human and non-human bodies as they participate in the co-production of atmospheres through ongoing processes of withdrawing and returning. And, in the circumstances, that felt like enough.

Notes

1 It is worth noting that for Morton an ambient poetics does not necessarily mean dismissing the concepts of atmosphere and ambience. Indeed, in *Ecology without Nature* (2007), Morton, drawing on Spitzer (1948), often uses these terms interchangeably. Where Morton, somewhat more recently (2015), *does* seem to make a distinction is between climate and weather. In some ways, climate is the ambient background becoming foregrounded in such a way that it intrudes into the worlds we inhabit, not least through different forms of atmospheric disturbance, violence, and trauma. Weather, in contrast, exists for Morton as a relatively localized lifeworld, a kind of circumscribed atmospheric phenomenon that adds tone to or colours the mood of our experience. It is worth noting that this distinction is made within the context of a wider claim about what is required for us to grasp the gravity of our planetary conditions. The claim here is that to focus on the localized experience of weather, like focusing on the localized experience of atmospheres or ambiences, prevents us from grasping the significance of a decentred ecology of the ambient.

2 See https://www.tate.org.uk/whats-on/tate-modern/hyundai-commission-anicka-yi/exhibition-guide

References

Adey, Peter. "Air/atmospheres of the megacity". *Theory, Culture & Society*, 30 (7/8) (2013): 291–308.

Adey, Peter, Laure Brayer, Damien Masson, Patrick Murphy, Paul Simpson, & Nicolas Tixier. "'Pour votre tranquillité': Ambiance, atmosphere, and surveillance". *Geoforum*, 49 (2013): 299–309.

Ahmed, Sara. *Queer phenomenology*. Durham, NC: Duke University Press, 2006.

Alaimo, Stacy. *Bodily natures: Science, environment, and the material self*. Bloomington: Indiana University Press, 2010.

Alaimo, Stacy. *Exposed: Environmental politics and pleasure in post-human times*. Minneapolis: University of Minnesota Press, 2016.

Amoore, Louise. *Cloud ethics: Algorithms and the attributes of ourselves and others*. Durham: Duke University Press, 2020.

Anderson, Ben. "Affective Atmospheres". *Emotion, Space, and Society*, 2 (2) (2009): 77–81.

Anderson, Ben. "Neoliberal affects". *Progress in Human Geography*, 40 (6) (2016): 734–753.

Anderson, Ben and James Ash. "Atmospheric methods", edited by P. Vannini. In *Non-representational methodologies*, 24–51. London: Routledge, 2015.

Ash, James. "Rethinking affective atmospheres: Technology, perturbation and space times of the non-human". *Geoforum*, 49 (1) (2013): 20–28.

Ash, James, Ben Anderson, Rachel Gordon, Paul Langley. "Unit, vibration, tone: A post-phenomenological method for researching digital interfaces". *Cultural Geographies*, 25 (1) (2018): 165–181.

Ash, James, & Paul Simpson. "Geography and post-phenomenology". *Progress in Human Geography*, 40 (1) (2016): 48–66.

Böhme, Gernot and Jean-Paul Thibaud. *The aesthetics of atmospheres*. London: Routledge, 2016.

De la Cadena, Marisol. *Earth beings: Ecologies of practice across andean worlds*. Durham: Duke University Press, 2015.

Engelmann, Sasha. "Toward a poetics of air: Sequencing and surfacing breath". *Transactions of the Institute of British Geographers*, 40 (3) (2015): 430–444.
Engelmann, Sasha. *Sensing art in the atmosphere: Elemental lures and aerosolar practices*. London: Routledge, 2021.
Engelmann, Sasha, and Derek McCormack. "Elemental worlds: Specificities, exposures, alchemies". *Progress in Human Geography*, 45 (6) (2021): 1419–1439.
Fujikane, Candance. *Mapping abundance for a planetary future: Kanaka Maoli and critical settler cartographies in Hawai'i*. Durham: Duke University Press, 2021.
Furuhata, Yuriko. *Climatic media: Transpacific experiments in atmospheric control*. Durham: Duke University Press, 2022.
Griffero, Tonino. *Quasi-things: The paradigm of atmospheres*. Albany: SUNY Press, 2017.
Highmore, Ben. "Formations of feelings, constellations of things". *Cultural Studies Review*, 22 (1) (2016): 144–167.
Jue, Melody. *Wild blue media: Thinking through seawater*. Durham: Duke University Press, 2020.
Kinkaid, Eden. "Is post-phenomenology a critical geography? Subjectivity and difference in post-phenomenological geographies". *Progress in Human Geography*, 45 (2) (2021): 298–316.
Lee, Rachel. "Metabolic aesthetics: On the feminist scentscapes of Anicka Yi". *Food, Culture & Society*, 22 (5) (2019): 692–712.
Lorimer, Jamie, Timothy Hodgetts, & Maan Barua. "Animals' atmospheres". *Progress in Human Geography*, 43 (1) (2019): 26–45.
MacDowell, James. "Wes Anderson, tone and the quirky sensibility". *New Review of Film and Television Studies*, 10 (1) (2012): 6–27.
McCormack, Derek P. *Atmospheric things: On the allure of elemental envelopment*. Durham: Duke University Press, 2018.
McCullough, Malcolm. *Ambient commons: Attention in the age of embodied information*. Cambridge: MIT Press, 2013.
Morton, Timothy. *Ecology without nature: Rethinking environmental aesthetics*. Cambridge: Harvard University Press, 2007.
Morton, Timothy. *Hyperobjects: Philosophy and ecology after the end of the world*. Durham: Duke University Press, 2015.
Neimanis, Astrida. "The Sea and the Breathing". *E-flux architecture*. (May 2020). https://www.e- flux.com/architecture/oceans/331869/the-sea-and-the-breathing/
Neimanis, Astrida, & Jennifer Mae Hamilton. "Open Space Weathering". *Feminist Review*, 118 (1) (2018): 80–84.
Neimanis, Astrida & Rachel Loewen Walker. "Weathering: Climate change and the "thick time" of transcorporeality". *Hypatia*, 29 (3) (2014): 558–575.
Ngai, Sianne. *Ugly feelings*. Cambridge: Harvard University Press, 2005.
Peters, John Durham. *The marvelous clouds: Toward a philosophy of elemental media*. Chicago: University of Chicago Press, 2015.
Peterson, Marina. *Atmospheric noise: The indefinite urbanism of Los Angeles*. Durham: Duke University Press, 2021.
Philippopoulos-Mihalopoulos, Andreas. "Withdrawing from atmosphere: An ontology of air partitioning and affective engineering". *Environment and Planning D: Society and Space*, 34 (1) (2016): 150–167.
Riedel, Friedlind. "Atmosphere". In *Affective societies*, edited by Jan Slaby, Christian von Scheve, 85–95. London: Routledge, 2019.

Roquet, Paul. *Ambient media: Japanese atmospheres of the self.* Minneapolis: University of Minnesota Press, 2016.
Shapiro, Nicholas. "Attuning to the chemosphere: Domestic formaldehyde, bodily reasoning, and the chemical sublime". *Cultural Anthropology*, 30 (3) (2015): 368–393.
Sharpe, Christina. (2016). *In the wake: On blackness and being.* Durham: Duke University Press, 2016.
Spitzer, Leo. "Milieu and Ambience". In *Essays in Historical Semantics*. New York: Russell and Russell, 1948/1968.
Stanley, Eric. *Violent atmospheres: Structuring antagonism and the trans/queer ungovernable*. Durham: Duke University Press, 2021.
Starosielski, Nicole. "The Elements of Media Studies". *Media+ Environment 1.1* (2019): https://doi.org/10.1525/001c.10780.
Thibaud, Jean-Paul. "The sensory fabric of urban ambiances". *The Senses and Society*, 6 (2) (2011): 203–215.
Thibaud, Jean-Paul. "The backstage of urban ambiances: When atmospheres pervade everyday experience". *Emotion, Space and Society*, 15 (1) (2015): 39–46.
Whitehead, Alfred North. *Process and reality*. New York: The Free Press, 1978.
Williams, Raymond. *Marxism and literature*. Oxford: Oxford University Press, 1977.
Williams, Raymond. *Politics and letters: Interviews with new left review.* New York: Verso Books, 1981.

10 Ambiance as a concept of francophone cultural geography

Challenges and prospects[1]

Rainer Kazig and Damien Masson

Introduction

The notions of ambiance and atmosphere have been used for several decades in the humanities and social sciences to address people's potential for sensitivity. They highlight a multi-sensory relationship with the environment and that a person's state of mind, soul, or even *body* (Thomas et al., 2014) can evolve with the sensory context in which they find themselves.[2] In the French-speaking world, the notion of ambiance finds its place mainly within the field of architectural and urban research, while the notion of atmosphere is clearly positioned within cultural geography in Anglo-Saxon and German-speaking countries. French-speaking cultural geography instead makes very little reference to the notion of ambiance. In the third edition of Paul Claval's *Géographie culturelle* – a reference book in French-speaking cultural geography – the term 'ambiance' does not appear in the book's index (Claval, 2012). A simple search using the search engine for the journal *Géographie et cultures* reveals the notion's scant presence. Admittedly, in 2023, we find more than a hundred articles that use the term in their text, but there are only eight citations of the key authors who contributed to building this concept in French-speaking spatial research (four for Jean-François Augoyard, five for Jean-Paul Thibaud, and one for Pascal Amphoux) in all the indexed online content of the journal – a very low number of references to the term.

On the basis of this observation, the aim of our contribution is to discuss the prospects and challenges for giving greater consideration to the notion of ambiance in francophone cultural geography, where it is much less present than atmosphere in anglophone and germanophone cultural geography. We will begin by presenting the notion and its uses in the field of francophone architectural research. We will then develop two avenues for bringing together the field of ambiances and francophone cultural geography. The first outlines three new research perspectives that emerge when the two fields of research meet, while the second shows how ambiances can become an object of cultural geography. Finally, this chapter concludes with a methodological invitation.

DOI: 10.4324/9781032655307-16

The concept of ambiance and the field of architectural and urban ambiances

In the French language, the notion of ambiance has two inseparable levels of understanding that concern both the phenomenal, sensory dimensions of atmospheric phenomena and their apprehension by individuals. The definition in the Le Robert dictionary (*Le nouveau petit Robert*, 2002, p. 77) captures this very well, referring to 'the material or moral atmosphere that surrounds a person.' Ambiance surrounds bodies, qualifies the environment, and forms a climate; it is eminently geographical,[3] even if this notion is not central to the vocabulary of French-speaking geography and does not constitute one of its concepts. Ambiance is also an individual and social construct, conditioned by the presence of individuals who in turn experience and describe it.

Research into the notion of architectural and urban ambiance as a critique of sectoral approaches that take the sensory in the architectural field into account (thermal, acoustic, and lighting, among others) has been developing over the last thirty years. In this time, research into the notion of *architectural and urban ambiance* has focused on understanding ambiances at the crossroads of their phenomenality and their lived dimension.[4] Jean-Paul Thibaud's approach, which is particularly heuristic for our analysis, consists of understanding ambiance in terms of its potential to articulate the uses of the body and the lived space, which, for him, manifests itself in the following four dimensions: its indivisibility, its immediacy, its omnipresence, and its diffusion (Thibaud, 2002). These dimensions emphasize the comprehensive, immanent, and subjective nature of ambiance, its capacity to touch the body before consciousness and to convey affect and meaning.

These ways of understanding ambiances yield a localist and situated apprehension of them. Ambiance or ambiances are never ideal or immaterial. They are material and immanent, described as *architectural and urban*, always related to situations and understandable at the interface of three spaces – or levels – of manifestation: sensory, social, and built (Amphoux, 1994). Second, we understand that ambiance lies at the interface of the cognitive and the pre-cognitive (Thibaud, 2002): it is both expressible and a sensory-motor vector. Ambiance is not something external to a subject who can grasp it and dissect it to comment on it; it is a materiality as much as an experience, forming a filter through which we feel, perceive, and act.

Thus, research on ambiances cannot be limited to characterizing them by naming their qualities. Above all, it enables the creation of a mode of attention specific to the facts being studied. From being an object, ambiance becomes a mode of approach. As perceiving the ambiance is tantamount to perceiving the situation through the ambiance, the latter can be understood as an intermediary, like a 'dispositif,' to use Foucault's vocabulary (1993 [1975]), which would then be understood as being sensory, that is to say, a condition of possibility for the advent of the experience of the situation as much as a character of the situation. Taken in conjunction with this notion, the intermediary dimension of ambiance becomes manifest; it is part of the phenomenal world while at the same time placing individuals in relation to it. The consequence, then, is a potential practice of research *on* ambiances, which is also likely to be researched *through* ambiances. By this we mean a specific mode of approach to space at the meeting point of the three registers of interest mentioned above, thereby situating the question of ambiance at the intersection of three spatialities: social, sensory, and built. The consequence is a necessarily interdisciplinary apprehension of sensorily situated issues for these works, which seek never to 'disintegrate' these dimensions. However, the cultural theme does not emerge from these dimensions. Moreover, we have just shown that ambiance is a spatial concept, that it conveys meanings and affects, and that it is intimately linked to situated bodies: all qualities that make it a concept of cultural geography. So how can we go about accounting for it?

Three research perspectives at the crossroads of ambiances and cultural geography

Similarly to what we have just said about the notion of ambiance, the notion of culture can have different statuses in cultural geography; it can be an object of analysis as well as a research perspective in human geography (Claval, Staszak, 2008; Crang, 2009). In fact, for a geography that analyses cultural phenomena, ambiance can easily be taken into consideration as a dimension of the phenomenon being analysed. In this sense, it can be integrated into the analysis of numerous research objects such as commercial or tourist areas, which we will develop in the last section of this chapter. For the moment, let us start by looking at the intersection between ambiance and cultural geography, as research perspectives to be deployed in three ways that are qualified as the *emergence of spatial representations*, *differentiated dispositions,* and *micro-macro articulation.*

The emergence of spatial representations

One of the concerns of cultural approaches in the humanities and social sciences is the analysis of the social construction of the world, to which the notion of ambiance can contribute. Following the classification of cultural approaches given by the sociologist Andreas Reckwitz (2008), our proposal can be anchored in the perspective of 'cultural contingency.' Research that follows this perspective aims to make explicit the symbolic orders, cultural codes, and systems of representations that underpin human practices and their emergence. Within this framework, the geographical approach is particularly interested in the construction of spatial representations. There are two approaches to conceptualizing this 'cultural contingency': culture as discourse and culture as a set of practices. The first approach can be summarized as a semiotic or post-structuralist cultural geography, which was important for the development of the new cultural geography (Crang, 2009). The second approach, which we call the praxeological approach to culture – and which we are interested in crossing with the notion of ambiance – deals with the system of meanings through ordinary practices. It is based on the idea that representations are not an established and stable reality, that they instead emerge, are reproduced, are modified, and are called into question by everyday practices and incorporations in which the sensory dimension plays an important role (Wylie, 2007, p. 164 *ff*; Harrison, 2000). The challenge of this first proposal is to articulate the theme of ambiances in conjunction with practices in the constitution of representations and to present the resulting research perspectives. Three modalities specific to the question of ambiances can be mobilized to this end.

First, we have the modality of *poly-sensory openness*, which makes it possible to consider on a pragmatic level the question of the emergence of the sensory dimension within the sphere of the cognitive, by not focusing on an *a priori* sense. The differentiated consideration of the sensory dimension is undoubtedly the most obvious characteristic of a praxeological approach to the culture that the notion of ambiance enables. By not giving primacy to one sense, it lets us question the role of the sensory dimension of practices, considered as a whole, in the emergence of spatial representations. It also keeps open the question of which sensory modality should be favoured where, traditionally,[5] the visual modality has predominated in the production of geographical representations. However, when a geographical object as fundamentally visual as the landscape is penetrated by readings through the other senses,[6] we understand the interest and diversity of the geographies and new spatial categories that thinking through ambiances produces.

Second, there is the modality of the *involvement of the body*, which reminds us of the importance of the course of action to the constitution of experience. This raises the question of the role of ordinary practices in the emergence of representations. The theme of mobility and associated practices provides good examples that illustrate this line of inquiry. Each form of mobility involves a specific sensory encounter with the materiality and sociality of the city, which can translate into systematically different experiences and emotions, and thus feed into representations of the city in different ways. Without being explicitly anchored in a cultural geography perspective, Masson (2012) shows how specific representations of urban space emerge during ordinary journeys through the city by train.[7] Generally speaking, taking ambiances into consideration leads to a research perspective that aims to render explicit the role of situated practices in the emergence of spatial representations, obviously without limiting itself to mobility practices or representations of urban space.

Third, there is the *diffuse materiality* of places and its role in reminding us that space is not neutral, nor is the sole support of practices, rather that it contributes to the formation of our individual dispositions to the world. This dimension of ambiance has two implications for a praxeological approach to the cultural. The first is based on the observation that a person's dispositions – or, in other words, forms of attention – vary during the course of their activities. It raises the issue of the ability to determine and qualify moments that, during a specific activity, 'affect' the spatial representation in question. The challenge, then, is to describe and understand the 'ecological' rationales of these moments. The second implication concerns the importance accorded to diffuse materiality in francophone research on ambiances. This is characterized by a sustained attention to the material conditions at which experiences unfold, and this has been made possible by major methodological developments (Grosjean, Thibaud, 2001). Therefore, a praxeological approach to the cultural based on the notion of ambiance must aim to analyse explicitly the material conditions that contribute to the emergence of spatial representations.

To sum up, this initial proposal of elements that contributes to the foundation of a sensory praxeology of the cultural stems from an understanding of the articulation of sensory, cognitive, and emotional availability, of bodily engagement, and of the materiality of situations in the process of constituting representations. Such a perspective requires that we extend the existing work in the field of ambiances by placing greater emphasis on the conscious dimension of their experience. In a certain sense, this use of the notion of ambiance within a cultural geography[8] aimed at clarifying the forms of emergence of spatial representations amounts to reversing the existing perspective of research on ambiances. In the field of architectural research, it is used in particular to question the sensory effects of material constructions on users or inhabitants, and it ultimately aims to qualify volatile phenomena often considered to be indescribable and ineffable. Here, by contrast, it is used to understand and render explicit the emergence of representations through a sensory encounter with the world. This use of the notion of ambiance reinforces the positioning of cultural geography as a discipline within a broader field of cultural research that aims to understand the contingency of the social world.

Differentiated arrangements

The second crossover between ambiances and cultural geography consists in pointing out that cultural geography can contribute ideas and issues to ambiance research by culturalizing it, something that has already been initiated by Kazig (2013). Culturalizing ambiance research here means making it sensitive to cultural differences in the lived experience of ambiances. Until now, most research on ambiances[9] has been based on the analysis of situations, with little consideration of the question of sharing – or rather of not sharing – the ambiance under consideration – with the exception of a recent piece of work on this theme (Thibaud, Duarte, 2013).

This lack of sensitivity to perceptual differences can be understood by considering the history of the development of the notion of ambiance, which is strongly rooted in phenomenology. Its specific method consists of characterizing sensory phenomena before any cultural relativization (Raab *et al.*, 2008, p. 12 *ff.*). In addition, references to the ecological perspective of perception, in particular the work of Gibson (1979), and to ethnomethodology (Thibaud, 2002), contribute, alongside phenomenology, to little room being left for considering cultural differentiation.

There is, however, an openness to the theme of differentiation, at least on the part of the theorist behind this notion in the German-speaking world. Hermann Schmitz explicitly mentions the idea that biography plays a role in one's openness to ambiances (Werhahn, 2011). To render the thesis of the existence of differences in dispositions to ambiances more explicit on a theoretical level, we draw on Bourdieu's theory of practice, which can be included in the category of praxeological theories of the culture mentioned above.

At the heart of Schmitz's approach is the idea of actors forming socialized bodies. Bourdieu emphasizes that practices are always generated by actors who involve their socialized bodies in a specific way. The notions of practical sense, *habitus* and *corporeal hexis,* clarify this understanding of a socialized corporeality. The notion of a practical sense is understood as a mastery of know-how based on the creation of practices and producing, in the field concerned, an action considered to be normal (Bourdieu, 1977; Reckwitz, 2006). This practical knowledge is not reflexive, instead incorporated and based on experience. Bourdieu applied his theory of practice in particular to the theme of social inequalities, showing how, in different fields, relations of domination are expressed in unequal forms of *habitus* and are reflected in an equally unequal repertoire of available practices. Practice theory can also be used – as is particularly the case in the uses made of it by German-speaking researchers – in the form of a general cultural theory of the social sciences to analyse the emergence and influence of practices in different fields of practice (Hillebrandt, 2009, pp. 388 *ff*; Reckwitz, 2006, p. 309). It is in this sense that Bourdieu's theory of culture can also serve as a basis for research into ambiances. Following on from the above, the question arises for culturalized research on ambiances regarding the extent to which, and within what framework, culturally transmitted dispositions come into play when people deal with ambiances.

This perspective of culturalizing research on ambiances opens it up to a different approach to the issues that have already been broached and developed. For example, it lets us take a different approach to the question of accessibility and the opening up of public spaces to all. This theme has already been opened up by the non-neutralization of the perceiving subject, as proposed in the work of Rachel Thomas, who approaches the question of disability by taking the variability of its forms and their effects on the experience of space into account (Thomas, 2005). As an extension of this movement, the geoculturalization, in particular, of research into ambiances makes it possible to understand how developments, public space policies, and certain dominant practices are likely to affect users with distinct habits and dispositions in different ways, for whom the same spaces may appear to be more or less welcoming, intimate, or practicable, for example. In this sense, this opening up is in line with certain aspects of Rachel Thomas's work. It is also in line with certain themes developed by the recent French-speaking social and cultural geography of the body, which Di Méo (2010) outlines in some detail.

A challenge: micro-macro articulation

Raising the question of the culturalization of research on ambiances also represents the first step in a more consistent interrogation of the capacity of work on ambiances to fit into a broader, more comprehensive scheme, specifically the social and political context in which the situations under study are embedded. As mentioned above, French language work in the field of ambiances has its roots in phenomenology and has developed mainly in schools of architecture. One of its main aims is to produce a critique of constructed situations and the social and sensory experiences to which they give rise. This paradigm of ambiances, which aims to shed light on the interactions between constructed, sensory, and social spaces, serves a rationale for aiding the design of the space produced, and to this end, it takes into account situations and their ambiances *here and now*, or in other word, necessarily in situ and in the present time. While such a mode of attention is necessary for the deployment of situated, sensitive phenomenologies, it also produces certain difficulties in terms of its capacity to articulate what is observable with certain conditions of possibility that originate 'outside' the place where the situations in question manifest themselves. In this way, the ambiance approach puts in brackets the historical, societal, political, ideological, legal, and other conditions prior to the emergence of the situations under consideration to focus as closely as possible on their modes (sensory, interactional, and material) of actualization. Such a choice is not without its tensions, as we may sometimes observe the risk that this kind of work removes the political dimension of the ambiance in favour of research into the sole question of comfort, for example, by avoiding a position on the question of inequalities.[10] Above all, these criticisms give way to a fundamental observation that we may make about the limits of the architectural and urban ambiance approach to the study and analysis of situations in which social and political structures are prevalent. Let us take the example of a recent study that examines the effects, in terms of ambiance, of policies to make public space safer and more peaceful (Thomas et al., 2014). This research has made it possible to account for certain methodological limitations of the ambiance-based approach, in particular for taking into account the power issues expressed above, but also concerning the articulation of these structural modalities together with our more contingent field observations. In fact, the question of going beyond the analysis of micro-situations arises with a view to a more general approach and asks us to confront the same challenges that certain micro-sociological approaches have faced in other times. Thus, this proposal, which starts from within research on ambiances, aims to reflect on the articulation of the ambiance-based approach within the broader issues considered in the social sciences. Finally, in addition to contributing to research in the field of ambiances, the theme of culturalization is also of interest to cultural geography because it indicates one of its potential contributions to the 'humanist geographies' that are sometimes accused – without foundation, as Bailly and Scariati (2004) show – of lacking operational capacity, or even relevance.

Ambiances as an object of analysis in francophone cultural geography

A second way of considering ambiances in cultural geography involves looking at them as an object of cultural geography, and not just as an approach. Cultural geography is particularly interested in areas of consumption and leisure and in the problem of security, themes in which the sensory dimension is of primary importance. The creation and marketing of new consumer spaces nowadays almost systematically involves the creation of ambiances to justify the sensory specificity of these places and to make them attractive. This use of ambiances can be understood as a dimension of the 'aestheticization of the world' that, according to Lipovetsky and Serroy (2013), characterizes contemporary capitalism. An analysis and critique of these developments would gain in relevance by explicitly taking their sensory dimension into account. Gernot Böhme, one of the founders of German-language research on ambiances, indicated the role to be played by research on ambiances in criticizing this development (Böhme, 2001). Cultural geography can help to respond to this call by deploying new field research in which ambiance represents a useful notion alongside others – such as that of the body (Volvey, 2000; Di Méo, 2010; Barthe-Deloizy, 2011) or that of affect (Feildel, 2013) – to analyse jointly the lived experience and the sensory dimension of consumer and leisure spaces. Furthermore, with its explicit interest in the joint consideration of the sensory environment and bodily affects, the notion of ambiance seems particularly appropriate for analysing the spatial dimension of the sensitive in a situated manner. In this sense, work on the surveillance and securing of communal spaces (Adey *et al.* 2013) shows that the use of ambiances enables an understanding of the effects of urban security measures on the feeling of safety. Ultimately, this second line of inquiry is a reminder that considering ambiances is always useful for understanding the role of the material context (understood in a broad sense, including human presence in particular) in lived experience and the construction of situated representations.

Encouraging methodological innovation

Finally, the inclusion of the notion of ambiance in francophone cultural geography encourages methodological extensions for the study of micro-spatial phenomena. One of the specific features of French-language research on ambiances is the development of empirical approaches that help us grasp the lived, material, and phenomenal (at a sensory level) dimensions of a space jointly and in great detail (Grosjean, Thibaud 2001). In actual terms, this means combining several methods to shift the focus of observation from the lived experience to a characterization of the constructed and sensory dimensions of the places being studied. The results of these multiple approaches help contextualize the question of how people experience space. Nevertheless, these methods, with their convincing results, remain marginal within the social sciences and cultural studies. Therefore, we can only invite French-speaking cultural geographers to take up and develop this innovative methodological approach to consolidate the work in this field, which is one of the few to take an interest in the micro-dimensions of experienced space and its material and sensory aspects.

Notes

1 This text was originally written following a paper given at the conference 'Géographies et Cultures: le tourment culturel' held at Cerisy-la-Salle in 2014 and published in French in the journal *Géographie et cultures* (Kazig and Masson, 2015). It is translated here in an abbreviated, slightly updated version. The original version of the text starts from the observation of a relative absence (already in 2014!) of the notion of ambiance in francophone cultural geography and proposes avenues for its integration into cultural geography. This translated and abbreviated version presents this particular situation and the avenues it opens up to a non-French-speaking readership. The French text, including the quotes, was translated by Damien Masson and Rainer Kazig. We wish to thank Margret Vince for proofreading and correcting the final English version of the text.

2 While the term 'ambiance' dominated francophone academic discourse until recently and the term 'atmosphère' was hardly used, the term 'atmosphere' almost consistently determines anglophone and germanophone academic discourse.

3 This is evidenced by the entry 'ambiance' in the 'Dictionnaire de Géographie, de l'espace et des sociétés' (Lévy and Lussault, 2003).

4 We refer here to a construction of this notion developed within the UMR CNRS Ambiances Architectures Urbanités, based in particular on the work of the CRESSON team, which is one of the two founding teams of this research group.

5 We could even say 'culturally' in respect of the Western world since the *Quattrocento*, as Augoyard (1991) shows.

6 These give rise to soundscapes, thermal landscapes, and olfactory landscapes, as well as those that are harder to represent, such as tactile or gustatory versions, and finally those that arise from the very question of action, such as walkscapes and taskscapes, among others.

7 Certain 'dynamic visual effects' bear witness to this process of creating representations of the space travelled through the intermediary of movement and the material and social framework formed by the train. See, for example, the effects of disaffection, openness, landscape, and geometric abstraction (Masson, 2009, 108–124).

8 We could perhaps speak of an 'ambiantization' of the cultural, echoing what we refer to below as a culturalization of ambiances.

9 Those from the CRESSON team at the Ambiances, Architectures, Urbanités Laboratory, to say the least.
10 This is borne out by the many debates – particularly with Brazilian colleagues – that have taken place during seminars organized by the international ambiances network (http://www.ambiances.net).

Bibliography

Adey, Peter, Laure Brayer, Damien Masson, Patrick Murphy, Paul Simpson, and Nicolas Tixier. "«Pour votre tranquillité » : ambiance, atmosphere, and surveillance". *Geoforum*, 49 (2013): 299–309.

Amphoux, Pascal. "Environnement, milieu et paysage sonores". In *Les faces cachées de l'urbain*, edited by Michel Bassand and Jean-Philippe Leresche, 159–176. Bern: Peter Lang, 1994.

Augoyard, Jean-François. "La vue est-elle souveraine dans l'esthétique paysagère?" *Le Débat*, 65, (1991): 51–59.

Augoyard, Jean-François. "L'environnement sensible et les ambiances architecturales". *L'Espace géographique*, 24 no. 4 (1995): 302–318.

Bailly, Antoine S., Renato Scariati, "L'humanisme en géographie". In *Les concepts de la géographie humaine*, edited by Antoine S. Bailly, 213–222. Paris: Armand Colin, 2004.

Barthe-Deloizy, Francine. "Le corps peut-il être "un objet" du savoir géographique?". *Géographie et cultures*, 80 (2011): 229–247.

Böhme, Gernot. *Aisthetik. Vorlesungen über Ästhetik als allgemeine Wahrnehmungslehre*. Munich: Fink, 2001.

Bourdieu, Pierre. *Outline of a Theory of Practice*. Translated by Richard Nice. *Cambridge Studies in Social and Cultural Anthropology*. Cambridge: Cambridge University Press, 1977.

Claval, Paul. *Géographie culturelle. Une nouvelle approche des sociétés et milieux*. Paris: Armand Colin, 2012.

Claval, Paul, Jean-François Staszak. "Où en est la géographie culturelle?: introduction". *Annales de géographie*, 660–661 (2008): 3–7.

Crang, Mike. "Cultural Geography". In *The dictionary of human geography* (5th edition), edited by Derek Gregory et al., 129–133. Chichester: Wiley/Blackwell, 2009.

Di Méo, Guy. "Subjectivité, socialité, spatialité: le corps, cet impensé de la géographie". *Annales de géographie*, no. 675 (2010): 466–491.

Feildel, Benoit, "Vers un urbanisme affectif. Pour une prise en compte de la dimension sensible en aménagement et en urbanisme", *Norois*, 227 (2013): 55–68.

Foucault, Michel. *Discipline and punish: The birth of the prison*. Translated by Alan Sheridan. New York: Pantheon Books, 1978 [1975].

Gibson, James Jerome. *The ecological approach to visual perception*. Boston: Houghton Mifflin, 1979.

Grosjean, Michèle, Jean-Paul Thibaud et al. *L'espace urbain en méthodes*. Marseille: Parenthèses, 2001.

Harrison, Paul, 2000. "Making sense: embodiment and the sensibilities of the everyday". *Environment and Planning D: Society and Space*, 18 no. 4 (2000): 497–517.

Hillebrandt, Frank. "Praxistheorie". In *Handbuch soziologische Theorien*, edited by Georg Kneer and Markus Schroer. Wiesbaden: VS Verlag für Sozialwissenschaften, 2009.

Kazig, Rainer. "Vers une culturalisation de la recherche sur les ambiances". In Jean-Paul Thibaud, Cristiane Rose Duarte et al. *Ambiances urbaines en partage*, 37–48. Lausanne: MétisPresses, 2013.

Kazig, Rainer, Damien Masson. "L'ambiance comme concept de la géographie culturelle francophone: défis et perspectives". *Géographie et Cultures*, 93–94 (2015): 215–232.

Le nouveau petit Robert: dictionnaire de la langue française, text revised and edited under the supervision of Josette Rey-Debove and Alain Rey. Paris: Dictionnaires le Robert, 2002.

Lévy, Jacques, Michel Lussault. *Dictionnaire de la géographie et de l'espace des sociétés*. Paris: Belin, 2003.

Lipovetsky, Gilles, Jean Serroy. *L'esthétisation du monde. Vivre à l'âge du capitalisme artiste*. Paris: Éditions Gallimard, 2013.

Masson, Damien. "Expériences sensibles de la mobilité urbaine". *Les Cahiers de la recherche architecturale et urbaine*, 26/27 (2012): 136–43.

Raab, Jürgen et al. *Phänomenologie und Soziologie. Theoretische Positionen, aktuelle Problemfelder und empirische Umsetzungen*. Wiesbaden: VS Verlag für Sozialwissenschften, 2008.

Reckwitz, Andreas. *Die Transformation der Kulturtheorien. Zur Entwicklung eines Theorieprogramms*. Weilerswist: Velbrück Wissenschaft, 2006.

Reckwitz, Andreas. *Unscharfe Grenzen. Perspektiven der Kultursoziologie*. Bielefeld: Transcript, 2008.

Thibaud, Jean-Paul. "L'horizon des ambiances urbaines". *Communications*, 73 (2002): 185–201.

Thibaud, Jean-Paul, Cristiane Rose Duarte et al. *Ambiances urbaines en partage*. Lausanne: MétisPresses, 2013.

Thomas, Rachel. *Les trajectoires de l'accessibilité*. Bernin: À la Croisée, 2005.

Thomas, Rachel, Grégoire Chelkoff, Sandra Fiori, Damien Masson et al. *Les* énigmes *sensibles des mobilités urbaines contemporaines*. Grenoble: rapport de recherche ANR MUSE, programme « Espace et Territoire », 2014.

Volvey, Anne. "L'espace vu du corps". In *Logiques de l'espace, esprit des lieux*, edited by Jacques Lévy and Michel Lussault, 319–332. Paris: Belin, 2000.

Werhahn, Hans et al. *Neue Phänomenologie. Hermann Schmitz im Gespräch*. Freiburg: Alber, 2011.

Wylie, John, *Landscape*. London, New York: Routledge, 2007.

11 Affective inflationism and atmospheric competence[1]

Tonino Griffero

Welcome to inflationism: atmospheres and (even) quasi-things

If we take a look at a famous sequence taken from Marcel Carne's *Hotel du Nord* (1938),[2] it is easy to see that the word "atmosphere" sounded pedantic and exotic in the daily vocabulary of that time. The fact that the word "atmosphere" is now in force in scientific field enough to justify talk of an "atmospheric turn"[3] and that atmospherization even appears to have become our common condition[4] is surely a consequence of the new sensibility spread by the so-called and broader "affective turn" in Humanities.

But there are several ways to discuss affect. There are "affect theories", based on an anonymous ontological process dynamic blind to local affect contexts, and "affect studies", focused either on the reduction of affects to human-social relationships or on exploring the emotional sphere in order to activistically create better affective relationships. Although my atmospherological approach from the outset[5] partly meets both ways, it is quite different. Focusing on atmospheres, in fact, means looking carefully at how situations and places can produce affects, without making excessive metaphysical claims but also, at least initially, without reducing affects neither to quantitative data and statistics nor to a continuation of politics with other means. The aim of my approach is to stick as much as possible to a descriptive phenomenological stance, which in my case stems in particular from Hermann Schmitz's New Phenomenology.[6]

There have so far been attempts not so much at analysing in depth but at merely "translating"[7] Schmitz's neophenomenological theory of feelings. This effort to translate it would provide a kind of "northwest passage", capable both of synthesizing the instances of cognitive and sensory emotivism and of embracing a motivational and intentional theory[8] of affectivity as a ubiquitous dimension of existence: for this reason, I also intend to "urbanize Schmitz's province", so to speak.[9] However, I do not share the hermeneutic *coté* often implicit in such urbanizations and summarized in the invitation to move from the letter to the spirit of the original text. For me, it is rather a matter of promoting an updating of the neophenomenological approach and a problematization of it that doesn't reduce it to a sort of philosophical idiolect but also, obviously, does not coincide with its dogmatic acceptance.

DOI: 10.4324/9781032655307-17

A first step of this urbanization, while pursuing the highest possible fidelity to concrete daily experience, amounts to embracing a real inflationism that is both ontological and phenomenological. Ontological since it should be able to also include vague entities[10] like quasi-things and, among them, atmospheres;[11] phenomenological in so far as it should know how to pluralize the typology of atmospheric feelings as well as the forms in which they are perceived. The neophenomenological setup of my project, which I have called "pathic aesthetics"[12], presupposes thus defining feelings as atmospheres: which means that they are not segregated in the (fictional) container of the psyche, but poured out into lived and predimensional (i.e. non-geometric-local) space and that, as semi-demonic powers felt-bodily involving the person, they exercise (sometimes uncontrolled) authority over her.

According to this project, there are three ideal types of atmospheres. On the basis of their decreasing objectivity (as well as intensity) and increasing objectual focalization,[13] atmospheres can be, first of all, *prototypic*, that is objective, external, unintentional, sometimes lacking a precise name and relating to space of vastness, therefore broadly equivalent to the *Stimmungen* of the German philosophical tradition. But there are also *derivative-relational* atmospheres, i.e. objective, external, intentionally produced, related to directional space, and implying an "in-between" between subject and object. And there are even quite *spurious* atmospheres that are subjective, projective, and even related to single objects and local spaces.

Provided this wide range of atmospheric feelings, whose extreme poles are, as seen, groundless quasi-objective moods and merely projective individual's moods, one may wonder what atmospheres can really do. While, in fact, the objects of thought and judgement presuppose the autonomous initiative of the conscience-bearer and can, so to speak, multiply at will and even be completely made up,[14] the perception of an (atmospheric) feeling concerns a restricted and historically invariable number of affects by which a person quite susceptible to them is practically and irresistibly grasped. But what are the power and authority of this (sometimes) irresistible atmospheric affect?

This (almost prescriptive) power[15] is not only social and/or cultural.[16] Consider the ingressive and prototypic atmosphere that you feel when entering a space pervaded by an affect that is completely different from your subjective mood and for this reason exerts an authority over your mood, whether by deciding to adapt or resist to it. Well, a similar affect can also be produced outside of social scenarios. To make a simple example, when the sky darkens because of a fleeting cloud and turns gloomy, thus tonalizing the surrounding environment in a specific way, it exerts an affective power that cannot be explained as an unconscious projection of a purely subjective emotion. Contrary to what is often claimed by sociological approaches to collective emotions, the power and authority of an atmospheric feeling cannot be reduced to the need or desire to correspond to social-emotional norms or social practices, because, even though it has certainly powerful social (inclusive, excluding, etc.) effects, the segmentation of reality generated by the atmospheric experience cannot be sufficiently explained by social conventions (even if introjected).

Now, a (physicalist and/or culturalist) reductionist approach is obliged to explain the undoubted experience of living in a sentimentally impregnated space as a subjective and at most intersubjective projection on a physical-objective constellation previously[17] deprived of any affective significance. Such a view would surely stigmatize the neophenomenological perspective, seeing it as an undue theologization, thus overestimating the undeniable influence exercised on Schmitz by Rudolf Otto's conception of the numinous, and as a crypto-metaphysicization of the simply contextual-cultural character of the felt authority.

This objection is for me seriously inadequate to prototypic atmospheres: in fact, as quasi-thing, they grasp the percipient without being really owned by her. While having an intermittent and quasi-thingly life, in the sense that they come and go, without there being any point in asking what they did in the meantime (as one would ask instead for a thing proper),[18] prototypic atmospheric affects are much more involving than social norms, thoughts, or even factual evidences.[19] In short, even without any physical coercion, a prototypic atmosphere unfolds a set of environmental affordances[20] or, in neophenomenological terms, motor suggestions and synaesthetic characters understood as felt-bodily bridge-qualities. The unmissable and precise felt-bodily (and sometimes even physical) resonance of these affordances in the perceiver attests to the indisputable authority of that atmosphere.[21]

And by the way, in spite of the already mentioned analogy between the prototypic atmosphere and Otto's idea of the holy as the numinous, it would be really reductive to see atmospherology only as a chapter of theological emotivism. In fact, the numinous is at best just the example of a specifically religious atmosphere, as such surely just a little more demonic than that created by the eros,[22] by the *genius loci*[23] and perhaps even by the Kantian *Gewissen*, which works almost as a courtroom. All these atmospheres, however, could be called "divine" not because they refer to transcendence, but only because they exercise a serious and absolute authority over those who feel them.[24] Indeed, as we have already mentioned, both those who obey them and those who resist them, precisely by so doing, demonstrate their quasi-objective reality.

The power and authority of atmospheres

What further demonstrates the quasi-objectivity of atmospheres and their different degree of authority is the elementary but emblematic example of contrasting feelings.[25] A cheerful person who meets people wrapped in an atmosphere of true and serious sadness will feel the authority of this sadness, respects it (not just for social etiquette) and, even if not infected by it, will mitigate her own joy,[26] whose atmospheric irradiation force is evidently more limited. This clearly means that some atmospheric feelings legitimately claim to colonize the surrounding space more than others.[27] In turn, a sad person may feel entitled to abandon herself to her sadness, to reject any consolatory attempt,[28] to punish "with disregard anything that might cheer him up",[29] and even to stigmatize the joy of others, which she evidently perceives as offensive: this further proves the existence (and sometimes coexistence) of atmospheres equipped with different power and authority.[30]

But it would be a mistake to think that this kind of authority is only exerted by social atmospheres or, at most, by inanimate situations still somehow related to human psychicity (as Tellenbach claims).[31] In fact, just like sadness, shame, the ideal of justice, a building, an artistic masterpiece, gratuitous generosity, love, a landscape (even if devoid of human traces), the silence of an autumn morning and even music as an ideal (unreservedly condemning a bad performance, for example!) can very well be emotionally authoritative and binding. And this happens not only in a synthonic form, as when one feels oppressed by a grey and sultry morning, but also in a dystonic form, as when a beautiful landscape, exactly because of its beauty, sharpens someone's (previous) sadness. To use Wilhelm Raabe's literary example, a sad person takes nature's beauty as mockery, as an insult, and starts to loathe all seven days of creation.[32]

More precisely, a prototypic atmosphere has an absolute authority only when it grasps and, at least initially, generates a relationship so tangling as not to allow the percipient to take any position towards what grabs her and to mobilize the critical reserves provided by her level of personal emancipation.[33] When, for example, I break a rule that (for various reasons) I do not feel superior to and am pervaded by an atmosphere of shame,[34] this feeling captures me completely and generates a felt-bodily resonance that is totally specific and often rationally uncontrollable. Building on Schmitz's felt-bodily "alphabet" and theory of felt-bodily communication,[35] one could guess that in such cases there is a unilaterally antagonistic incorporation, due to which one of the partners involved in the felt-bodily communication[36] is in a sense "sucked in" her partner's prevailing felt-bodily narrowness. An atmosphere's authority, of course, also depends on the perceiver's felt-bodily disposition and biography. However, it is not something completely relative, no more than the same language spoken in a relatively different way by different people. It should be stressed that a lovely sunny day, for example, may even sharpen the sadness of those who suffer, and not because they perceive it as sad, but exactly because they feel the delight-induced sunshine as foreign and irritating.[37]

A few *caveats* are still required here.

A Even if atmospheres are affects permeating a lived and predimensional space, one must never forget also the role of the local-metrical space with its boundaries, as exemplified by the fact that sometimes you no longer feel the oppressive atmosphere of a certain room as soon as you leave it and go outside.

B Even if the prototypic atmosphere exerts authority, one must never forget that even a powerful social feeling like charisma, able sometimes to mesmerize large masses, is so fragile that it can be instantly dissolved by a simple *faux pas* (the king is naked!),[38] whose felt-bodily resonance consists in changing the previous antagonistic one-sided incorporation. In less serious cases, this incorporation becomes alternate and the mimesis of the charismatic person weakens; in the most serious cases, it remains unilateral but turns inside out and the masses slaughter the leaders they idolized before.

C Finally, the prototypic atmosphere is cognitively impenetrable at first. The funeral's sadness, for example, is not really mitigated simply because we know that every biological organism must die. One should yet not forget that, if the beautiful evening sky turns out to be caused by pollution, it loses (at least a little) its atmospheric beauty later. So, precise knowledge and location of what generates an atmosphere often trigger a relatively non-affective distancing, exactly as when one is not really persuaded by a speech whose rhetorical structure is too obvious: like money, an (atmospheric) affect ceases to function (better: to involve affectively, not to be observed) at the very moment when one ceases to believe in it! But things are perhaps more complicated than that, because sometimes a situation radiates an atmosphere *despite* what one knows (an atmosphere might please and gratify, for example, even if it is only accidental), while, at other times, it radiates an atmosphere *only if* one knows something (a place has an *aura*, for example, only if one knows that an important person live there, etc.). Therefore, the relationship between the cognitive and the affective is a very contentious issue and can never be unilaterally resolved. What can be said is that an atmosphere is not affected by what one knows (for example that its generator is just an illusion) only at first, when she perceives the first and involuntary impression that its appearance suggests.

The very next question is: what is the most powerful atmosphere? Apparently, not one that is latent, but one so blatant as to oppose the mood of the percipient who enters the space tonalized by it and re-orientate her emotionally (it's what we might call an ingressive experience in a discrepant atmosphere). When the outside atmosphere instead confirms the perceiver's emotional protention (it's what we might call a syntonic atmosphere),[39] the situation may even appear non-atmospheric, whereas, actually, it is rather permeated by an impersonal and anonymous atmosphere. Except for the use of a generic "evaluation" Jung (1921, 544) is right in saying that "even an 'indifferent' sensation possesses a 'feeling tone', namely, that of indifference, which again expresses a certain valuation. Hence feeling is also a kind of judging." And that is because a situation can never be separated from the atmosphere that necessarily enhances its (also emotional) meaning.[40]

Apparent non-atmosphericness is thus far from powerless.[41] It is simply due to the failure of a diffused feeling to achieve the perceptual threshold required and to its remaining on the (Gestaltic) background of a different qualitative figure.[42] As an intransitive-unfocused condition of perceptions and therefore of specifically transitive-focused feelings, it exercises its power in all its intensity allusively[43] rather than through explicit pressure. Sure, if one conceives of them as non-actual and only partially conscious states, atmospheres turn into *Stimmungen*, i.e. less intense, focused, and fleeting feelings.[44] This is close to what Boehme defines the quasi-objective "atmospheric" as opposed to the "atmosphere" (in the strict sense).[45] Thus conceived, latent atmospheres are nothing but those existential or background feelings whose non-thematic nature should not be confused with an affectless one. So, the most influential and powerful atmospheres, against all odds, may be, according to Heidegger, those

> to which we pay no heed at all, the attunements [*Stimmungen*] we least observe, those attunements which attune us in such a way that we feel as though there is no attunement there at all, as though we were not attuned in any way at all.[46]

If one focuses then on the feeling's progress instead of stopping at the first impression, the most powerful atmospheres might be the mixed ones,[47] which by virtue of their non-saturating and long-term effect penetrate deeper and leave room for the perceiver's probable imaginative additions.[48]

The strange case of the atmosphere you recognize without feeling it

The question of the power and authority of atmospheres also brings us to consider the well-known experience[49] of objectively recognizing and "registering" an atmosphere but without personally feeling it or being felt-bodily involved by it.[50] In other words, one must take into account the ontological-phenomenological distinction between a quasi-thingly atmosphere and its (only potential) capturing power. Is the former really a fully non-affective atmosphere, as Schmitz says, and Moritz Geiger explains by drawing a distinction between an observing attitude (*betrachtende Einstellung*) and an absorbing attitude (*aufnehmende Einstellung*)?[51] Or is it, more simply, a low-intensity atmosphere, which for the most diverse reasons is not engaging but without being purely informative, given that no perception is ever a mere sociosemiotic "reading" of situation data but always a somewhat capturing felt-bodily communication? The low-intensity option has the merit of recognizing that the body feeling is never "opaque" but always transparently refers to the world of which it highlights some aspects.[52] Moreover, it obviously proves that even the most neutral-cognitive observation implies at least a minimal affective component, thus embracing Heidegger's idea of a continuous albeit not always conscious affective situatedness, where background feelings alone, thanks to their motivational, hedonic, and axiologic link to the world,[53] enable any specific relationship with the world, with ourselves and with others.

Let us take Rainer Maria Rilke's Malte as an example. He is in the middle of a festive crowd, whose "faces were full of the light that came from the show-booths, bubbled from their mouths like matter from open sores [...] People stopped me and laughed, and I felt that I should laugh too, but I couldn't" (Rilke 1945, 45). I believe that only a low-intensity and incipient empathic share of the others' joy can explain and justify Malte's disappointment, thus making him feel a discrepant atmosphere.[54] Even sadism[55] and *Schadenfreude*, actually, are such only because the subject feels or empathizes (at least partially), respectively, the pain and sadness of her victims! However, Schmitz's dissociation between neutral feeling and felt-bodily involving feeling, proven by the uncertainty concerning the right gestural-postural conduct (which instead is instinctive in the case of a real and effective felt-bodily resonance), should at least be nuanced[56] but certainly not eliminated, if only because, if too radical a dichotomy between living a feeling and feeling as such actually risks reifying the affective too much, the identity between emotional involvement and atmospheric perception risks instead of making inexplicable that distance from feelings (not only as attitude-position posterior to the initial emotional passivity but also sometimes even contemporary to it) which is also admitted.[57] Narrowing the too drastic gap between the cognitive and the affective, it can be assumed that the distinction between these two dimensions is always only gradual.[58] It follows that every atmospheric perception is merely a possibility[59] (based on the subject's felt- bodily disposition and resonance-capability) within an affective *continuum*[60] that finds its extreme (and very rare) cases in emotional "fusion" and neutral distancing.

Over time

The affective power of an atmosphere also depends on the too often unrecognized temporal dynamic of feeling, starting from the way atmospheric feelings – especially "derivative-relational" ones – arise. These feelings are then deeply influenced (surely more than "prototypic" ones) by the changing in-between that "connects" perceiver and percept. The in-between of derivative atmospheres is, by the way, very different from that of prototypic atmospheres, which should be antidualistically conceived not as the outcome of two autonomous poles but rather of a relation prior to its relata.[61] Anyway, because of their in-between, the derivative atmospheres also raise (otherwise impossible) doubts as to whether one should attribute these feelings to the subject or rather to the environment.[62] If you pay attention to perceptual experience after the first impression,[63] you will note that the perceiver, after necessarily adopting the atmosphere at first, is then able to take her own stance through her personal emancipation. She can therefore a) surrender to that feeling, b) constructively embrace its radiating effect, c) oppose this effect, for example when she better identifies the objects and places into which the atmospheric feeling condenses (perhaps improperly), or simply escapes from it, or lastly d) specifically mix acceptance and resistance.[64]

But there's more, because due to rituals and cultural conventions, a given atmosphere – for a person and even for a society – can maintain and strengthen[65] or reduce and even lose its power. In both cases, this is due to variations of the perceptual field, affected by factors such as the distance of the perceived, the brightness of the environment, the speed with which one approaches the perceived, some scale shift, a change of personal mood, the unveiling of a sensory illusion, etc. But it is also due, more simply, to the perceiver's physiological and felt-bodily conditions. This increase or decrease in power, more generally, also depends on the subject's sensible-cognitive awareness, that is, additional and divergent information on the perceived, further purely idiosyncratic experiences, focusing over time on misleading feelings' anchoring points and fields of condensation, etc. Not to mention the perceiver's sociological and/or variable aesthetic competence in detecting the atmospheric potential of a certain situation. This multifactorial long-term atmospheric change – which has been underestimated until now – could be compared to the feeling of being emotionally involved by a piece of music (a score),[66] during which, in fact, one can be engaged by previously overlooked sub-atmospheres that may even be in conflict with the original atmosphere.

Hence, for example, a sense of disappointment with respect to an affective expectation, the demystification of a manipulative attunement, the setting aside of a trivial emotional *cliché*, etc. The best solution is for me to make the already suggested three-way distinction of atmospheres even more flexible. In fact, by changing the perspective and felt-bodily incorporation, a prototypic atmosphere may also become a derivative-relational one, such as when, for example, an atmosphere breaks free of its anchoring point (which is the source of its radiation) and materializes into random or at least improper things and situations. Likewise, a spurious atmosphere may increasingly (albeit deceptively) get rid of its original idiosyncratic attitude and become (almost) as powerful as the prototypic one.

It is worth remembering, though, that making the types and perceptions of atmospheres more flexible and fluid does not at all justify neglecting an anti-introjectionist approach. This flexibility should not be overestimated, indeed, to the point of confusing the diversity of affective reactions to an atmospheric feeling (of the way the perceiver "filters" it) with the diversity of the atmosphere perceived, which instead – precisely because it is relatively the same – allows for the actual coexistence (and intersubjective communicability)[67] of relatively different moods.[68] To take just one example: the impressive entrance hall of a major banking institution will express an aggressive atmosphere of power for those who venture there in search of a loan, while expressing, on the contrary, a quiet atmosphere of proud belonging, not even clearly felt, for an employee who has developed a strong *esprit de corps*. And yet what generates both polar opposing atmospheric moods (conscious aversion and overwhelming awe or unnoticed sense of well-being and pride) is precisely the same spatial-emotional atmospheric quality of intimidating vastness.

One should also choose the more brilliant literary example of Lawrence's *The Rainbow*, where the "same" atmosphere of the Lincoln Cathedral is experienced in a very different way by two characters of the novel: both are overcome with wonder and awe and clearly feel the powerful *genius loci*, but whereas for Will "'before' and 'after' were folded together, all was contained in oneness", resulting in a "timeless ecstasy", and the church "was all, this was everything", Anna is "but silenced rather than tuned to the place", so that Will's "passion in the cathedral at first awed her, then made her angry". More precisely, she feels that, compared to the "pillars upwards [...] there was the sky outside", and this

> *open sky was no blue vault, no dark dome hung with many twinkling lamps, but a space where stars were wheeling in freedom, with freedom above them always higher. The cathedral roused her too. But she would never consent to the knitting of all the leaping stone in a great roof that closed her in, and beyond which was nothing, nothing, it was the ultimate confine [...] Her soul too was carried forward to the altar, to the threshold of Eternity, in reverence and fear and joy [...] But even in the dazed swoon of the cathedral, she claimed another right. The altar was barren, its lights gone out. God burned no more in that bush. It was dead matter lying there. She claimed the right to freedom above her, higher than the roof. She had always a sense of being roofed in.*

The widespread atmosphere is clearly the "same" (wonder and awe), but she "wanted to get out of this fixed, leaping, forward-travelling movement", for example by catching "at little things, which saved her from being swept forward headlong in the tide of passion that leaps on into the Infinite in a great mass, triumphant and flinging its own course". More, by means of a resistance based on indulging on the details (which act here as really discrepant and disturbing sub-atmospheres), she maliciously succeeds in "spoiling his passionate intercourse with the cathedral" and destroying

> the passion he had [...] Strive as he would, he could not keep the cathedral wonderful to him. He was disillusioned. That which had been his absolute, containing all heaven and earth, was become to him as to her, a shapely heap of dead matter-but dead, dead.
>
> (Lawrence 1955, 146–148)

Atmospheric competence and provisional atmospheric morality

When a person's narrow-pole absorbs into itself its passive partner within a unilaterally antagonistic incorporation, we talk about charisma. A charismatic pole or person can rely on a power that sounds completely irrational only because it transcends the categories of legal and traditional power. Especially in an era in which a ubiquitous late-capitalistic economy atmospherizes its own power and sphere of influence through a media-economic anonymity[69] and a diffused aestheticization based on stage values, it is necessary to deeply reflect on how to object to those who are opposed to a neophenomenological atmospherology. They see in it not only an undue hypostatization and reification of the affective realm, but also an emotionalistic stance, an irresponsible arousalism that would end up confining people to being nothing more than blind passengers of freely floating atmospheres.[70]

I don't want to reply to these points using some late-romantic and vitalistic *cliché*, some Frankfurtian critical-theoretical argument, or advocating the person's full capacity for self-determination (which is usually exaggerated by personalistic theories). I believe it is much better to question the meaning and role of the often-used but little-explored "atmospheric competence". It is surely not enough to simply repeat with Bollnow, Heidegger, and Schmitz that the perceiver is responsible, after an initial involving phase, of saying yes or no to a *Stimmung* or atmosphere,[71] for example shaping it by means of keeping a certain "posture"[72] (which contradicts a bit the previous claim of a deep affective involving). Nor is it enough to say that one successfully fights against a mood only by opposing a contrary and more powerful mood to it, thus somewhat using a mechanical energetics whose intentional governance contradicts the pathicity elsewhere highlighted. I surely welcome Schmitz's recent renunciation[73] of the too provocative earlier claim according to which, feelings (and therefore atmospheres) are "no more subjective than highways [...], only less easy to fixate".[74] However, it is for me worthwhile to defend the neophenomenological anti-introjectionism against attempts at trivializing this radical de-psychologization of the affective sphere. These attempts, actually, explain this externalization by means of social reasons and norms,[75] cultural practices,[76] or even by showing their illusory nature. In their opinion, "perceiving atmospheres" would be tantamount to simply "perceiving atmospherically", where the subject would be unaware of projecting her own feeling externally and therefore she then deceivingly perceives it as foreign.[77]

I reject any – even only implicit – hydraulic-projectivistic explanation of outside feelings in terms of psychic filling of the extra-psychic world, that is, a softer and more academic version of atmospherology,[78] which it is not more true only because it is an easily shared view. Instead, I adopt the undoubtedly provocative and counterintuitive claim that atmospheric feelings, especially prototypic ones, of course, are objective and quasi-thingly powers (or even daemons!) that are irreducible both to the model of self-determined *homo clausus* embraced by the bimillenary Western dualism and to the resulting causalistic-prognostic epistemic "obsession". But this compels me to spend a few words explaining the triple function of an "atmospheric competence".

A) First of all, this competence is the skill of generating atmospheres, regardless of whether one chooses a hard version of the theory, according to which, people can intentionally and consciously produce atmospheres,[79] or a softer one, according to which, people could only set the generative conditions in which atmospheres appear.[80] When one stages an atmosphere, he certainly could end up being governed, like in any other human activity, by a heterogony of ends or by the usual gap between a project and its material implementation: a failure that explains perhaps the production of unwanted or very volatile atmospheric feelings but surely not their full absence. But the hard version seems unquestionable to me, and many are in fact busy researching and investigating what Böhme calls generators of atmosphere.[81] And they do that despite the traditional view which claims, but in a way that is not always consistent, that an intentionally produced atmosphere should be impossible or at least (which is not the same thing) deleterious. According to this glorious tradition, when producer and user aim at creating and experiencing the feelings they do want, they would a) produce nothing more than a sentimentalist dilettantism only focused on the feeling experienced ("internal concentration") rather than on the axiological qualities inherent in what radiates that feeling ("external concentration").[82] Moreover, an affective planning b) would be impossible, because, like *Stimmungen*, atmospheres belong to a preconscious sphere that is prior to the subject/world dualism and that you therefore couldn't produce but only "awaken".[83] Every intentional staging then c) risks to lead to an instrumental emotional hygiene that is seriously unethical in perverting life's earnestness (such as when one falls in love with own feeling so as to taste it!)[84], enjoys her inner capitulation through a superficial resonance,[85] or, more trivially, gives life to sentimental kitsch and demagogic situation.[86] Finally, d) it reveals a fundamentally rhetorical-propagandistic nature, based on a technique of impression (*Eindruckstechnik*) designed to transform segmented situations into impressive ones (*Plakatsituationen*), as such characterized by accentuated and artificial pathic-semantic contours.[87]

Let's take a look at Schmitz's perspective, now. He is certainly right in reminding us that the world of media and advertising generates serial atmospheres and spreads several false illusions. The first illusion is that everything could be produced, even affective life.[88] The second illusion is that single things could ecstatically radiate atmospheres right away (as Böhme actually claims), thus perpetuating the wrong prejudice of singularism, according to which, one first of all encounters (perceives) single objects rather than chaotic-multiple situations. And, lastly, the third illusion is that an atmosphere could be perfectly explained in dualistic-informational terms, i.e. as a message that a transmitter – an aesthetic worker in this case – sends to a receiver clearly separated from herself.

However, fully accepting Schmitz's and now Begout's[89] strong argument against the idea that atmospheres can be intentionally produced would mean to contradictorily ignore that the felt-bodily communication – which Schmitz himself regards as pervasive – necessarily also applies to the relationship between aesthetic workers and users. Not to mention that it would also be impossible to justify Schmitz's claim that in certain privileged closed situations (dwelling, garden, church, tea house, etc.) atmospheres might be "cultivated" in a non-manipulative way. But, above all, fully subscribing to Schmitz's warning would amount to say that entire spheres of social and cultural life are inexplicable and to surreptitiously introduce in a phenomenological approach programmatically supposed to be descriptive of some normative-axiological parameters, at least existential (like authenticity!) or aesthetic ones (like the rejection of kitsch).

But this requires some clarifications now. First of all, are intentionally produced atmospheres really *fake* atmospheres (i.e. non-atmospheres) or simply *bad* atmospheres (i.e. manipulative)? The first option counts on a third-person *adaequatio* parameter, as such very far from a first-person and neophenomenological approach and from the affective sphere in general, where, if really felt, a feeling cannot ever be false or only illusory (compared with what?).[90] The second one presupposes an apparent dualism between a manipulative and unethical agent and a culpably manipulated and hetero-managed receiver.[91] But this dualism fails to recognize that a certain atmosphere (of fear, for example) could be created by complete political chaos or stem from many different sources, as well as that, given the holistic nature of every situation, it is enormously difficult to distinguish between subject and object. One should not ignore, therefore, the fact that the manipulated person is always co-responsible for her seemingly irresistible involvement. Moreover, the bad atmospheres, that is the manipulative ones, simply do not have absolute authority, and they are almost never, unfortunately, the most artificial and even the totalitarian.[92] The thing is that no one is ever really involved in feelings that one clearly sees as manipulative.

It is obvious from the above that my intention is then to de-axiologize the notion of atmosphere, which means that an atmosphere is really such as soon as it is felt by someone, regardless of whether it is positive or negative, spontaneous or induced. I'd like to break the illusion, by the way, that the affective sphere could be fully transparent (let alone in a cognitive way), that the so-called "suspension of disbelief" could apply outside of the artistic sphere and, lastly, that there are somewhat quantitative (and non-introspective) criteria for affective intensity. Anyway, let me conclude by saying something about atmospheric competence (or intelligence), understood as the ability to b) feel atmospheres, c) understand them, and possibly d) distance oneself from them. What has often been referred to as atmospheric "instinct" or "flair"[93] here should be understood as a skill that can also be improved through exercise: that is, the ability to critically examine the atmospheres one feels. That would enable us to benefit from a "provisional atmospheric morality" (to paraphrase Descartes). There are other three main points that I would like to highlight.

B) A "good" atmospheric competence should also be able to distinguish between "toxic" and "benign" atmospheres, while being aware of walking on thin ice, where aisthesis and ethics mingle. It must be pointed out that toxic atmospheres, which cannot however be reduced to non-atmospheres, are not only those arousing stress and distress but also the dissuasive-sedative ones. Through them many *dispositifs* (in Foucault's sense) aim at defusing any social contradiction with the help both of artificial-conformistic attunements[94] and of the inhibiting effects resulting from the alarmist demand, today become obsessive, to regulate every fragment of everyday experience, not least through the alibi of privacy and political correctness.[95]

C) A good atmospheric competence, however, also consists in accepting the fact that, due to the lack in our post-traditional societies of a paradigmatic-normative place of atmospheric awareness, that is of a situation that may act as a paradigm of every other atmospheric experience, one should rather learn to have as many and different atmospheres as possible and thus to allow the resulting experiences interact with each other. This could give rise to a well-being that, exactly as happens in democracy, depends on a division of powers (affective in this case) that relativizes their impact.

D) Lastly, a good atmospheric competence should favour and foster those atmospheres where, as happens with a *trompe l'oeil*, an early pathic-immersive step may and should be followed by an emersion phase. In this respect, not to mention here the case of protests offering real counter-atmospheres, an example of atmospheres that require both immersion and emersion – which might sound at least in part like the "the dreamer calling out to himself in the midst of the illusory dream world, but without disturbing it, 'It is a dream, I will dream on'" (Nietzsche 1999, 25) –, i.e. are powerfully and influentially contagious without ever being oppressive and coercive, could be easily found in aesthetic experience, especially in contemporary art:[96] unlike the populist, as such hypnotic-somnambulistic, atmosphere, in fact, the critical-artistic one generates through its provocatory and irritating impact cognitive and affective discontinuities that always make a critical distance possible as well as empower whoever deeply experiences them.

Provided then that I am unsatisfied by the *affectus non nisi parendo vincitur* solution, I look around very carefully, but I do not see other ways of immunizing oneself (even only partially) against today's pervasive atmospherization. The usual ways proposed to do so are romantic-moralistic or cognitive-naturalistic, but they are naive either way. The former overestimates personal freedom and the critical power of merely cognitive scepticism; the latter is unrealistically ascetic in requiring a big distance from the affective world, thus degrading human beings to neutral observers of life: a view from nowhere that, as lack of felt-bodily resonance, becomes a real symptom of a psychopathological crisis. A pathic aesthetics, on the contrary, "subversively" promotes an ontological and phenomenological inflationism that even includes quasi-things, and, being moderately critical towards today's predominant stage values, limits itself to proposing a "provisional atmospheric morality". It cannot do much more than this – atmospheres management – and could be useful to adopt this duly revised and detheologized version of Niebuhr's formula: that atmospheric competence may "give us grace to accept with serenity the atmospheres that cannot be changed, courage to change the atmospheres which should be changed (or softened), and the wisdom to distinguish the one from the other".

Many other issues, of course, deserve to be further discussed and explored. To give an idea of that: can the theoretical boom in atmospheres be explained by the growing awareness of the risks of manipulation and unliveability of the (physical and psychic) atmosphere, as Peter Sloterdijk thinks? Can an atmosphere be produced and controlled intentionally, or is it something that floats in space totally independently of human beings? Is it a metaphorical-linguistic effect or a real entity of the external world? Is it a "between" in the sense that it is the outcome of the subject-object relationship, or is it the pre-dualistic background from which subject and object only emerge later? Is it something we perceive thetically, or only the condition of possibility for more focused perceptions? Is it cognitively penetrable, and thus influenced by what we know and, in general, by cultural socialization? Is it an "entity" that only supervenes on its material generators, or is it totally independent of them? Is it only given in first-person experience, analogous to a mood (*Stimmung*), or is it something that we can remember, even counterfactually project and inter-subjectively communicate (and could be understood by others) without significant loss of meaning? Is it a type of perception subject to change over time? Are there different types of atmosphere, depending on the level of independence from the percipient, the type of generator from which it arises, the privileged sensory channel, the level of synaestheticity involved, even the syntactic structure with which one attempts to express it linguistically? Is it possible to totally convert the traditional categories of aesthetics (beautiful, sublime, grace, ugly, etc.) into exquisitely atmospheric concepts? And so on.

One will say, at his point, that for many of these problems I have not provided a real answer or have only suggested a provisional one. But this is not a bad thing: being a philosopher means, for me, also asking the questions that, for now or even in principle, cannot be answered.[97]

Notes

1 This is a modified and extended version of Griffero (2021, 85–103).
2 Edmonde (Louis Jouvet): "I need a change of atmosphere, and you are my atmosphere!" Raymonde (Arletty): "This is the first time I've ever been called atmosphere... Atmosphere! Is that so? Alright then, go by yourself to La Varenne. Have fun fishing and enjoy the atmosphere!"
3 Griffero (2019a, 2020a).
4 Latour (2003).

5 See Griffero (2014).
6 For a very short introduction cf. Schmitz (2003, 2019, 2021).
7 Slaby (2008, 342, fn.29).
8 In the non-mentalistic sense of Merleau-Ponty's "operative intentionality", of course.
9 In analogy with what Gadamer did with Heidegger, according to Habermas.
10 Which are such *de re*, not due to epistemic deficits.
11 I have developed Schmitz's view (suggested from 1978 on) in Griffero (2017a).
12 Cf. Griffero (2020b).
13 Further criteria may be greater or lesser intensity, awareness, etc.
14 See Haensel (1946, 38–41, 52–53).
15 Schmitz (2008b).
16 As stated by Blume/Demmerling (2007, 126) und Demmerling (2011, 48). See Griffero (2014b).
17 Archetypically, starting from Democritus' atomistic materialism.
18 See Griffero (2017a, 1–18).
19 Think of the alarmist climate promoted by the new media, especially breaking news (Milev 2012, 301). Cf. also Schmitz (2003, 315).
20 To use Gibson's notion (see Gibson 1986) in a broader and not only visual-pragmatic sense. See now Griffero (2022b).
21 See Schmitz (2008a, 77; 2008b, 8) and Griffero (2014).
22 Rappe (1995, 312–323).
23 See Schmitz (1977, 133–134) and Griffero (2020b, 137–149).
24 Schmitz (1977, 91).
25 Schmitz (2003, 47–48; 2008b, 9). One could explain the contrast between atmosphere and individual feeling by distinguishing between perceiving affects of others or collective emotions and understanding a situational significance (Slaby 2008, 341). But how can one also explain the contrast with inanimate situations, such as when a wonderful day makes me sad? On the vexed issue of atmosphere as a (possibly) collective feeling see Griffero (2021, 2022a) and Trigg (2022).
26 Bollnow (1956[3], 111) calls it a respectful distance.
27 Schmitz (2008b, 9; 2019, 96).
28 Schmitz (2012, 57).
29 Musil (1995, 1261).
30 For alternative explanations of contrasting feelings see Hauskeller (1995, 23) and Demmerling (2011, 47).
31 Tellenbach (1968, 47–48).
32 Quoted by Bollnow (1956[3], 56). For further discussion and other examples see Griffero (2021, 63–66).
33 Schmitz (2002, 169–170; 2016, 224).
34 See Griffero (2017a, 79–92). Personal shame is deeply atmospheric as a form of affectivity that condemns those who are subject to it in a centripetal form. However, the same also holds for vicarious shame which we feel (non-empathetically) for those who "should" be ashamed, powerfully acts as an atmospheric and quasi-thingly feeling.
35 See especially Schmitz (1965, 2011).
36 Which is a specific relationship that lies between a felt body and any of its "encountered" beings or forms, even if inanimate. For a quick overview of Schmitz's theory of *leibliche Kommunikation*, see Griffero (2017b).
37 For these inverted correspondances cf. also Seel (1996, 101–102).
38 See Böhme (2007, 289–290).
39 For a more detailed phenomenology of see Griffero (2014, 129–141, and, lastly, 2021, 37–66).
40 Language, which according to Schmitz is an example of a non-atmospheric shared situation Schmitz (2014, 55; see also 10 und 50), seems to me non-atmospheric

only if it is considered abstractly – certainly not when it is emotionally experienced, for example when one hears its sound away from home or when (in literature as well as in ordinary language) we realize that its expressions are largely untranslatable. As Wigley (1998, 26) says, even "the very rejection of an atmosphere constructs a particular atmosphere".

41 See also Ratcliffe (2008, 148, 178).

42 About non-thetic function of atmospheres see, for example, Thibaud (2003, 293), Bockemühl (2002, 221) and Minkowski (1936, 234).

43 Think of eroticism or the "porous" connection typical of non-oppressive love. See Schmitz (2008b, 11–12).

44 This is hardly surprising, since "due to their vastness as atmospheres [...] all feelings [are] moods [*Stimmungen*]" (Schmitz 2016, 237). Cf. also Julmi (2015, 57).

45 See Böhme (2001, 59–62).

46 Heidegger (1995, 68).

47 Mendelssohn (1761, 157).

48 Cf. Julmi (2015, 102). What is love, for example, might at first seem "an unpleasant and disturbing restlessness" (Schmitz 2003, 53).

49 It is not really a special case, as Demmerling claims (2011, 50).

50 Schmitz (2016, 241).

51 Geiger (1911, 27f.).

52 Slaby (2008, 333).

53 Slaby (2008, 333).

54 "When we fight against penetration of an objective feeling, it has already taken hold of us" (Baensch 1923, 9).

55 Ben-Ze'ev (2001, 126).

56 Those who reject Schmit's distinction as an exaggeration of a localized phenomenon claim that a *feeling* (the feeling's content) and *to feel* (the act of feeling), which can only be distinguished for heuristic reasons, are actually the same thing. Their only apparent distinction could be explained through the distinction between the only affective and only perceptual use of an evaluative notion (thus Slaby 2008, 339–334).

57 See Slaby (2017, 239–240, 243–244).

58 Slaby (2008, 335–336).

59 For example, according to the felt-bodily dynamic: a cyclothymic person's oscillation tendency and a schizothymic's splitting tendency should foster their involvement (this is all the more true in the case of uneasiness: cf. Sonntag 2013, 178), whereas a bathmothymic impulse to compactness would foster the mere statement of what affectively happens. Can a correlation be made between a certain atmosphere and a specific felt-bodily island? See Griffero (2016, 2017c, 2019b, 20–25).

60 See Slaby (2008, 335–336) and especially Schouten (2011^2, 203).

61 To define the "in-between" as a relation between two autonomous beings is, in fact, ontologically quite different from defining it simply as a condition of possibility of their "emergence".

62 Böhme (1995, 22).

63 Which is a co-action with no reaction time like a jump scare but irreducible to stimulus-response because it implies retentions and protentions.

64 Cf. Schmitz (2011, 89, 95; 2014, 86; 2016, 223).

65 Julmi (2015, 196).

66 See Frese (1995).

67 Böhme (2001, 49).

68 As even Schmitz (2014, 55) recognises.

69 You can think of neutral concepts such as "apparatus", "market", "governance" and influential acronyms such as SPREAD, PIL, etc. Cf. Heibach (2010, 11),

Schmitz (2008b, 15–16), Hennecke (2008, 57–58). For a less pessimistic conception of the atmosphericity of aesthetic capitalism see Böhme (2017a).

70 Soentgen (1998, 117).

71 See, for example, Bollnow (1956^3, 132).

72 Bollnow (1956^3, 154–161).

73 Schmitz (2014, 9).

74 Schmitz (1969, 87).

75 Preusker (2014, 132).

76 Feelings would seem to be outside and objective only because they are shared and subject to public discussion by those they deeply influence.

77 Rauh (2012, 252).

78 See, for example, Slaby (2008, 339) or Bulka (2015).

79 Böhme (1995, 39).

80 Böhme (1995, 199–200). See also Grossheim/Kluck/Nörenberg (2014, 65).

81 Böhme (2001, 87ff.; 2017b, 92–94: moving impressions, synaesthesia, social characters, etc.).

82 Exemplarily in the aesthetic field: see Geiger (1911, 37). Contemplating a landscape only in order to evoke feelings is completely different from letting oneself go to its values and radiances (Geiger 1928).

83 "All making conscious means destroying, altering in each case, whereas in awakening an attunement we are concerned to let this attunement be as it is, as this attunement" (Heidegger 19950, 65).

84 Lersch (1962, 313).

85 As resonant appeals the emotions (feelings or affects in the broader sense) can vary according to intensity and degree of excitement. Beyond this, they can be genuine or false. False feelings have nothing to do with myself [...] Sentimentality, which is out to be touched and misses no opportunity to enjoy inner capitulation, scarcely permits true feelings to arise. For true feeling is an appeal bound through and through to what is the case. Not the resonance alone, with its qualities of mood, like cheerful, gloomy, joyful, irritated, sad, etc., governs our state, but the resonant appeal of some particular quality. In false feeling there is no appeal in an objective context, and only the resonance itself absorbs us.

(Plessner 1970, 128)

86 Bollnow (1956^3 59, 132, 140, 152–153).

87 Schmitz (1998; 1999, 242–245, 335; 2002, 169; 2003, 243–261; 2005, 266).

88 It is almost a *contradictio in adjecto*, given that a feeling is "expression of a reality [...] and this reality needs resistance in order to be felt" (Haensel 1946, 127).

89 For an approach that excludes the need to know how atmospheres are designed (Böhme), their direct producibility (Sloterdijk's "self-climatisation") and even the search for favourable conditions for the emergence of more appropriate atmospheres (Bollnow) see Begout (2019, 377–403). His thesis – *ambiance* is too transient, medial and authoritarian (and therefore passively felt) for man to dispose of it at will, and any desire to control it would end up reifying it, depriving it of its involuntary, mersif et enveloppant character – is partly shareable, of course. But if it becomes the precondition for any *ambiance* (atmosphere), it basically raises the same perplexities raised by Schmitz's thesis, beginning with the fact that Begout himself acknowledges that always and since time immemorial men have sought (evidently not always in vain) to control or modify *ambiances* in their favour. That the *ambiance* is only authentically involving if it erases all traces of its own laborious intentional genesis, and that the final effect of a voluntary production is (sometimes) an affective tone different from the planned one, does not in principle seem to exclude the generability of atmosphere – first of all, because the unwanted *ambience* is in any case generated. Moreover, in what way, from the

first-person perspective of the percipient, would authentic *ambiances*, because they are produced by man involuntarily and without any competence, differ from the *fantômes d'expérience* aroused by intentionally produced ones? The lapidary thesis that activity cannot produce what is passive, after all, in no way excludes active production (by someone) being experienced passively (by others). See also Griffero (2020c).

90 One can, indeed, have doubts about the object of her feeling but not about her own feeling.

91 Heibach (2012, 263).

92 Landweer (1999, 145).

93 Tellenbach (1968, 49).

94 Schouten (2011^2, 103).

95 For this critique of our alarmed society see Sloterdijk (2011–16, especially the third volume (*Foams)*.

96 Schouten (2011^2, 106).

97 Marquard (1989, 62).

Bibliography

Baensch, Otto. *Kunst und Gefühl.* "Logos". 12, 1 (1923), 1–28.

Begout, Bruce. *Le concept d'ambiance.* Paris: Seuil, 2019.

Ben-Ze'ev, Aaron. *The subtlety of emotions.* Cambridge: MIT Press, 2001.

Blume, Anne & Christophe Demmerling. "Gefühle als Atmosphären? Zur Gefühlstheorie von Hermann Schmitz". In *Gefühle – Struktur und Funktion*, edited by Hilge Landweer, 113–133. Berlin: Akademie Verlag, 2007.

Bockemühl, Michael. "Atmosphären sehen. Ästhetische Wahrnehmung als Praxis". In *Neue Ästhetik. Das Atmosphärische und die Kunst*, edited by Ziad Mahayni, 203–222. Munich: Fink, 2002.

Böhme, Gernot. *Atmosphäre. Essays zur neuen Ästhetik.* Frankfurt a. M.: Suhrkamp, 1995.

Böhme, Gernot. *Aisthetik: Vorlesungen über Ästhetik als allgemeine Wahrnehmungslehre.* München: Fink, 2001.

Böhme, Gernot. "Atmosphären in zwischenmenschlicher Kommunikation". In *Atmosphären im Alltag. Über ihre Erzeugung und Wirkung*, edited by Stephane Debus and Roland Posner, 281–293. Bonn: Psychiatrie-Verlag, 2007.

Böhme, Gernot. *Atmospheric architectures. The aesthetics of felt spaces.* London: Bloomsbury, 2017a.

Böhme, Gernot. *Critique of aesthetic capitalism.* Milan: Mimesis International, 2017b.

Bollnow, Otto Friedrich. *Das Wesen der Stimmungen.* Frankfurt a.M.: Klostermann, 1941/1956.

Bulka, Thomas. *Stimmung, Emotion, Atmosphäre. Phänomenologische Untersuchungen zur Struktur der menschlichen Affektivität.* Münster: mentis, 2015.

Demmerling, Christophe. "Gefühle, Sprache und Intersubjektivität. Überlegungen zum Atmosphärenbegriff der Neuen Phänomenologie". In *Gefühle als Atmosphären. Neue Phänomenologie und philosophische Emotionstheorie,* edited by Kerstin Andermann and Undine Eberlein, 43–55. Berlin: Akademie Verlag, 2011.

Frese, Jürgen. "Gefühls-Partituren". In *Leib und Gefühl. Beiträge zur Anthropologie,* edited by Michael Großheim, 45–70. Berlin: Akademie Verlag, 1995.

Geiger, Moritz. "Zum Problem der Stimmungseinfühlung". *Zeitschrift für Ästhetik und allgemeine Kunstwissenschaft,* 6 (1911), 1–42.

Geiger, Moritz. *Zugänge zur Ästhetik*. Leipzig: Der Neue Geist, 1928.

Gibson, James J. *The ecological approach to visual perception*: London: Routledge, 1986.

Griffero, Tonino. *Atmospheres. Aesthetics of emotional spaces* (2010). London-New York: Routledge, 2014.

Griffero, Tonino. "Atmospheres and felt-bodily resonances". *Studi di estetica,* XLIV, 1, (2016): 1–41.

Griffero, Tonino. *Quasi-Things. The paradigm of atmospheres*. Albany: SUNY Press, 2013.

Griffero, Tonino. "Felt-bodily communication: A neophenomenological approach to embodied affects". *Studi di estetica*, XLV, 8/2, (2017a): 71–86.

Griffero, Tonino. "Felt-bodily resonances. Towards a pathic aesthetics". *Yearbook for Eastern and Western Philosophy* (Embodiment. Phenomenology East/West), 2, (2017b): 149–164.

Griffero, Tonino. "Is there such a thing as an 'atmospheric turn'? Instead of an introduction". In *Atmosphere and aesthetics: A plural perspective*, edited by Tonino Griffero and Marco Tedeschini, 11–62. Basingstoke: Palgrave Macmillan, 2019a.

Griffero, Tonino. "The invasion of felt-bodily atmospheres: Between pathic aesthetics and psychopathology". In *Psychopathology and atmospheres. Neither inside nor outside*, edited by Gianni Francesetti & Tonino Griffero, 6–34. Newcastle upon Tyne: Cambridge Scholar, 2019b.

Griffero, Tonino. "Emotional atmospheres". In *The Routledge Handbook of phenomenology of emotion*, Thomas Szanto & Hilge Landweer, 262–274. Abingdon and New York: Routledge, 2020a.

Griffero, Tonino. *Places, affordances, atmospheres: A pathic aesthetics*. London and New York: Routledge, 2020b.

Griffero, Tonino. "Was kann eine Gefühlsatmosphäre tun? Atmosphären zwischen Immersion und Emersion". In *Die Macht der Atmosphären*, edited by Barbara Wolf & Christian Julmi, 77–96. Freiburg-München: Alber, 2020c.

Griffero, Tonino. *The atmospheric "we". Moods and collective feelings*. Milan: Mimesis International, 2021.

Griffero, Tonino. "Are atmospheres shared feelings?" In *Atmospheres and shared emotions*, edited by Dylan Trigg, 17–39. London & New York: Routledge, 2022a.

Griffero, Tonino. "They are there to be perceived: Affordances and atmospheres". In *Affordances in everyday life*, edited by Zakaria Djebbara, 85–95. Cham: Springer, 2022b.

Grossheim, Michael, Steffen Kluck, Henning Nörenberg. *Kollektive Lebensgefühle. Zur Phänomenologie von Gemeinschaften*. Rostocker Phænomenologische Manuskripte, Vol. 20. Rostock: Rostock University, 2014.

Haensel, Karl. *Das Wesen der Gefühle*. Überlingen: Wulff, 1946.

Hauskeller, Michael. *Atmosphären erleben. Philosophische Untersuchungen zur Sinneswahrnehmung*. Berlin: Akademie Verlag, 1995.

Heibach, Christiane. "Manipulative Atmosphären. Zwischen unmittelbarem Erleben und medialer Konstruktion". In *Atmosphären. Dimensionen eines diffusen Phänomens* edited by Christiane Heibach, 261–282. München: Fink, 2012.

Heidegger, Martin. *The fundamental concepts of metaphysics. World, finitude, solitude* (1929–30). Bloomington and Indianapolis: Indiana University Press, 1995.

Hennecke, Hans Jochen. "Macht, Legitimität und Ordnung". In *Zur Legitimierbarkeit von Macht*, edited by Hans Jürgen Wendel & Steffen Kluck, 55–81. Freiburg-Munich: Alber, 2008.

Julmi, Christian. *Atmosphären in Organisationen. Wie Gefühle das Zusammenleben in Organisationen beherrschen.* Bochum/Freiburg: projektverlag, 2015.
Jung, Carl Gustav. *Psychological types or the psychology of individuation.* New York: Pantheon Books, 1921.
Landweer, Hilge. *Scham und Macht. Phänomenologische Untersuchungen zur Sozialität eines Gefühls.* Tübingen: Mohr Siebeck, 1999.
Latour, Bruno. "Atmosphere, Atmosphere". In *Olafur eliasson: The weather project,* edited by Susan May. London: Tate Gallery, 2003.
Lawrence, David Herbert. *The Rainbow* (1915). Ware: Wordsworth Classics, 1955.
Lersch, Philipp. *Aufbau der Person* (1938). Munich: Barth, 1962.
Marquard, Odo. *Aesthetica und Anaesthetica. Philosophische Überlegungen.* Paderborn: Schöningh, 1989.
Mendelssohn, Moses. "Rhapsodie, oder Zusätze zu den Briefen über die Empfindungen" (1761). In *Ästhetische Schriften,* edited by Moses Mendelssohn. Hamburg: Meiner, 2005.
Milev, Yana. (2012). "Design governance und breaking news: das Mediendesign der permanenten Katastrophe". In *Atmosphären. Dimensionen eines diffusen Phänomens,* edited by Christiane Heibach, 285–303. Munich: Fink, 2012.
Minkowski, Eugène. *Vers une cosmologie: fragments philosophiques.* Paris: Aubier Montaigne, 1936.
Musil, Robert. *The Man without Qualities.* II. *Into the millennium. From the posthumous papers* (1930, 1933, 1943). New York: Knopf, 1995.
Nietzsche, Friedrich. *The birth of tragedy and other writings.* New York: Cambridge University Press, 1999.
Plessner, Helmuth. *Laughing and crying. A study of the limits of human behavior* (1941). Evanston: Northwestern University Press, 1970.
Preusker, Johannes. *Die Gemeinsamkeit der Leiber. Eine sprachkritische Interexistenzialanalyse der Leibphänomenologie von Hermann Schmitz und Thomas Fuchs.* Frankfurt a.M.: Lang, 2014.
Rappe, Guido. *Archaische Leiberfahrung. Der Leib in der frühgriechischen Philosophie und in außereuropäischen Kulturen.* Berlin: Akademie Verlag, 1995.
Ratcliffe, Matthew. *Feelings of being. Phenomenology, psychiatry and the sense of reality.* Oxford-New York: Oxford University Press, 2008.
Rauh, Andreas. *Die besondere Atmosphäre. Ästhetische Feldforschungen.* Bielefeld: Transcript, 2012.
Rilke, Rainer Maria. *Malte Laurids Briggs and other prose* (1906–26). New York: Norton, 1945.
Schmitz, Hermann. (1965). *System der Philosophie.* II.1, Der Leib. Bonn: Bouvier, 1965.
Schmitz, Hermann. *System der Philosophie,* II. 2, *Der Gefühlsraum.* Bonn: Bouvier, 1969.
Schmitz, Hermann. *System der Philosophie,* III. 4., *Das Göttliche und der Raum.* Bonn: Bouvier, 1977.
Schmitz, Hermann. *System der Philosophie,* III. 5., Die Wahrnehmung. Bonn: Bouvier, 1978.
Schmitz, Hermann. "Situationen und Atmosphären. Zur Ästhetik und Ontologie bei Gernot Böhme". In *Naturerkenntnis und Natursein* edited by Michael Hauskeller, 176–190. Frankfurt a.M.: Suhrkamp, 1998.
Schmitz, Hermann. *Adolf Hitler in der Geschichte.* Bonn: Bouvier, 1999.

Schmitz, Hermann. *Begriffene Erfahrung. Beiträge zur antireduktionistischen Phänomenologie.* with G. Marx, A. Moldzio. Rostock: Ingo Koch Verlag, 2002.

Schmitz, Hermann. *Was ist Phänomenologie?* Rostock: Ingo Koch Verlag, 2003.

Schmitz, Hermann. *Situationen und Konstellationen. Wider die Ideologie totaler Vernetzung.* Freiburg/Munich: Alber, 2005.

Schmitz, Hermann. *Logische Untersuchungen.* Freiburg/Munich: Alber, 2008a.

Schmitz, Hermann. "Die Legitimierbarkeit der Macht". In *Zur Legitimierbarkeit von Macht*, edited by Hans Jürgen Wendel & Steffen Kluck, 5–19. Freiburg/Munich: Alber, 2008b.

Schmitz, Hermann. *Der Leib.* Berlin/Boston: De Gruyter, 2011.

Schmitz, Hermann. *Das Reich der Normen.* Freiburg/Munich: Alber, 2012.

Schmitz, Hermann. *Atmosphären.* Freiburg/Munich: Alber, 2014.

Schmitz, Hermann. *Ausgrabungen zum wirklichen Leben. Eine Bilanz.* Freiburg/Munich: Alber, 2016.

Schmitz, Hermann. *New phenomenology. A brief introduction* (2009). Milan: Mimesis International, 2019.

Schmitz, Hermann. *Sich selbst verstehen. Eine Lesebuch.* Freiburg/Munich: Alber, 2021.

Schouten, Sabine. *Sinnliches Spüren. Wahrnehmung und Erzeugung von Atmosphären im Theater.* Berlin: Theater der Zeit, 2011.

Seel, Martin. *Eine Ästhetik der Natur.* Frankfurt a.M.: Suhrkamp, 1996.

Slaby, Jan. *Gefühl und Weltbezug. Die menschliche Affektivität im Kontext einer neo-existentialistischen Konzeption von Personalität.* Paderborn: mentis, 2008.

Slaby, Jan. *Möglichkeitsraum und Möglichkeitssinn. Bausteine einer phänomenologischen Gefühlstheorie.* In *Körperskandale. Zum Konzept der gespürten Leiblichkeit*, edited by Stefan Volke & Steffen Kluck, 220–248. Freiburg/Munich: Alber, 2017.

Sloterdijk, Peter. *Spheres* (1998–2004), 3 vol. Los Angeles: Semiotext(e), 2011–2016.

Soentgen, Jens. *Die verdeckte Wirklichkeit. Einführung in die Neue Phänomenologie von Hermann Schmitz.* Bonn: Bouvier, 1998.

Sonntag, Jan. *Demenz und Atmosphäre. Musiktherapie als ästhetische Arbeit.* Frankfurt a. M.: Mabuse-Verlag, 2013.

Tellenbach, Hubertus. *Geschmack und Atmosphäre. Medien menschlichen Elementarkontaktes.* Salzburg: Otto Müller Verlag, 1968.

Thibaud, Jean-Paul. (2003). "Die sinnliche Umwelt von Städten. Zum Verständnis urbaner Atmosphären". In *Die Kunst der Wahrnehmung. Beiträge zu einer Philosophie der sinnlichen Erkenntnis*, edited by Michael Hauskeller, 280–297. Kusterdingen: Die Graue Edition, 2003.

Trigg, Dylan, ed. *Atmospheres and shared emotions.* London-New York: Routledge, 2022.

Wigley, Mark. "The architecture of atmosphere". *Daidalos*, 68, ("Constructing atmospheres") (1998): 18–27.

Urubupungá
20.093
MASP

Part 4

Ambiances, sensitivity, and culture

12 Towards a therapy of ambiances

Philippe Simay

Let us put forth the idea that ambiances played a therapeutic role in forming modern sensibilities and subjectivities during the metropolization of European societies from the late eighteenth to the mid-twentieth centuries.

Why are we talking about therapy here? And what were ambiances supposed to cure? The physical and psychological ills, real or alleged, that affected city dwellers, and which were caused by the city itself.[1] Since the late eighteenth century, the city has in fact been depicted as an artificial environment that summons frightening images of dark, miserable, trash-filled, miasmatic streets oozing with illness, vice, and madness. Such negative representations did not merely accompany urban growth. They also underlay most debates over the city and prompted a very diverse range of reactions to it. Two paradigmatic conceptions are particularly interesting to us here: the first, advanced by Jean-Jacques Rousseau, identified the city as an unnatural phenomenon that should be fled, while the second, held by Georg Simmel, Siegfried Kracauer, and Walter Benjamin, instead viewed the city as an environment to which we must adapt.

Rousseau, a strategy of distance

Rousseau is rightly considered one of the founders of the "urbaphobe" current; in fact, big cities were the target of some of his most pointed criticisms. In his *Confessions*, he describes his disappointment and repulsion upon arriving in Paris in June 1731: "I see nothing but dirty, smelly, little streets, dark, ugly houses, an atmosphere of filth, poverty… it affected me so deeply that it left me with a deep-seated sense of disgust while I lived in the city".[2] Even though Rousseau saw Paris as a place that concentrated knowledge, he remained averse to the city's ambiance, which was characterized by an assault on the senses and a congestion of so many people in such a small space that seemed to foul the air and the bodies of those who breathed it.

Such a physical corruption had a moral component as well. Rousseau was struck by the violence inherent to the social inequalities that arose from the division of labour and the monopolization of wealth. In *The New Heloise,* he wrote that Paris is "the city where the most sumptuous opulence and the

DOI: 10.4324/9781032655307-19

most deplorable misery both reign supreme".[3] He also denounced the vices of this nascent urban sociability, the predominance of what he called "suspect feelings": the clique mentality, obsessing with one's appearance and escaping into amusements, a cheap flattery and hypocrisy that replaced friendship with opportunism and genuine feelings with lies. For Rousseau, urban society both *forms* and *perverts* our modern sensibilities, which are thus far removed from our natural state of innocence.

In *Emile,* Rousseau said that big cities are "the chasm of human hope".[4] But since our natural state was thus lost once and for all, Rousseau did admit that the process of society's urbanization was nevertheless underway. And even if he wished that this were limited to the mid-sized cities that he preferred, he recognized its unavoidability.

Emile thus had to learn to live in this *anti-nature* environment of the city, making the most of its inhabitants and "living, if not like them, then at least with them". Her education should not lead her to flee the city forever, instead simply to prevent and counteract the harmful influence of a corrupted society. How? By combatting the city's deleterious effects through the rediscovery of a healthy, soothing nature.

What we now call a "need for nature" comes from Rousseau's own life. Following the condemnation of *Emile* and *The Social Contract* by the Parliament of Paris, he left Paris and took refuge in the Val-de-Travers in the canton of Neuchâtel, where he would live from 1762 to 1765. This is where the legendary image of the walking philosopher was formed, wearing his Armenian kaftan and clutching his pilgrim's walking stick, criss-crossing the gorges of the Areuse River, at the foot of a tree or the edge of a waterfall, fully absorbed in the greenery and in observing the nature around him.

The "Seventh Walk" in the *Reveries of the Solitary Walker* testifies to a strategy of distancing oneself from the city and returning to nature. The practice of walks through the forest and hikes through the mountains offers a pleasure that does not necessarily result from the bombardment of one's senses, as is the case with cities, rather in their being soothed by the silence of the woods, the sound of streams, and the caress of a breeze – in other words, by ambiances that heal the infirm sensibilities of modern city dwellers.

This therapeutic dimension of ambiances is particularly evident in the "Fifth Walk", in which Rousseau describes his recollection of a stay on Saint Peter's Island in the middle of Lake Bienne in Switzerland from September to October 1765. This is where he experienced the most complete happiness and the fullness of what he called the "sentiment of existence", one of Rousseau's most profound and complex concepts, with which he articulated a philosophy of perception through a phenomenology of states of consciousness.

"What does one enjoy in such a situation, says Rousseau? Nothing other than oneself and one's existence". On the one hand, Rousseau did not reduce feelings to sensation. The sentiment of existing is free of the passivity of organic affections. It is a "feeling oneself to exist", to which the mere interplay of protean sensations could obstruct access. On the other hand,

Rousseau clearly suggested that this feeling was not a pure awareness of oneself, because it is experienced as emotion and pleasure. The sensory is indeed present, but it cannot just express itself in any manner whatsoever.

Jean Starobinski very astutely observed this intermediate situation in which Rousseau situated himself. He seems to set aside everything that is not the "sentiment of existence", anything that could hinder or obscure it, as is the case with sensation. One must arrive at sensing things in a way that avoids "the iridescent diversity of sensuous experience". At the same time, this cannot be done without "surrounding objects".[5]

What Rousseau found on Saint Peter's Island was not simply "nature", instead a very specific type of ambiance that allows one to be stimulated by one's sensory environment, but without being overwhelmed by it. Rousseau described its characteristics clearly:

> As evening approached, I would come down from the island's summits and happily sit at the edge of a lake, in some hidden, sheltered spot. The sound of the waves and the agitation of the water captivated my senses and chased away any other agitation from my soul, plunging it into a delightful reverie, during which night-time would often creep up on me without my realizing it. The ebb and flow of this water, its continuous sound swelling incessantly into my ears and eyes, replaced the internal movements that the reverie extinguished in me and sufficed to make me feel my existence with pleasure, without having to think about it.[6]

The ambiance that allows the sentiment of self to arise is the same one by which what I think and feel indeed constitutes, here and now, *my* sensibility and *my* subjectivity; it is the gentle, regular, and continuous movement of the water. In other words, it is an acoustic ambiance, or, more precisely, a rhythm. The philosopher André Charrak commented on this connection between the rhythm of the water and the formation of the reverie: "The flow of imaginary affections is answered immediately by an ebbing, that is the sentiment's withdrawal upon its repetition, copied in the rhythm of the water and slowly generating a true state of openness to oneself within one's soul".[7]

Rousseau's life and work testify at length to the therapeutic dimension of ambiances as a way of healing the ills of the city. What Rousseau was fleeing in the city was not just its oppressive ambiances, the incessant, aggressive bombardment of stimulations, but also a semiotic landscape where everything formed a sign that reminded him of his unhappiness, his oppressors, and the shortcomings of a society to which he never managed to adapt. Because the ambiance of Lake Bienne consisted of rhythms instead of signs, it provided him with a path to recovery. The palindromic movement of the water soothes and cradles the imagination without anything in the outside world to trouble it. In the calm of the present, the sentiment of existence reconciles the love of self with that of the world. At least for a short while, because the city and its inhabitants would swiftly return to populate his reveries.

We can see that Rousseau's therapy was more of a balm than a vaccine. His recourse to nature – or, more precisely, a certain type of rhythmic ambiance in nature – temporarily mitigated the ills of the city, but it never cured him of them for good. For Rousseau, urban consciousness and a sensitivity to nature were inseparable aspects of a modern subjectivity that oscillated dialectically between these two extremes. Rousseau did not abandon the city; he merely left it momentarily to take refuge in what lay outside it. But in these constant gestures of distancing himself, Rousseau managed to grasp with an extraordinary prescience the modern city in its otherness, heterogeneity, and contradictions, as a universe saturated with too many nervous signs and stimulations from which one must learn to protect oneself.

Simmel, Benjamin, and Kracauer in the crucible of the metropolis

A different therapeutic relationship to ambiances is at work for Georg Simmel, Siegfried Kracauer, and Walter Benjamin. These three writers tried to understand the metropolis of the first third of the twentieth century not by fleeing it, as Rousseau had, instead by confronting it to grasp the formation of a modern sensibility on the basis of their own experience.

The sensorial approach to modernity shared by Simmel, Kracauer, and Benjamin consisted of taking into account how the material transformations of metropolises affect our sensory apparatus and transform the registers of individual and collective experience. It also sought to establish how the sensory apparatus adapts to, and eventually masters, an increasingly complex environment. Thus, the issue is one of skills, of an aptitude that requalifies the city dweller's relationship to the urban space.[8] The singularity of their thought resides in their demonstration that a city dweller's skills develop by means of ambiances, of places and objects within this apparently chaotic milieu that is the metropolis, and that urban pathologies also constitute resources that a city dweller may use. In other words, they envisioned the metropolis as a *pharmakon* – not just as a destructive force, but as also having the power to heal.

Simmel, Kracauer, and Benjamin did not invoke the concept of ambiance. Yet each one of them was interested in situations where ambiances are inseparable from the skills of perception and mobility that are indispensable to the workings of urban life: the comings and goings of cars and trams, the constant din of vehicles and their horns, the swarm of crowds filling the streets to their capacity, the omnipresence of advertising images, the flash of illuminated signs on facades, and the incessant cadence of machines.

Ambiance lies at the heart of their assessment of modern life, characterized by a hyperesthesia from an overabundance of nervous stimulations and shocks.[9] We may view shock as a constituent trait of metropolitan ambiances. It is an ordinary characteristic of the urban environment that is incorporated into the realm of the perceptible that affects city dwellers and lends a particular tonality to their experience.

Shock was analysed as a pathological element of metropolitan ambiances. For Simmel, it was tantamount to a rape of the psyche. Kracauer said that it placed individuals in a state of permanent inattention. Benjamin added that it dispelled collective memory, dulled the senses and encouraged neurasthenia, rendered experiences incommunicable, created psychomotor disorders, and more.

And yet, in these same analyses, Simmel, Kracauer, and Benjamin also emphasize the city dweller's capacity for adaptation and, at times, autonomy.

In his 1939 essay, *The Work of Art in the Age of Mechanical Reproduction*, Benjamin emphasized "the need to expose oneself to shocks represents humans' adaptation to the perils that threaten them".[10] According to Benjamin, controlling these shocks came not from protecting oneself, but instead from exposing oneself to them. Paradoxically, reliving the experience of the shock enables the city dweller to overcome its traumatizing effects. Benjamin revealed here the pharmacological aspect of shock as both a poison and a remedy, on which one must become drunk to rid oneself of it and thus allow for an urban self-therapy.

We should specify at this point that this voluntary exposure to stimulations is directly dependent on the affective tonality of ambiances. A shock is never sought out for its own sake. The same shock can be perceived as either traumatizing or exciting. Ambiances configure city dwellers' sensory impressions and thus stimulate their openness to the spaces where they find themselves.

This is why these three writers paid extensive attention to places of consumption and entertainment (trade fairs, department stores, railway stations, restaurants, hotels, movie theatres, and amusement parks), which – for all the obvious commercial reasons – rely on the attractiveness of their ambiances.[11]

Simmel published a piece on the 1896 Berlin Trade Exhibition, in which he described the ambiance of a vast commercial hall, its visitors electrified by what he called the "showcase quality of the objects",[12] meaning the aesthetic supplement to the merchandise, in anticipation of a critique of the cosmeticization and spectacularization of market capitalism.

Kracauer was the first to analyse the appearance of urban entertainments as a response to the modern need for stimuli. The detective novel, dance reviews, carnivals, and cinema all share ambiances in which the new "thrill-seeking culture" sought by the masses manifested itself. These ambiances draw the masses to places that metabolize the experience of shock. City dwellers relive some of the difficulties encountered in their daily lives in a more playful and hypertrophic form. Thus, such ambiances fulfil the city dwellers' need for *adaptation* and *conformation* to a rapidly changing context. They represent a kind of therapeutic device that people use to acclimatize themselves to the rigors of city life.

Obviously, the theory of a therapeutic aspect to the ambiances of mass entertainment is only relevant if it reaches past the conformation of city dweller's skills to the urban order and arrives at a reflection and critique. This is what Benjamin suggests in his analyses of cinema. For Benjamin, cinema is the art of urban modernity because it captures and represents the ambiances of a metropolis that audiences like to see in films. Cinema is also the medium that most thoroughly portrays the experience of metropolitan shock to which viewers expose themselves to become accustomed to it. And finally, cinema exploits a kind of attention that resembles the urban experience, which is one of distraction (*Zerstreuung*).[13] According to Benjamin, distraction is not an inability to concentrate, instead a redistribution of one's attention.

This decentred form of attention allows the viewer to gain a veritable expertise that has a clearly political dimension. Benjamin here stands directly opposite Kracauer's 1926 essay, *The Cult of Distraction*. Kracauer views cinema as a legitimate form of compensation for the masses, who are subjected to the mechanization of urban life and the rationalization of industrial and commercial labour; however, he agreed with the tenets of *Kulturkritik* in viewing distraction as passive and immobilizing. For Kracauer, cinema cannot be an emancipating force so long as it subjects viewers to unilateral, uncritical surface effects that only provide access to fragments of our social reality.

Unlike Kracauer, Benjamin argues that distraction in fact constitutes a form of active apprehension for grasping this fragmentary social reality and also acting on it, thus enacting a revolutionary reversal within an entertainment-driven society. Kracauer might have pointed out here to Benjamin that he had paid surprisingly little attention to the very ambiance of movie theatres, the goal of which is specifically to keep the audience at the periphery, such that an encounter between the film and the audience can never take place.[14] The divergence of opinion between Kracauer and Benjamin was not resolved, and it persists today in the differing assessments of mass culture and the forms of distraction that it generates.[15]

***

Rousseau on one side and Simmel, Kracauer, and Benjamin on the other represent the emergence of a modern sensibility tied to the growth of urban society, though in diametrically opposite ways. They share an investigation of the need for an urban therapy and to search for remedies to a city's ills, thereby putting a spotlight on the role of ambiances. Each one in his own way shows how ambiances allow us to work on ourselves, either by distancing ourselves from urban life or by conforming to it.

Notes

1 On "urbaphobe" currents and discourses, see Arnaud Baubérot and Florence Bourillon, *Urbaphobie. La détestation de la ville au XIXe et XXe siècles* ["Urbaphobia: Detesting the city in the 19th and 20th Centuries"] (Paris: Editions Bière, 2009); Joëlle Salomon Cavin and Bernard Marchand (ed.), *Antiurbain.*

Origines et conséquences de l'urbaphobie ["Anti-urban: Urbaphobia's origins and consequences"] (Rennes: PPUR, 2010).
2 Jean-Jacques Rousseau, *The Confessions*, trans. J.M. Cohen (London: Penguin Classics, 1953).
3 Jean-Jacques Rousseau, *The New Heloise*, trans. Philip Stewart and Jean Vaché (Hanover, NH: Dartmouth College Press, 1997).
4 Jean-Jacques Rousseau, *Emile,* translated by Allan Bloom (New York: Basic Books, 1979).
5 Jean Starobinski, *Jean-Jacques Rousseau, Transparency and Obstruction*, transl. Arthur Goldhammer, (Chicago: University of Chicago Press, 1988).
6 Jean-Jacques Rousseau, *Reveries of the Solitary Walker*, trans. Peter France (London: Penguin Classics: 1980).
7 André Charrak, *Rousseau. De l'empirisme à l'expérience* ["Rousseau: From Empiricism to Experience"] (Paris: Vrin, 2013), 163.
8 See S. Füzesséry and Ph. Simay, "Une théorie sensitive de la modernité" ["A sensorial theory of modernity"] in S. Füzesséry and Ph. Simay, *Le choc des métropoles: Simmel, Kracauer, Benjamin* ["The shock of metropolises: Simmel, Kracauer, and Benjamin"] (Paris: Editions de l'Eclat, 2008).
9 See *Le choc des métropoles.*
10 Walter Benjamin, *The Work of Art in the Age of Mechanical Reproduction*, trans. J.A. Underwood (London: Penguin Classics, 2008).
11 See Janet Ward, *Weimar Surfaces. Urban Visual Culture in 1920s Germany* (Berkeley: University of California Press, 2001).
12 G. Simmel, "The Berlin Trade Exhibition", originally published in *Die Zeit*, 25.7.1896, in G. Simmel, *Gesamtausgabe* ["Collected Works"], Volume 17 (Frankfurt am Main: Klaus Christian Köhnke, 2004), 33–36.
13 Frederic J. Schwartz, *Blind Spots. Critical Theory and the History of Arts in Twentieth-Century Germany* (New Haven, CT: Yale University Press, 2005), 37–101.
14 S. Kracauer, "Cult of Distraction" in *The Mass Ornament*, translated and edited by Thomas Levin (Cambridge, MA: Harvard University Press, 1988).
15 Yves Citton, *The Ecology of Attention*, transl. Barnaby Norman (Hoboken, NJ: John Wiley & Sons, 2017).

13 The future of our contemporary sensitivity

Anne Bossé

To discuss the future of our sensitivity, a researcher can use the history of the senses, a discipline that leads to the study of sensitivities (Gélard, 2016; 2017), to show the long-term transformations of uses of the senses, of means of attention, and the evolution of climatic perceptions. Besides the field of history, the study of the senses consists of multiple disciplines. As a subject per se or as a corollary interest, they inform a specific perspective that can open new paths of inquiry and expand our knowledge. The journal *Sensibilités. Histoire, critique et sciences sociales,*[1] which was founded in 2016, has thus adopted an editorial focus on "addressing the issue of sensibilities as a locus for scholarly inquiry into the ways in which the world is organized".

Why such an approach after more than a decade of work on sensitivities that some are tempted to label as a turning point in the realm of the sensory, emotional, and affective spheres? The environmental future of our planet surely urges a reassessment of the biological and physiological components of humanity; the rational actor is now also seen as a sensory being. For some authors, our environmental crisis is a crisis of our eco-sensibility, "a loss of sensitivity and knowledge of living things" (Morizot, 2021). On another level, the multiple spatialities of diasporas ask us to redefine cultural worlds, their powers, and the scales of collective or social configurations. We can also mention metropolitan urban spaces equipped with digital objects that hear, communicate, and memorize, thereby producing new sensitivities that may be cause for concern (consider governance by algorithm, for example). To demonstrate the new relevance of approaches based on sensitivities, we could mention many other disruptions that involve our bodies in multiple ways in a variety of "transitions". Here, we would like to specify three key elements for thinking about the transformations of our contemporary sensory condition.

A critique from and by the body

The development of the neurosciences has invigorated branches of psychology focusing on the search for happiness and the workings of affects and emotions. The humanities and social sciences try to nurture demonstrations of a historically and socially constructed nature and of the situational nature

DOI: 10.4324/9781032655307-20

of our emotions, and some have sought to consider what the best-understood activities of the human brain can teach us. For example, Joël Candau questions biological hierarchies by exploring the relationship between sensory stimuli and identity construction (Candau, 2013). Work is also being conducted on cognitive phenomena by adapting a scientific approach to the social sciences as a way to think about collective intelligence, learning, and trust (Lahire, Rosental (ed.), 2008). Contemporary sensory ruptures, characterized by the use of technology to create an augmented human experience, by the spread of "ubiquitous" experiences, and by different kinds of friction between the world of social networks and the realm of society have inspired work on modes of control and the governance of feelings. These changes, which are overturning our sensory perceptions, our way of being in time and space, and our ways of being present, have impacted inter-personal relationships and collective forms of gathering. For example, Jonathan Crary's book *24/7: Late Capitalism and the Ends of Sleep* is a renewed critique of our daily lives that attempts to show the effects of incessant sensory stimulation on the conditions for political action.

The body as object, which plays a central role in this research, remains a site of power (Sforzini, 2017), when the sensory assigns seats and roles. But it remains a field of infinite inquiry. Whether as a paroxysmic body in extreme situations, a competent, sensitive body, or a body that is the site of an "anodyne" production of feeling or reassessment (Chateauraynaud, 2015), the body – its inner life, its life within a milieu, its social flesh – is one of the keys to understanding the future of our contemporary sensitivity.

The body is deeply affected by the changes in climate, which may translate into new physical challenges (such as increasing heat waves), exceptional attacks on our well-being that require us to stay at a distance and confine our bodies, or everyday ones that generate chronic reactions and change our genetic makeup. The body seems at once vulnerable and raw, but also capable – both in relation to things over which we exercise full control and those that escape us through viral contamination, pollution, and radiation (Andrieu, 2017). The worldwide success of the miniseries *Chernobyl*[2] is a healthy indicator of contemporary interest in bodies and inherited environments. The narration of the events is framed by the deterioration of bodies in phases. The intention is clearly documentary (and scientifically quite precise, according to various commentators). In this series, the body becomes the site of a possible inhabitable future.

The effects on organisms of the steady increase in CO_2 in our atmosphere since the mid-nineteenth century or the presence of polymers in our bloodstream forces us to view our bodies as porous (Engelman, 2014), as incessantly immersed, exposed, and fundamentally transformed (over a long period of time) by industrial production and extraction (Zimmer, 2016). But the body, an organism that becomes a host also captures and reveals toxic substances such as formaldehyde, as Nicolas Shapiro's work on domestic spaces has shown (2015; 2020). Shapiro argues that the chemically wounded body, more so than a healthy one, is able "to transform itself *with* and to orient itself *because of* the molecular components in our household chemicals" (2020, 124). In this sense, the body allows us to grasp the compositions of common worlds. Invested anew by the pragmatist tradition, for example through hypersensitivity, the body allows individuals to trace, make visible, and bring to life infinitesimally small perceptions and to engage in forms of preliminary investigations of new threats and harmful exposures (Chateauraynaud, Debaz, 2017). The body receives, but it is also the source of a potential *care* of the self and of the living world as a whole.

Imagining an expansion of the sensitive

The current recognition of other ways of being in the world besides ours as humans is particularly appropriate for using a sensitivity-based approach. Work on non-humans, such as objects, animals, ghosts, and the dead, form part of this intellectual undertaking that seeks to de-anthropocentrize the humanities and the social sciences (Descola, 2005). Reimagining the diversity of connections between humans and non-humans allows for multiple relations and attachments to objects, and it argues for thinking about "sensory culture" to explore "the complexity of connections between the senses, material culture, and spaces (places, borders, and landscapes) that contain and produce different sensoriums" (Howes, Marcoux, 2006, p. 12). In the spirit of asking the humanities and the social sciences to reimagine themselves through a consideration of the world of the living as a whole, Edouardo Kohn's *How Forests Think* takes language outside its solely human attribution. Anna Tsing uses a mushroom – the matsutake – as a lesson on habitability in our contemporary world. Thus, an ethnographic immersion amongst the Runa D'Avila in the Ecuadorian Amazon or a multi-site research conducted between Oregon and Japan provides contrasting methodologies that characterize these things in new ways. The most important contributions of some of these works lie in their desire to propose alternatives to the dualisms of Western (scientific) thought, whether this involves thinking in terms of the connections between the living and the dead (Despret, 2017), focusing on the minor or silent mode of sensory life, abandoning certain persistent associations (such as between inert matter and living beings), or freeing oneself of preconceived representations. That is when the agency and mobility of objects, such as those that were pillaged in colonial times, allow us to consider their political or post-colonial relevance. Actual, practical experiences that involve contact with a multitude of beings, things, and milieus ground beliefs, fears, judgements, opinions, and reasons for acting. The whole of political life is thereby "affected" (Faure, Négrier, 2017).

More specifically, the growing interests among social scientists – and philosophers in particular – in the natural sciences, geosciences, animal and plant biology, and naturalist areas of knowledge, in general, are contributing to a renewal of descriptions of human life and ways of conducting research.

A better understanding of life gives rise to a decentring, a change in perspective, and a questioning of human singularity or exceptionalism. This desire to know (oneself) is more concerned with scale (the history of the present as well as prehistory or geological time), spatial layers (from the troposphere to below ground), and bacteria and large wild animals. For Baptiste Morizot (2021), the practice of researching animal tracking (among wolves and bears) reintroduces the sensory in observing these tracks, and it also leads us to reconsider our knowledge of *Homo sapiens* and inquiry. While work written from "the point of view of..."[3] uses a theoretical and epistemological framework, many researchers also look to other than human worlds to renew how we can write and describe our investigations. For example, Mathieu Duperrex revisits the sensory narration of ground in *Voyages en sol incertain* using animals and plants as narrators, such as the group of "sentinels", to describe the interrelations and hybridizations of Anthropocene landscapes in the Rhône and Mississippi Deltas. These new alliances or multi-specific arrangements can help us observe the state of our ecosystems and perceive warning signs (Keck, 2020). For example, the sensitivity of a lichen as a bio-indicator of air quality is leading to a scientific renewal, this time in terms of participatory science (Gramaglia, Dauphin, 2017). Be it in learning how to see, in enriching words, imagination and representations, especially cartographic ones by repopulating maps (Aït-Touati, Arènes, Grégoire, 2019), or in assigning a personality to non-humans, more and more areas of inquiry are being made possible through this expansion of our notion of sensitivity.

Documenting the city as a privileged environment for experiencing one's contemporary sensitivities

The notion of a sensitive city has, for some time, now occasioned a striking encounter between the sensitization of urban experience, the fabrication and design of urban spaces, and the use of explicit strategies to create ambiances and control feelings (Thibaud, 2010; 2018). The spread of artificial fragrances marketed by the perfume industry comes with an awareness of the olfactory dimension of how we relate to spaces and the issues that this raises for urban development. Shimmering ornamental ponds and other features of sensorially aware urban planning where urban bodies "free" themselves go hand in hand with a security-driven urban development that results in the design of "anti-urban" furniture. "Shock", the term used to describe metropolises in the nineteenth century, hardly applies to the critical observation and description of the metropolitan condition of the twenty-first century (Füzessery and Simay, 2008), because it is so marked by ambivalence. For example, does the omnipresence of indignation in public spaces, driven by reactions to terrorism, reassure us as to the democratic value of these spaces, or does it nurture urban mistrust? (Gayet-Viaud, 2015). But these two periods seem to share a need to return to a description of reality that encourages the hybridization of literature and the social sciences through a comparable practice of inquiry. (Demanze, 2019). The attention we pay to the future of our contemporary sensitivity must enlarge the scope of interactions, redraw the contours of urban civic consciousness, all the while acknowledging that the movement of individuals brings about transfers of things tangible and intangible, as well as the intercultural encounters that characterize a number of contemporary urban sites, for which ambiances and sensitivities are an operative concept, even in political terms. Consider the Sunday occupations by Filipina women along Chater Road in Hong Kong that became a climate protest (Law, 2001), the Buddhist community picnics in the Bois de Vincennes, or the Turkish immigrants and Berliners of Turkish origin who congregate in Berlin's Tiergarten Park, all events that suggest the possibility of being at home far away from home.

The Covid-19 pandemic has reawakened our interest in thinking about urban environments as sensory experiences. Urban history as a relationship between urban planning and health, urban forms and pandemics, and citizen behaviours and a public (hygienic) order has moved from being a backdrop to front and centre stage. And the healthcare crisis has accelerated, or simply updated, Anthropocene phenomena already at work. The desire for breathability, well-being, being in "nature", and environmental and ecological inequalities have become part of our concerns about transitions and mobility, urban microclimates, and risk management. More attention is being paid to reinventing the city as an ecological space (Blanc & Paddeu, 2018; Blanc, 2021) and new sensitivities have begun to crop up and become more visible. The proliferation of the term "wild", which we can view as a tool for qualifying the relationship between humanity and its surroundings, thus seems to express new urban sensitivities, whether to the presence of animals, the virtues of plants, what nurtures us, the different states of water, or waste. What transformations of the urban experience can these emerging sensitivities bring about? Should we see them as a possibility for rethinking urban hospitality (Stavo-Debauge, Deleixhe and Carlier Louise, 2018) and ways of being together in public spaces? Can this lead us to envision other ways of inhabiting?

The social sciences must fully address the issue of sensitivities because they participate in how the world is constructed and organized. They are not a minor aspect of our society's workings, and there are multiple scales and objects for understanding and documenting them. Here we have dwelled on the body and what affects it, the multiple sensory relations in inhabited environments, and the fabrication of the city as the fabrication of sensitivities.

Notes

1 A semi-annual review created in 2016 and published by Anamosa.
2 A British-American miniseries of five episodes, written by Craig Mazin and directed by Johan Renck.
3 Consider bacteria for Julie Beauté (2022), and lichens for Karine Prévot (2018).

Bibliography

Aït-Touati, Frédérique, Alexandra Arènes, and Axelle Grégoire. *Terra forma. Manuel de cartographies potentielles*. Paris: B42, 2019.

Andrieu, Bernard. *Se fondre dans la nature, Figures de la cosmose. Cosmotique 1*. Montreal: Liber, 2017.

Beauté, Julie. "Bactéries alchimistes: de l'agentivité microbienne aux matières vivantes". *Tracés. Revue de Sciences humaines* 40 (2021): posted online on 03 February 2022, consulted on 23 February 2022. URL: https://journals.openedition.org/traces/12340; https://doi.org/10.4000/traces.12340

Blanc, Nathalie. "Impossible sauvage urbain". In *Textes et contextes* 16–2 (2021): posted online on 10 December 2021, consulted on 3 January 2022. URL: https://hal.archives-ouvertes.fr/hal-03482690/document

Blanc, Nathalie, and Flaminia Paddeu. "L'environnementalisme ordinaire. Transformer l'espace public métropolitain à bas bruit". Paris: Association Espaces Temps.net, 2018.

Candau, Joël. "Une théorie sensorielle de l'identité". In P.- L. Colon, ed. *Ethnographier les sens*, 201–234. Paris: Éditions Petra, 2013.

Chateauraynaud, Francis. "L'emprise comme expérience. Enquêtes pragmatiques et théories du pouvoir." *SociologieS* Dossier Pragmatisme et sciences sociales: explorations, enquêtes, expérimentations (2015): posted online on 23 February 2015, consulted on 17 July 2018. URL: https://journals.openedition.org/sociologies/4931

Chateauraynaud, Francis, and Josquin Debaz. *Aux bords de l'irréversible. Sociologie pragmatique des transformations*. Paris: Éditions Pétra, 2017.

Demanze, Laurent. *Le nouvel âge de l'enquête*. Paris: Éditions Corti, 2019.

Descola, Philippe. *Par-delà nature et culture*. Paris: Gallimard, 2005.

Despret, Vinciane. *Au bonheur des morts. Récits de ceux qui restent*. Paris: Editions de la Découverte, 2017.

Duperrex, Mathieu. *Voyages en sol incertain. Enquête dans les deltas du Rhône et du Mississippi*. Marseille: Éditions Wildproject, 2019.

Faure, Alain, and Emmanuel Négrier, ed. *La politique à l'épreuve des émotions*. Rennes: Presses Universitaires de Rennes, 2017.

Engelman, Sasha. "Toward a poetics of air: sequencing and surfacing breath". *Geography and the Environment*, Oxford University Centre for the Environment, St Hilda's College, 2014. URL: https://static1.squarespace.com/static/52faae3ee4b00527f300d31f/t/5ea9603b6ade3c4d2c1eba6f/1588158525996/Engelmann_PoeticsofAir.pdf

Füzessery, Stéphane, Simay, Philippe, ed. *Le choc des métropoles. Simmel, Kracauer, Benjamin et la Großstadt*. Paris: Editions de l'Eclat, 2008.

Gayet-Viaud, Carole. "Les espaces publics démocratiques à l'épreuve du terrorisme". *Métropolitiques*, 20 novembre 2015. URL: https://www.metropolitiques.eu/Les-espaces-publics-democratiques.html

Gélard, Marie-Luce. "L'anthropologie sensorielle en France". *L'Homme* 217 (2016): posted online on 24 February 2018. URL: https://journals.openedition.org/lhomme/28868; DOI: 10.4000/lhomme.28868

Gélard, Marie-Luce. *Les sens en mots. Entretiens avec Joël Candau*, Alain Corbin, David Howes, François Laplantine, David Le Breton, Georges Vigarello. Paris: Editions Pétra, 2017.

Gramaglia, Christelle, and Charles-Enzo Dauphin. "Toucher la pollution industrielle du doigt grâce aux lichens". *Techniques & Culture* Supplements to Vol. No. 68

(2017): posted online on 18 December 2017, consulted on 24 January 2018. URL:// journals.openedition.org/tc/8610

Howes, David, and Jean-Sébastien Marcoux. "Introduction à la culture sensible". *Anthropologie et Sociétés* 30(3) (2006): 7–17. https://doi.org/10.7202/014922ar

Keck, Frédéric. *Les sentinelles des pandémies. Chasseurs de virus et observateurs d'oiseaux aux frontières de la Chine.* Brussels: Zones sensibles, 2020.

Kohn, Eduardo. *How Forests Think: Towards an Anthropology Beyond the Human.* Berkeley: University of California Press, 2013.

Lahire, Bernard, and Claude Rosental, ed. *La cognition au prisme des sciences sociales.* Paris: Editions des Archives contemporaines, 2008.

Law, Lisa. "Home Cooking: Filipino Women and Geographies of the Senses in Hong Kong". *Ecumene* (2001): 264–283. https://doi.org/10.1177/096746080100800302

Morizot, Baptiste. *Sur la piste animale.* Arles: Babel, 2021 [2018].

Prévot, Karine. "Sommes-nous des lichens? Une perspective végétale sur l'individu". *Critique* 3 (2018): 204–213.

Sforzini, Anne. *Michel Foucault, une pensée du corps.* Paris: Presses Universitaires de France, 2017.

Shapiro, Nicholas. "Attuning to the Chemosphere: Domestic Formaldehyde, Bodily Reasoning, and the Chemical Sublime". *Cultural Anthropology* 30(3) (2015): 368–393. https://doi.org/10.14506/ca30.3.02

Shapiro, Nicholas. "Persistent Ephemeral Pollutants". In *Being Material,* edited by Marie-Pier Boucher, Stefan Helmreich, Leila W Kinney, Skylar Tibbits, Rebecca Uchill, and Evan Ziporyn, pp. 154–161. Cambridge, MA: MIT Press, 2019.

Stavo-Debauge, Joan, Martin Deleixhe, and Louise Carlier. "HospitalitéS. L'urgence politique et l'appauvrissement des concepts". *SociologieS* (2018). URL: https:// journals.openedition.org/sociologies/6785

Thibaud, Jean-Paul. "The City as a Sensory Experience". In *Ecologies urbaines: états des savoirs et perspectives,* edited by O. Coutard and J.- P. Lévy, 198–213. Paris: Economica-Anthropos, 2010.

Thibaud, Jean-Paul. "Vers une écologie ambiante de l'urbain". *Environnement urbain/Urban environment* Volume 13 (2018): posted online on 4 July 2018, consulted on 13 January 2020. URL: https://journals.opendedition.org/eue/2135

Tsing, Anna Lowenhaupt. *The Mushroom at the End of the World: On the Possibility of Life in Capitalist Ruins.* Princeton: Princeton University Press, 2015.

Zimmer, Alexis. *Brouillards toxiques. Vallée de la Meuse;* 1930. *Contre-enquête.* Brussels: Zones sensibles, 2016.

14 Sonic ambiances, intellectual ambiances

Thoreau and Ives

Pauline Nadrigny

This chapter investigates the atmospheric character of Concord, Massachusetts' intellectual environment, as well as the emergence of a specific aesthetic, namely a musical one. Two figures are mentioned here: the philosopher, poet, and naturalist Henry David Thoreau and the composer Charles Ives. For Thoreau, a bond is forged between the concrete experience of a certain environment (the North American woods), more particularly, its sound environment, and a way of thinking. This connection leads to a musical oeuvre inspired by Transcendentalism, that of Charles Ives.

Where do we find ourselves?

In his essay *Experience*, Ralph Waldo Emerson set forth one of the founding principles of what came to be known as the Transcendentalist movement, which holds that our encounters with a specific environment – both in everyday and exploratory aspects – are essential to the development of our conceptual frameworks. Emerson believed that categories, the ideas that structure our thought, are encountered in a particular field of experience in the truest sense of the word, and not just actualized there. The Emersonian categories of illusion, temperament, succession, surface, surprise, reality, and subjectivity – which he refers to as "the lords of life" – are less rigid than the Kantian transcendental framework. They "are on our way", a way that we have to discover, to feel, to explore fully. However, the essay *Experience*, written in 1844, was conceived by Emerson in the same year that Thoreau stayed on his land and built his famous cabin near Walden Pond.

This concurrence is important and invites us to search Thoreau's work itself for a relation between the concrete experience of an environment and the constitution of a philosophical thought. In relating his own experience in the woods of Walden, and later in the forests of Maine, Thoreau intended to start from something both simple and common: a dimension of experience that is more authentic, wilder, and certainly universal, but, by the same token, unique and critical. Looking back on Walden, on what happened during this time when Thoreau lived, worked, observed, listened, and walked

DOI: 10.4324/9781032655307-21

in the New England woods, means working in the material of thought: a thought that, in Emerson's words, is impregnated with the physical environment in which it exists.

This process brings up the question of borders. The point is to describe the woods as a closed environment (as in the "walled-in" metaphor) and also as a permeable reality. While retaining a sense of consistency, Walden remains a diffuse and relatively diversified entity. Therefore, Thoreau's situation in the woods is neither totally wild nor totally civilized. "The writer's sentences must at each point come to an edge",[1] writes Stanley Cavell in *Senses of Walden*. We must therefore resist the temptation – to which one has too often given in – to *wilden* Thoreau's experience, and to reduce his narrative to the glorification of a plainly untamed environment – "the Wild". One does not return to the Wild as a primitive state; it is always impure, ambiguous. To perceive this ambiguity, one must start from the perceptible woods, through the eye and ear, as well as through the contact of one's feet with the ground. In philosophical works, the forest is often depicted as a form of thought (for instance, Descartes' maze forest, which represents the difficult path to knowledge). However, it is not an atmosphere, rather an allegory. Thoreau's forest is, to the contrary, an actual place of experience, at once affective and practical, within and through which the individual finds himself. The expression "I found myself doing" is thus recurrent in Thoreau, echoing Emerson's contemporary question in *Experience*: "Where do we find ourselves?" This question is not so much answered as echoed, in an *ad vitam* repetition.

Into the sonic forest

First, there is the Walden Forest, visited by the people of Concord, located on Emerson's land, its pond serving as an ice reservoir for the village and where the whistle of the train intrudes frequently. There is something pastoral about it, especially when echoes of cowboys and village bells resound. It is not so much the forest as "the woods". Solitude is never complete, and one aspect of Thoreau's narrative in *Walden* accounts for this peripheral situation; it is not so much *off* the edge as *on* it. There is also the Maine forest that Thoreau would describe in his later journals, a primary forest, dangerous and dense, where civilization sometimes fades away, although never completely. This forest, which Thoreau describes also as a naturalist, is "wild, damp, and shaggy"; "the surface of the ground in the Maine woods is everywhere spongy and saturated with moisture".[2] It muffles the sounds that echo in the woods of Walden.

In these depictions, Thoreau always avoids abstract representation. The atmosphere of the woods is described in a hyper-aesthetic way, as sounds reveal a specific awareness and acute sensitivity. Indeed, an entire chapter of *Walden* is devoted to sounds[3]. Thoreau invites his reader to listen with him, to put an end to the characteristic deafness of our practical and limited relationship to our environment. For the forest is a *sonic* environment, in the three meanings of his word: sounds are *produced* in it (by the living creatures that cross it, by the nearby town), it *produces* sounds as a botanical whole (rustling leaves, falling branches), and it is also a *medium* for sound diffusion, through a set of diffractions in which the trunks and the leaves play an essential role. It is an envelope, but it is also a medium that remains porous to the external world. As a sonic ambiance, the forest is thus carefully listened to by Thoreau, and described with a remarkable acuity, as in the following description of the sound of a simple bough falling to the ground:

> The sumach (*Rhus glabra*) grew luxuriantly about the house, pushing up through the embankment which I had made, and growing five or six feet the first season. [...] The large buds, suddenly pushing out late in the spring from dry sticks which had seemed to be dead, developed themselves as by magic into graceful green and tender boughs, an inch in diameter; and sometimes, as I sat at my window, so heedlessly did they grow and tax their weak joints, I heard a fresh and tender bough suddenly fall like a fan to the ground, when there was not a breath of air stirring, broken off by its own weight.[4]

It is a matter of "being always on the alert"– and hearing is the sense of alertness par excellence – but while freeing ourselves from a sense of utility and our daily concerns. The attention aims at the phenomenon for itself, to enjoy its pure manifestation and to describe its development. The Walden cabin serves as a focal point, a "clearing" from which one can pay attention to the sounds of the surroundings. In Thoreau's later writings, sound keeps its full significance in his naturalistic descriptions. The sounds of the wind in the treetops, the reflections on the river that runs through the Maine woods, and the call of the moose mimicked by the Native American guiding the author in his excursions all blend together. The water and the damp woods thus become a theatre, a resonating place for the sounds that occur there. It gives these sounds a substance, a texture, and a dramatic impact. Thus, on the Chesuncook:

> Once, when Joe had called again, and we were listening for moose, we heard, come faintly echoing, or creeping from far, through the moss-clad aisles, a dull, dry, rushing sound, with a solid core to it, yet as if half smothered under the grasp of the luxuriant and fungus-like forest, like the shutting of a door in some distant entry of the damp and shaggy wilderness. If we had not been there, no mortal had heard it. When we asked Joe in a whisper what it was, he answered, — "Tree fall".[5]

The tuning of the world: harmony and anarchy

Why does sound atmosphere occupy such an important place in Thoreau's perception of the woods? Probably because for him, hearing is not just about sounds, but always about a certain harmony, a tuning of the world – something that would inspire Raymond Murray Schafer more than a century later in his book *The Soundscape*[6]. Hence, the importance of the water surface plays both a unifying role in the sound atmosphere (as it provides a sound-bearing surface) and a source for subtle variations (depending on distance, agitation on the surface, and atmospheric conditions). For Thoreau, the forest is not just an acoustic world, but also a musical one, in the sense that there is harmony between the various elements that make up this environment: animals, plants, and climatic phenomena. This harmony expresses a bond between humans and the natural world they observe, the most evident manifestation of which is the aesthetic pleasure gained from the sound. In *Walden* especially, the influence of Emerson and his essay *Nature* is noticeable in the use of certain themes that already sound like *topoi* in the Emersonian text: thus, the "silent music" of nature, this "universal lyre", or the image of the "aeolian harp" taken from Novalis, on which all phenomenon resonates and through which everything concurs in a great harmony. In this respect, it is interesting to study the echo effect in Thoreau's writings. Such a phenomenon is not a simple reflection of sound. It reveals a musical composition at work in the woods, where each element of the ecosystem – trunks, needles, water surfaces – plays its role, composing the diversity in unity.

> All sound heard at the greatest possible distance produces one and the same effect, a vibration of the universal lyre, just as the intervening atmosphere makes a distant ridge of earth interesting to our eyes by the azure tint it imparts to it. There came to me in this case a melody which the air had strained, and which had conversed with every leaf and needle of the wood, that portion of the sound which the elements had taken up and modulated and echoed from vale to vale. The echo is, to some extent, an original sound, and therein is the magic and charm of it. It is not merely a repetition of what was worth repeating in the bell, but partly the voice of the wood; the same trivial words and notes sung by a wood-nymph.[7]

But these ideas of expression and harmony of the world are quite different from the Pythagorean view. The goal is to consider a concrete ecosystem in terms of composition and to produce a carefully orchestrated account of that composition. This must be understood in a very literal way; in fact, pine needles do play the role of prisms, reflecting the high-pitched sounds of the woods. And tree trunks effectively diffuse owl calls over long distances. Thoreau's description of the echo is not merely a poetic image; it is based on physical observation. This attentive listening goes on with birdsongs: from the call of the whippoorwill to the call of the rooster and of the owls. These sounds, as well as the frog song, are rendered with onomatopoeias in a transcription that is both comical and comprehensive. A poetic imaginary (that of mourners, for example) is linked to an attentive listening, a sound transcription that also involves walking actively into the woods, inscribing Thoreau's steps in it. Indeed, ambiance is often constituted by strolling, what Thoreau calls "sauntering".[8]

In Thoreau's work, philosophical and even metaphysical reflection emerge from a sensitive understanding of a particular atmosphere and its careful observation. We are dealing with a tangible nature. This idea leads to a reassessment of noise compared to the musical sounds that our culture creates, cultivates, and enjoys. The thesis of a musical nature therefore carries a critical perspective regarding the cultural forms of musical composition. True music can be heard around the cabin in Walden, and not in the concert halls of the upper classes.

> One will lose no music by not attending the oratorios and operas. The really inspiring melodies are cheap and universal, and are as audible to the poor man's son as to the rich man's. Listening to the harmonies of the universe is not allied to dissipation.

The democratic claim that music is audible to the poor as well as the rich relies on another, metaphysical claim: the universal nature of harmony, to which every individual can be sensitive, and to which he or she participates through an inscription in the sound environment. Consider the song of the cowboys that Thoreau mentions in the same chapter, for example.

Alongside the criticism of music in its traditional form, the return to natural harmony in its purely sonic expression also involves a criticism of another type of sound, exemplified by the whistle of the locomotive that breaks through the woods. On this point, Thoreau's descriptions urge us to think about what a sound ambiance is and what differs from it. It is interesting to compare the sounds of the surroundings of Walden and that of the engine's whistle. What we call natural sounds invite us to stay alert and to listen. These sounds surround us, both diffuse and unstable. To the contrary, the engine's whistle reaches the author up to his cabin; it imposes listening. It is an invasive sound that fits immediately into the type of signal-response network that Thoreau aimed to put in abeyance. If the echo is a phenomenon in which everything contributes to the transmission of sound, the engine's whistle violently penetrates the environment of Walden, like a shout. The farmer cannot ignore it and must respond to it. It disrupts and disturbs the ambiance; it is a reminder of another world, that of a civilization whose relationship to nature is not what Thoreau seeks to recreate in his retreat. In "Sounds", the railroad becomes the mythical figure of *fatum*, the "Atropos", whose signal sound imposed on nature is not cyclical, but mechanical. The harmony of the world becomes a worldwide trade.

Thoreau's response to this invasion is first of all a naturalization of the technical sound: the sound of the whistle is transformed in the text into the scream of the hawk, a remediation that incorporates the parasite signal into the woods. This solution allows us to avoid a misunderstanding. Thoreau's writing portrays an ambiance, not a system. Besides the topos of harmony, there is also a form of sonic anarchism in Thoreau. In fact, in *Walden*, the woods cannot be reduced to a biological environment, to what we would now call an ecosystem, and this is particularly clear in terms of sound.

Thoreau's sound descriptions do not hold the environment to be a soundscape either, despite the use of the concepts of "concert" or "harmony". The description is organized into several temporal and focal points. "Sometimes" a bough falls to the ground. This opening to the uncertain, to the punctual, makes the sound ambiance of the forest differ from the industrial sound environment, which is characterized by a sterile regularity. Hence, the fact that, in the ambiance of the woods, one does not exactly *find* oneself. At least, if one does, it is in the very concrete sense of a specific situation, where one meditates, among other themes, on the history of the American people, its problematic destiny, on an awakening of thought to the absolute through experience. This absolute – reality itself – is not subject to a dogma, to a definitive statement. On this point, Thoreau remains a disciple of Emerson; the universal is expressed by the courage of the individual. The representation of an ambiance in which numerous sound individuals express themselves links Thoreau's aesthetic considerations to an ethical and political thought: individualism, which the author of *Walden* inherits from Emerson, is always situated somewhere. It is not solipsistic; it instead welcomes the universal and finds itself by the call to the common and the ordinary experience.

Composing ambiances: Charles Ives

This knotting between singularity and totality continued, *de proche en proche*, in the field of musical composition, nearly four decades later. In fact, the Thoreauvian critique of conventional music in favour of the concrete sonorities of his environment does resonate in the field of musical composition. Before John Cage – the author of *Silence* never fails to return to Thoreau – Charles Ives was one of the first to be inspired by this Concordian ambiance. We will not describe here the very particular personality of Charles Ives, nor will we recount the life of a man who, in a very Emersonian way, moved between music composition and a career as an insurance company director[9]. We will note one significant fact: from his father, Ives inherited a musical culture that was both military and Christian, a sense of sound experimentation, as well as a deep interest in Transcendentalist thought. This interest is most evident in two works: the *Piano Sonata No. 2*, titled *Concord, Massachusetts 1840–1860,* which was composed in several stages between 1904 and 1940, and an orchestral piece composed between 1914 and 1916, titled *Three Places in New England.* The first is composed of four movements, each named after figures of Transcendentalism: Emerson, Hawthorne, the Alcotts, and Thoreau. As Philippe Albèra writes in an introductory paper to Ives' *Essays before a Sonata*, "transcendental thought certainly provides" Ives with a "conceptual framework for his work and a legitimation for the originality of his aesthetical choices".[10] In the *Concord Sonata,* one notices in particular Ives' typical use of a varied material made of learned musical forms, marches, popular songs, and religious hymns, which the musicologist J. Peter Burkholder catalogued in his book *All Made of Tunes.*[11] There is also a sensitivity to German Romanticism that emerges from some harmonic patterns. However, just as Thoreau did not gather the many sounds of Walden into a limited soundscape, Ives does not attempt to synthesize these various materials. Rather, he proceeds through tonal and temporal polyrhythms and heterophony. This type of composition can be traced back to Ives' childhood experiments with his father, who organized a gathering of different military bands at a crossroads where independent voices mingled. As Albèra notes:

> Dissonances, rhythmic inequalities, overlapping musical layers and polyrhythms do not appear in his music for the sake of musical structure, as the development of a learned language pushed beyond its limits, but as extensions, manipulations, or grafts from basic materials.[12]

The resulting musical entity is therefore somewhat ambiguous for being woven from heterogeneous elements. Like the patchwork quilts that were patiently sewn together in New England households, unity is obtained through a synthetic effect that does not result directly from the articulation of individual voices – and John Cage, composing the immense musical patchwork of *Apartment House 1776*, would draw inspiration from this process.

The orchestral piece *Three Places in New England,* especially in the movement entitled *The Housatonic at Stockbridge*, also displays a Transcendentalist aesthetic, this time in the concrete symphonic transposition of a natural sound environment. The piece is indeed inspired by a stroll that Ives took with his wife along a pond, as the sound of different music, including hymns, drifted over the water from a church on the other side. The hymn and the natural soundscape of a summer evening blend into an irreducibly complex atmosphere wherein the listener wanders. As Albèra writes, "in such music, neither time nor space are homogeneous and pre-established elements; they become part of the composing process".[13] Here again, the influence of Ives' father is decisive to his composition. As David Toop points out in *Ocean of Sounds*:

> In the late nineteenth century, Charles Ives's father, George, used to play musical instruments over the pond in Danbury, Connecticut. He was, Frank R. Rossiter wrote in *Charles Ives & His America*, "captivated by the tone quality of the echo that returned to him across the water" [...].[14]

Hence, this piece is not a symbolic figuration, but an accurate reflection of a sound walk in New England; like Thoreau's descriptions, it is at once naturalistic and poetic. In his *Essay*, Ives also presents a highly developed analysis of how to represent a sound environment in the manner of a visual landscape, taking into account the effects of distance and transformations due to atmospheric changes. Here, the pond, which the musical writing reproduces through the undulating movement of the strings, acts as a bridge, a resonating surface, while at the same time maintaining the strange quality of sound interference. Indeed, it is a characteristic of ambiences to be irreducibly varied, to embrace this kind of strangeness, to be not merely directional (like background music specifically designed to enhance production, for example), but open to various affective tunings. "Where do we find ourselves?": in Stockbridge, and this can imply a multiplicity of existential arrangements.

Referring to Thoreau, Ives wrote that he was able to "weave perfect transcendental symphonies" out of a certain natural environment. And his writing indeed has a legacy in Charles Ives' polyphonies. He also drew inspiration from an ambient environment, the sonic reality of New England, to produce a Transcendentalist symphony. Why then Transcendentalist? Because, in Ives as in Thoreau, musical and literary composition can carry ethical and political issues that deal with individual independence, with the ability to combine self-reliance and faith in the universal, and with the ability to find meaning in the ordinary, in the commons. Ives thought precisely of this atmosphere – simple human lives, a simplicity that is nevertheless highly demanding and difficult to portray –when he wondered, at the beginning of his *Concord Sonata*:

> It is darker, the poet's flute is heard over the pond, and Walden hears the swan song of this 'Day' – and one faintly echoes it. Is it a transcendental tune from Concord?[15]

Notes

1 Stanley Cavell, *Senses of Walden, An Expanded Edition* (Chicago: The University of Chicago Press, 1972), 44.

2 Henry David Thoreau, Chesuncook (1858), in *The Writings of Henry David Thoreau,* vol. III (Boston: Houghton, Mifflin, The Riverside Press edition, 1892), 206.

3 A detailed study of sounds in *Walden* (and in Thoreau's thought) has been carried out by other scholars, such as Roxana Oltean, who considers the relationship between transcendentalism and technological modernity from the standpoint of the sonic world. Cf. Roxana Oltean, "Language . . . Without Metaphor: Soundscapes and Worldly Engagements in Henry David Thoreau's *Walden", JAAAS, Journal of the Austrian Association for American Studies* 1, no. 2 (2020): 267–281. We also find a close analysis of the listening experience in Thoreau's work by Jeff Todd Titon, who stresses its ecological significance. See Jeff Todd Titon, "Thoreau's ear", *Sound Studies* 1, no. 1 (2015): 144–154. See also Kenneth W. Rhoads, "Thoreau: The Ear and the Music", *American Literature* 46, no. 3 (1974): 313–328.

4 Henry David Thoreau, *Walden* (Princeton, NJ: Princeton University Press, 1971), 114.

5 Henry David Thoreau, Chesuncook (1858), *in The Writings of Henry David Thoreau,* vol. III (Boston: Houghton, Mifflin, The Riverside Press edition, 1892), 139.

6 R. Murray Schafer, *The Soundscape. Our Sonic Environment and the Tuning of the World* (Rochester: Destiny Books, 1977).
7 Henry David Thoreau, *Walden,* 123.
8 Henry David Thoreau, *Journal*, in *The Writings of Henry David Thoreau* (Princeton, NJ: Princeton University Press), 83.
9 See Sidney Cowell, *Charles Ives and His Music* (Oxford: Oxford University Press, 1955); Helen R. Sive, *Music's Connecticut Yankee: An Introduction to the Life and Music of Charles Ives* (New York: Atheneum, 1977); Geoffrey Block, *Charles Ives: A Bio-Bibliography* (New York: Greenwood Press, 1988); J. Peter Burkholder, ed., *Charles Ives and His World. The Bard Music Festival* (Princeton, NJ: Princeton University Press, 1996).
10 Philippe Albèra and Vincent Barras, "Avant-propos", in Ives Charles, *Essais avant une sonate et autres textes*, trans. C. Russi, V. Barras, V. Aliberti, D. Collins and S. Ji, Genève (Paris: Éditions Contrechamps, 2016) online at URL: <http://books.openedition.org.ezpaarse.univ-paris1.fr/contrechamps/2303>. Our translation.
11 J. Peter Burkholder, *All Made of Tunes: Charles Ives and the Uses of Musical Borrowing* (New Haven, CT: Yale University Press, 1995).
12 Philippe Albèra and Vincent Barras, "Avant-propos".
13 Philippe Albèra and Vincent Barras, "Avant-propos".
14 David Toop, *Ocean of Sound, Ambiant Sound and Radical Listening in the Age of Communication* (London: Serpent's Tail, 2001): 244. See Frank R. Rossiter, *Charles Ives & His America* (London: Victor Gollancz, 1976).
15 Charles Ives, Note for the *Second Piano Sonata* ("Concord, Mass: 1840–1860") – 1904–1915.

15 From urban ambiances to cinema

A few reflections based on Siegfried Kracauer

Nia Perivolaropoulou

Kracauer, the city, and film

The terms "ambiance" and "atmosphere" have long been used in the discourse of film and literary criticism. Critics will sometimes describe an atmosphere of trouble, anxious or joyous, and at times they attempt to describe it in a way to make readers feel it. Yet, as an aesthetic category in the field of film studies, these terms have, to my knowledge, only begun to appear relatively recently, even though the terms *Stimmung* and "mood", which stem from a much older tradition, can be found in the very first theoretical works on film.[1] As an important aesthetic category in nineteenth century thought, *Stimmung* made its debut appearance in about 1910, in the first attempts to theorize film, both in Europe and the United States.[2] The use of the concept represented a discursive strategy for bridging the gap between film and the visual arts, especially painting, and thus for granting film a sense of cultural legitimacy.

In the twenties and thirties, film, which had come to be embraced by all levels of society, became a mode of mass distraction, especially in cities, at the same time that the idea of "film as art"[3] had begun to take shape. Siegfried Kracauer's position on this topic stood out within this theoretical and cultural context. His analysis of photography and film led him – and Walter Benjamin – to view means of mechanical reproduction as both consequences and drivers of the process of modernization, the apparition of which had rocked people's sensibilities and led to new modes of perception. Their theoretical approach represents a rupture with the tradition of aesthetics and a search for new modes of conceptualization.[4]

So, it should not come as any surprise that the terms "atmosphere" or *Stimmung* appeared neither in Kracauer's writings on film in the 1920s and 1930s nor anywhere else – other than in their customary meaning. As yet, in his major work *Theory of Film*, published in 1960 in the United States, the occurrences of "atmosphere" and "mood", though quite rare, are, upon closer inspection, revealing.[5] As they had been used for a long time in commentaries on film music and in the context of production and film criticism, these references – as one might expect – appear most frequently in the chapter devoted to music.

DOI: 10.4324/9781032655307-22

However, in the chapter that deals with fantasy, Kracauer, in citing a statement made by Carl Dreyer during the filming of *Vampyr*, dwelled on a remarkable case that demonstrates the effects of the narrative elements on the creation of atmosphere in film:

> Imagine that we are sitting in an ordinary room. Suddenly we are told that there is a corpse behind the door. In an instant [the room we are sitting in is completely altered:] everything in it has taken on another look; the light, the atmosphere have changed, though they are physically the same. This is because *we* have changed, and the objects *are* as we conceive them. That is the effect I want to get in my film.[6]

This case, simple and extreme all at once, reminds us that an atmosphere thus created is an effect, and that it also invokes the subject of the perception and the spectator on the one hand, and the perceived film on the other.

It is in the first and apparently ordinary occurrence of the term "atmosphere" (as "mood") in *Theory of Film* that the watermark of Kracauer's concept of film and its relationship to the modern city begins to show through. It appears in a passage of the work that discusses the "flow of life". For Kracauer, this expression designates a mainly material continuum that "covers the stream of material situations and happenings with all that they intimate in terms of emotions, values, thoughts" as revealed when filmed by the camera.[7] The "flow of life" is included within the conceptual framework established by Kracauer to characterize the distinctive qualities of the medium of film, which, other than the "flow of life", are the same as for still photography. He elucidates the affinity of the two means of mechanical *reproduction* with physical reality through the notions of the "unstaged", the "fortuitous", "endlessness", and the "indeterminate".[8] The aesthetic potential of film is also consummately achieved through the revelation of unintentional spatial configurations, fortuitous figures and arrangements, and acoustic environments that elude meaning. The cinematographic approach to reality, which also exists in function of the director's attitude towards his/her material, is characterized by a tension, a constantly unstable balance between the director's tendency to shape the material (by staging, styling, editing, etc.) and the realism inherent to the camera lens (which he termed camera-realism), between what Kracauer called the "realistic tendency" and the "formative tendency" inherent to film.[9]

As a locus of chance able to generate countless stories, a city street is "a place where the flow of life is bound to assert itself".[10] Already in his first writings on film, the street represents the cinematographic subject par excellence for Kracauer. Over time and throughout his exploration of the medium's distinctive affinities, the notion of the street grew to encompass nearby spaces such as passageways, railway stations, bars, airports, and more. In the pages on the flow of life, Kracauer cites several examples of eminently cinematographic streets: the street in the films of D.W. Griffith, a director who inspired his thinking about changes in scale in both cinematographic narration and history; Karl Grune's film *Die Strasse* ("The Street"), which he saw in the early 1920s, and which proved a decisive experience for him, as it marked the starting point of his thinking about the affinities between film and the modern city.[11] He describes how streets emanate a constant sadness in De Sica's films, and lastly, he notes that "a *mood* of fin-de-siècle disenchantment and nostalgic longing for faraway countries lingers in crowded sailor hangouts"[12] in Louis Delluc's *Fièvre* and in Cavalcanti's *Sea Fever*.

The mention of these two films shot, respectively, in 1921 and 1927 in Marseille may seem arbitrary, but it does allude to the importance of this city to Kracauer, even before he stayed there in 1940–41, like so many other German exiles, anxiously awaiting a visa to be able to leave Europe. In fact, among his many texts about cities written between the wars, the ones about Marseille hold a special place and testify to an experience of considerable intensity. The first was in 1926, when Kracauer met his friends Walter Benjamin and Ernst Bloch there. Both Benjamin's and Kracauer's correspondence and the texts about this trip reveal a sense of deep anxiety provoked by the ambiance of some of the city's neighbourhoods.[13] Marseilles is the source of a sense of fear that, for Kracauer, often expressed itself in the form of spatial images. His description of the Place de l'Observance in the text titled "The Quadrangle" turns this site into a kind of inverted panopticon, where, as in Foucault's description of Bentham's Panopticon, "visibility is a trap".[14]

Kracauer's texts on the city are written in different registers, and, because of their frequent heterogeneity, they resist a clear typology. We may nevertheless draw certain distinctions from a heuristic standpoint. In his writings on the experience of being a flaneur, like those on Marseilles, the subjectivity that decodes the city's physiognomy is essential; in the urban miniatures, as defined by Philippe Despoix, the flaneur gives way to the narrator. As a form, the urban miniature is characterized by a concentration "on one sole phenomenon, one sole detail in the life of the street", which it tries to depict (*darstellen*) rather than interpret.[15]

Portraits of the city, texts on flaneurs, and urban miniatures taken all together sketch out a vast, composite landscape of urban modernity, one that would broaden and lend greater complexity to two of Kracauer's books: *Jacques Offenbach and the Paris of his* Time (1937) on the nineteenth century and *The Salaried Masses* (1930) on the twentieth. In this latter work, Kracauer does not merely measure "the social space in which employees live". Through the different writing processes he uses and the montage that stitches the parts of his study together, he sketches a portrait of Berlin in which, to use an expression by Hans Speier (himself the author of a classic study of employees), he portrays "the air they breathe",[16] the ambiance that surrounds them.

Urban and cinematic landscapes

In *Passagenwerk*, Walter Benjamin considers the appropriateness of using the term landscape "to designate an urbanized space", just as a new sensibility and new forms of perceptions had begun to characterize the subject of modernity. In his commentary on the principle of *flânerie* in Proust, he observed:

> This passage shows very clearly how the old Romantic sentiment for landscape dissolves and a new Romantic conception of landscape emerges – of a landscape that seems, rather, to be a cityscape, if it is true that the city is the properly sacred ground of flânerie.[17]

In fact, urban landscape is a translation of Benjamin's neologism *Stadtschaft*, or "cityscape", which he did not explore further, aside from the expression "the primordial landscape of consumption" (*Urlandschaft der Konsumption*), which he used to describe Parisian passageways.

Kracauer does not use the term "landscape" in his writings on the city. But, in 1931, he did publish a brief essay in the *Frankfurter Zeitung* titled *Berliner Landschaft* (*Berlin Landscape*), which he included in his 1964 book *Strassen in Berlin und anderswo* (*Streets in Berlin & Elsewhere*) with the title *Aus dem Fenster gesehen* (*Seen from the Window*).[18]

In an article on the city and the notion of landscape in Benjamin and Kracauer, Henrik Reeh first discusses this text by Kracauer before dwelling extensively on Benjamin's writings[19]. He notes that "Seen from the Window" extends the tradition of representing the city from a "bird's-eye view", as Kracauer insists first on the elevated perspective of his observation made from an apartment interior, through (the frame of) a window. Reeh's analysis essentially concerns language. He shows that the choice of words and the assemblage of terms referring to different registers of reality – natural, urban, and even industrial – reveal the intent to use the tradition of landscape theory to describe the urban space while systematically negating any identification of the city with nature. When taken out of their discursive context, the words with which Kracauer describes landscapes are used to describe the elements of an urban, industrial, and mechanized world, modern means of transportation, and the ornamental figures that emerge from them, such as the "swarm of shiny parallels" outlined by the rails emerging from the Charlottenburg railway station. The process culminates in a nocturnal description of the view from the window, which employs three registers of metaphors: agricultural, forest, and maritime. And as Reeh notes, it makes generic use of the term "landscape" to qualify the view of the city. He concludes that "the metaphor of landscape ultimately submits to the urban space, whose strangeness it helps describe".[20]

I will in turn dwell on this essay to consider it from a different perspective. It consists of five paragraphs, the first of which is a prologue, in which Kracauer distinguishes between two kinds of *images* of the city: intentional and unintentional images.[21] The former, as the result of an artistic intent, are often noted in guidebooks; the others, such as the masses of stones and webs of intertwining streets, are the result of chance. Such images of the city are "as loosely formed (*gestaltet*) as nature, and they resemble a landscape in that they assert themselves unconsciously".[22] The view from an apartment window, an everyday, ordinary situation that the following paragraph specifies, is expressed here as a series of unintentional images. The insistence on the observer's elevated point of view and the view from the window situate Kracauer's text within a literary tradition of forms of urban perception. It echoes the E.T.A. Hoffmann novella, *Des Vetters Eckfenster* (My cousin's corner window), "one of the earliest attempts to capture the street scene of a large city",[23] according to Walter Benjamin, who compared this novella to Edgar Allan Poe's *The Man of the Crowd*. He noted that Hoffmann's observer – unlike Poe's, who, drawn in by the crowds of the streets, loses himself in them – is perched above the street, and that the images he gleans constitute genre scenes. Instead of expressing urban chaos, they seek to put it at a distance, to organize it aesthetically in ways that neutralize it (Figure 15.1).

In the second paragraph, which describes the view from the window in the daytime, Kracauer maintains a distance. His description of the city merely renders the chaos and rhythm of the modern metropolis, but human beings are absent from it, and the explicit "I" from the previous paragraph has disappeared. The following paragraph, which depicts the same landscape at night, accentuates certain characteristics that were already present in the daytime description. The urban space seen from the window is merely a "field of lights" shining in the darkness. In the nocturnal city, volumes disappear, and the mass of buildings becomes a zone of indistinct shadows that blend into the dark background pierced by streaks of light, like the aggressive orange of a garage sign. This colour, like the ones described in "Luminous Advertisements", another text of Kracauer's where the description of the city appears interchangeable with one in a filmed sequence of the city, is a "light-colour" that contrasts with the "material-colours" of painting, to use a distinction drawn by Frieda Grafe.[24] The movement of the lights shreds urban darkness just as a car's headlights, a lighthouse in a port city, or projectors blazing through the night sky would in a scene in a film.

The perspective is blurred, the window frame cuts out a "field of lights", but it does not structure it in the same way as the frame of a painting. It is more like the framing of a film shot or a photograph. What lies outside the frame, the extent of the urban space, is implied but remains undefined. The image is a fragment that does not refer to a whole that can be represented.

Of course, Kracauer is questioning the very notion of landscape by linguistically transposing it into the urban space. More fundamentally, though, the text also gradually exposes the substitution of the *pictorial matrix* and of

Figure 15.1 Film stills from Helga Fanderl's film *Fireworks* (Super 8, 2000).

the *literary matrix* used in traditional descriptions of landscape by a *photo-cinematographic matrix*. And in a brief, final paragraph that serves as an epilogue, Kracauer concludes: "This landscape is Berlin, unstaged".[25]

Kracauer's process slides from a description of the city towards its recording on film (and vice-versa) in texts that, though of different genres, be they film critiques or urban writings, reflect the development of his thinking about the affinities between film and the modern city. Within this corpus, "Seen From the Window" distinguishes itself for its theoretical scope, one that the craft of the text deploys. It stitches together the concomitant effects of urbanization and photo-cinematographic recording on (the image of) the perception of one's surrounding space, thereby ultimately deconstructing the traditional notion of landscape.

In his 1913 essay "Philosophy of Landscape", which was published in the era of early cinema, and in which he reflected on the appearance of landscape in art, especially painting, and on the "awareness of landscape", Georg Simmel asserted that, to see a landscape, "our consciousness has to acquire a wholeness, a unity, over and above its component elements, without being tied to their specificity or mechanistically composed of them".[26] This mode of a subject's relationship to nature, which appeared in the West in modern times, "necessitated a tearing away from that unitary feeling of the whole of nature" that was characteristic of medieval religious sentiment, for example. As an individual, singular form defined by its limits, a landscape nonetheless entails nature as a whole. For Simmel, the *Stimmung* of the landscape is the main medium for grouping juxtaposed natural elements into a particular mode of unity. *Stimmung* constitutes the landscape, and at the same time, it is the "perceptible *unity*" for the subject contemplating it. And to the extent that it signifies "a general dimension [...] of this precise landscape", it grounds its individuality in an acknowledgement of nature as a totality (Figure 15.2).

From a Simmelian perspective, the notion of a fragment of nature is impossible. Now, the notion of the fragment is inherent to photography and flagrant in the case of the snapshot. Consider an individual's gesture in the street, caught on the fly by a photographer, or photographs of clouds taken in the 1910s and 20s, or countless film shots, such as the expanses of ice or water in the first exploration films from the early twentieth century, or views of a dark, hidden street corner on a night furtively illuminated by the lights of a car in a film noir movie or in certain experimental films. At times, the urban or natural landscapes shown in these photographs and views tend to become a surface, like the image of Berlin seen through the window frame in Kracauer's essay. The elements that traditionally structure a landscape, such as the horizon line, viewpoint, or perspective, disappear in an image where the environment is presented as an undefined continuum that spills out of the frame, which is perceived initially in synaesthetic terms, through its sensory attributes: texture, colours, and movement (Figure 15.3).

Figure 15.2 Film stills from footage shot in 1910–1913 by Herbert Ponting during the Terra Nova Expedition to Antarctica and used in his film *The Great White Silence* (Great Britain, 1924).

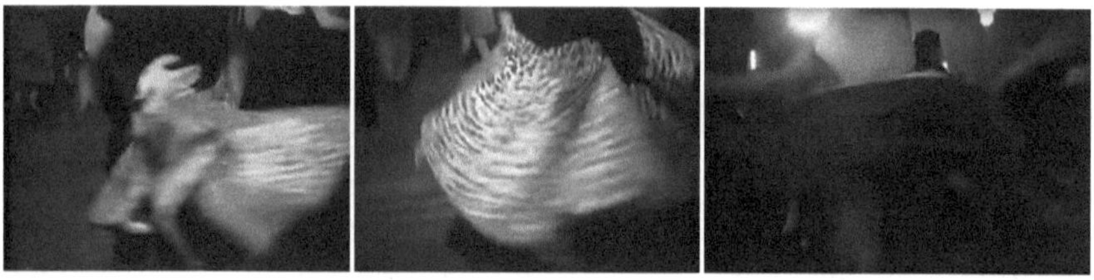

Figure 15.3 Film stills from Roger Tilton's film *Jazz Dance* (United States, 1954).

Urban ambiances: from the city to film

The affinities between the city and the photo-cinematographic medium that Kracauer addressed throughout his writings on cities and on film are inseparable from his theoretical approach to film. The central concept in *Theory of Film*, "material aesthetics", in which the view through the lens *and* the optical imprint on the film are the primary data of his thought, is the *camera-reality*, which designates this distinctive level of reality that the film reveals. This refers to a specific way of relating to reality that is both a revelation and a transformation. Such a theoretical approach opens the way to considering what happens to urban ambiances *simply for their being recorded on film*, aside or beyond any ambiance effects that are created using specific technical processes or through music.[27] It also invalidates any attempt at making systematic and general statements about ambiance effects inherent to the process of recording with a camera, and even more so, any attempts at formalization. Nevertheless, I do think it is possible at the end of this long detour to specify several aspects that seem particularly relevant to accounting for this and which thereby broach several issues worth considering.

- The focus on a detail or an object that a close-up draws from the film's diegetic context is not simply a change in scale. The possibility of dwelling on it intensifies the way in which this object or this detail can affect us, by its sensory qualities, by the set of associations that it provokes, or the memories that it summons. This may also be a question of the viewer's own memories or recollections of films he/she has seen.[28]

- The fragmentation inherent to the medium of film, while making objects unrecognizable, has the paradoxical power of transmitting the sensation of movements and rhythms, and to intensify their recognition. For example, this is the case of *Jazz Dance*, a film that Kracauer says initiates us "in the secrets of a material universe set afire by the dancers' frenzy," thus amplifying the ambiance of a jazz club, an emblematic locus of the modern city.[29]
- The exhibition of the perceptible qualities of the material environment that the inherently fragmentary photo-cinematographic image detaches, at least partially, from the narrative and semantic aspects of the film engenders another perception of the filmed reality at the same time that it contributes to the creation of a specifically cinematic mood.
- Beyond that, the relative autonomy of cinematic shots and their relative disconnect from the narration enable a perception of micro-ambiances that continue to exist throughout the film, at the same time that they help create a sense of atmosphere on another scale. Similarly, the atmosphere emanating from an object can persist. For Kracauer, there is only one fundamental principle for editing: in addition to serving the narration, it must also take an interest in "the objects represented so that they may appear in their suggestive indeterminacy".[30]
- Lastly, the means of articulating the different temporal strata of the past in an urban present exist in a homologous relationship with the multiple senses of time inherent to the film. In other words, cinematic recording retains an imprint of the heterogeneous multiplicity of a city's different layers of time. But time in a film is always dual, because there is a distance in time between the moment of recording and the moment of projection. The past-present of a film that becomes present upon its projection per se possesses an "atmospheric" potential.

Notes

1 The concept of atmosphere was introduced to film studies in roughly the 2000s. Its usage developed in German writings in the wake of Gernot Böhme's works, and more generally, within the context of new phenomenological approaches to film, works on empathy inspired by aesthetic and/or cognitivist theories, and works in the field of the theories of emotions.

2 One of the best-known approaches is in Vachel Lindsay, *The Art of the Moving Picture* (New York: Random House, 2000) (originally published in 1915). Herbert Tannenbaum was one of the first to introduce it in the German-speaking world in his article "Lichtspielbühnen und Theaterbühnen" ["Movie Stages and Theatre Stages"] in 1912, which was republished in Helmut Diederichs (ed.), *Geschichte der Filmtheorie* ["A History of Film Theory"] (Frankfurt am Main: Suhrkamp, 2004).

3 Emblematic in this regard is the title of Rudolf Arnheim's classic, *Film als Kunst*, originally published in 1932, republished in English as *Film as Art* (Berkeley, CA: University of California Press, 2006).

4 In 1926 Kracauer sketched out the concept of *distraction*, which Benjamin picked up and explored further; see Siegfried Kracauer, "Cult of Distraction" in *The Mass Ornament*, translated and edited by Thomas Levin (Cambridge, MA: Harvard University Press, 1995), 323–328.
5 Siegfried Kracauer, *Theory of Film. The Redemption of Physical Reality,* introduction by Miriam Bratu Hansen (Princeton: Princeton University Press, 1997). I must specify that the corresponding terms in the German translation, reviewed by Kracauer himself, were *Atmosphäre* and *Stimmung*.
6 *Theory of Film*, 90.
7 *Theory of Film*, 71.
8 The concepts "cobbled together" by Kracauer and the absence of the customary aesthetic categories in theoretical approaches to film may confuse the reader of *Theory of Film*. On the specificities and issues of Kracauerian conceptualization, see Philippe Despoix and Nia Perivolaropoulou, "Une théorie intempestive du cinema" ["An untimely theory of film"], in the French edition, *Théorie du film, La rédemption de la réalité matérielle* (Paris: Flammarion, 2010), VII-XXIX.
9 A whole set of terms is used to describe the medium's qualities and this tension characteristic of film: the camera-eye, camera-realism, and, finally, camera-reality.
10 *Theory of Film*, 72.
11 On the importance of Karl Grune's film, see my article "Du flâneur au spectateur. Modernité, grande ville et cinéma chez Siegfried Kracauer" [From flaneur to spectator: modernity, the big city, and film in Siegfried Kracauer"], in Stéphane Füzessery, Philippe Simay (ed.), *Le choc des métropoles. Simmel, Kracauer, Benjamin*, ["The shock of the metropolis: Simmel, Kracauer, Benjamin"] (Paris/Tel-Aviv: Éditions de l'éclat, 2008), 125–148.
12 *Theory of Film*, 73 (emphasis added).
13 On this subject, I refer to the beautiful volume by Christine Breton and Sylvain Maestraggi, which also includes translations into French of the texts and letters by the three authors: *Mais de quoi ont-ils eu si peur ? Walter Benjamin, Ernst Bloch et Siegfried Kracauer à Marseille le 8 septembre 1926* ['What are they so afraid of? Walter Benjamin, Ernst Bloch and Siegfried Kracauer in Marseille on 8 September 1926"] (Marseille: Éditions Commune, 2016).
14 Michel Foucault, *Discipline and Punish*, (New York: Vintage, 1995), 200. "Das Karree" was included in the collection *Strassen in Berlin und anderswo* ["Streets in Berlin and Elswhere"] (Frankfurt a.M.: Suhrkamp, 1964), 26–27.
15 Philippe Despoix, "La miniature urbaine comme genre. Kracauer entre ethnographie urbaine et heuristique du cinema" ["The Urban Miniature as a Genre: Kracauer between Urban Ethnography and Film Heuristics"], in Nia Perivolaropoulou and Philippe Despoix (ed.), *Culture de masse et modernité. Siegfried Kracauer, sociologue, critique,* écrivain ["Mass Culture and Modernity: Siegfried Kracauer: Sociologist, Critic, and Writer"] (Paris: Éditions de la Maison des sciences de l'homme, 2001), 165.
16 Cited by Inka Mülder-Bach, "Introduction", in Siegfried Kracauer, *The Salaried Masses. Duty and Distraction in Weimar Germany* (London-New York: Verso, 1998), 17.
17 Walter Benjamin, *The Arcades Project*, Rolf Tiedemann (ed.) (Cambridge, MA: Harvard University Press, 1999), 420–421.
18 Siegfried Kracauer, "Aus dem Fenster gesehen" ["Seen from the Window"], in *Strassen in Berlin und anderswo*, 51–53.
19 Henrik Reeh, "'Un paysage hanté, intense comme l'opium'. Ville et notion de paysage selon Siegfried Kracauer et Walter Benjamin" ["'A Haunted Landscape as Intense as Opium': the City and the Notion of Landscape in the works of Siegfried Kracauer and Walter Benjamin"], *Revue d'esthétique*, No. 39, "Autres

sites, nouveaux paysages" ["Other Places, New Landscapes"] (2001): 13–26. We should note that neither one ever uses the composite form *Stadtlandschaft*.

20 Henrik Reeh, "'Un paysage hanté, intense comme l'opium'. Ville et notion de paysage selon Siegfried Kracauer et Walter Benjamin", p. 16.

21 The ambiguity of the word image – a configuration of an actual space, image(s) of the perception one may have of this, and images one may make out of it (artistic or literary) – is crucial to Kracauer's description. It is therefore especially regrettable that the word disappears in the French translation published as *Rues de Berlin et d'ailleurs*.

22 "Aus dem Fenster gesehen", in *Strassen in Berlin und anderswo*, 51 (our translation).

23 Walter Benjamin, *Charles Baudelaire: A Lyric Poet in the Era of High Capitalism*, trans. Harry Zohn (London: Verso Books, 1997), 129.

24 Frieda Grafe, "Befreiende Farbe" ["Liberating Colours"], in *Filmfarben* ["Film Colours"] (Berlin: Brinkmann & Bose, 2002), 77.

25 "Aus dem Fenster gesehen", in *Strassen in Berlin und anderswo*, 53 (our translation).

26 Georg Simmel, "The Philosophy of the Landscape", *Theory, Culture & Society*, vol. 24, no. 7–8 (2007): 21.

27 Most studies of film or genres focus either on the mood produced by cinematic processes such as lighting or colours, film music, or the impact that the conditions for the reception of films may have.

28 German has an almost untranslatable term to designate the way in which an object appears to us, what emanates from it, and how it affects us: *Anmutung*. The concept is present in the discussions of *Stimmung*, atmospheres, and ambiances.

29 Roger Tilton, *Jazz Dance*, United States, 1954; see *Theory of Film*, 43, 49, 54.

30 *Theory of Film*, 71.

Part 5

Ambiances, architecture, and place

16 Towards a modal conception of architectural ambiances

Grégoire Chelkoff

When we move through a work of architecture, it gets us to see, hear, and feel at the same time. Our body and senses come into play in a way that is both obvious and indescribable. Once we exceed an exclusively visual experience, one of the first effects of the concept of ambiance is that it completely expands the sensory world of architecture beyond the scope of our retinas. When sounds, movement, air, or heat-related elements are converted into ambiance (which until that point are barely experienced as "poetic" in architecture), can they acquire an aesthetic character and allow us to sketch out specific imaginary worlds? In considering these dimensions carefully, shouldn't we say that the aesthetic experience of architecture exceeds the strict framework of a visible built structure, and that it concerns one's entire sensory, active experience of space? In such a case, it is no longer a contemplative or receptive aesthetic, but instead an experience embedded in our everyday gestures as we inhabit the world that in some way renders us present within it. This constitutes core of the notion of ambiance, and it heavily involves the architecture and space that we inhabit. What kind of sensory experience and what ways of inhabiting shall we prepare when we design our projects?

It is precise because architecture is an art that addresses all the senses that the notion of ambiance acquires a very particular significance within it. It is also undoubtedly a way to ensure that architecture remains open to the multiple phenomena that concern it, to the conditions for inhabiting that it prepares, the imaginary worlds and representations that it generates, and, lastly, to forms of aesthetic experience, both unexpected and ordinary. Thus, approaching architecture in terms of ambiance means perceiving it as a culture that is sensitive to the inhabited environment, and this requires a modal – instead of a causal – approach to the physical and sensorial dimensions that one decides to address. This becomes an interrogation of architecture and urban planning according to an approach that privileges the dimensions of *feeling,* through which we experience "our being with the world in a pathic mode".[1] One of the major issues of this modal approach to ambiances in architecture is the renewal of our way of conceiving of the sensory, as well as our way of working with the sensory[2] and the senses while, at the same time, technical knowledge of natural lighting, acoustics,

DOI: 10.4324/9781032655307-24

and thermic properties is developing constantly. We must explore the implications of these ideas to the aesthetic, technical, and design-related thinking in architecture.

It is true that, when we consider the knowledge and tools that engineering, the humanities, and social sciences summon and use, we may ask, what contributions and demands does architecture make in relation to the concept of ambiance?

The notion of ambiance,[3] though not called such, begins to appear in architecture alongside the lighting (natural lighting), acoustic, and thermal issues that have begun to play a bigger role in construction and through the evolution of technology. The domains of engineering knowledge have led to a mastery that Rayner Banham called a "well-tempered environment".[4] Some architects began to integrate these technological dimensions in their projects extensively, in manner that was primarily functional, and at times an aesthetic.

The convergence of engineering and the social sciences,[5] initially on the subject of acoustic and thermal comfort, is a recent one. In their criticism of the design of a universal comfort as the result of constant technological progress, the social sciences also addressed the sensorial dimensions through the development of more localized modal approaches that assigned greater importance to context and on-site analysis,[6] and which were distinct from the more distant and global structural approaches. This convergence made ambiance an interdisciplinary field of research[7] that seeks the further enmeshing of all relevant physical, sensorial, and social questions. At the same time, the interweaving of these dimensions renders the articulation of these different fields of knowledge exciting, though difficult, and any theorization becomes extremely delicate, if not artificial.[8] Characterizing the contours and qualities of such an elusive object approaches the impossible, and those who attempt a complete definition of ambiance at times leave one sceptical. Others instead see it less as a notion or precise object to be defined than a way to transform architectural practice through processes of negotiation.[9]

The questions we will broach here reaffirm a theoretical position that gives full priority to the means and phenomenality of an experience and of usages, as well as to the envisioning of the relevant technical and sensory aspects. As a result, considering the conditions for experiencing built forms and spaces as completely and precisely as possible is one of the cornerstones of an approach to architectural ambiances that we have thus defined as *modal*.

Architecture and sensory practices

It is not so much a question of a spectacular usage of ambiance-related effects or very sophisticated technical ideas as it is an attempt to stimulate aesthetic experiences in terms of the *creation of ambiance* for functions that are as ordinary as they are anodyne. Standing on one's balcony, walking along a wall, going up or down a staircase, and crossing an expanse are experiences involving issues of ambiance that can be legitimately integrated into architecture. These experiences may be considered in function of light exposure, sonorousness, or climate, from the perspective of the person walking, or from that of the person perceiving. On each occasion, the link between the environmental conditions and the situated means of action specifies the architectural ambiance and renders it perceptible as the basis for an aesthetic experience and a medium for imagining how we inhabit. Architecture, with its spatial and technical means, can thus integrate certain aesthetic dimensions into current spatial practices and reveal dynamics for imagining how we inhabit. This characterizes it as the art of shared ambiance.

This act of imagination, which renders habitation possible, is fertilized by the ensemble of sensory registers that come into play in everyday life. It takes the form of narratives and memories, as well as the ordinary practices and experiences of spaces. Improving our understanding of this and of the processes of its formation may help modes of design that privilege a formalism that tends to impose its views or effects, or a functionalism that is often overly reductive of human variations of inhabiting.

We can see that this modal approach to architectural ambiances fundamentally questions how one works with the senses in a project, especially in terms of the details. With regard to a holistic approach to ambiance (developed using major categories or types, such as ambiances for railway stations, hospitals, and public spaces,[10] among others), the challenge resides in identifying and designing the elements able to solicit people's imaginations on the basis of the situated experience.

The sensory and its subtle variations

We should first of all grant the sensory its fair due. In the name of intelligibility, the body and the senses have generally been thought of as secondary and considered misleading. They were assigned to a different sphere by our tendency to construct using abstraction, to the detriment of sound and image. It is true that the flow of sensations, the event, rattles the linear, more controllable order of causality.[11] This overly variable flow is thus sacrificed for its resistance to the determination of the exclusive order of the concept. As a result, rationalist or formalist thinking has concerned itself little with feeling, with what happens to the body. Architecture theories sometimes have a hard time resisting this general trend towards desensitization. Whenever the sensory is invoked, it is done so in the name of art and sensation, and it is most often reduced to a question of sight, to the author's own perception.

Architecture's relationship to the sensory world should be explored with greater precision. The passage from the world of solids arranged in space to a world consisting essentially of variable flows is indeed disconcerting. In taking inspiration from a phenomenological method, architects, who by their very nature study environments, should more often place within parentheses the "natural world" of the objects they encounter to grasp the structure of our feelings more fully. We may not always be aware of these flows attached to the objects we know (a dapple of light is considered an opening, and a sound is named for the object that produces it), but we cannot deny that they permeate our everyday, ordinary experience. Paradoxically, this flow is constant and so "natural" that it disappears from our field of awareness.

Imagination, poetry, and phenomenology are thus essential for qualifying certain processes for the emergence of the sensory, as if we were describing an aesthetic intention. Unlike a pre-determined object designed outside any context in which it then appears, the object expresses itself through a phenomenality that inserts it into a tactile, thermal, acoustic, or light-related network, depending on the context. Hence, the plurality of appearances or sensory presences of a work of architecture – depending on the circumstances, on lighting, sound, air, or heat and also depending on the means of usage – merits attention and can stimulate our designing imagination as well as our awareness of our everyday reality. "If we multiplied images and considered them in terms of light and sound, hot and cold, we would generate an ontology that is much slower, but definitely more sure-footed than one relying on geometric images".[12] Privileging geometric intuition tends to reduce other structures that are less stable but just as illustrative. Thus, hearing[13] substitutes new images for habitual objects. Sounds can stimulate our imaginations especially when the visual world yields less information. "Some doors left ajar laugh wildly… melodies rise from between the cobblestones," wrote Balzac[14] in describing the old, dark streets of Paris. Literary imagination suggests a kind of fusion between the object and what it triggers in our senses, here specifically in the realm of sound.

Thus, with regard to an architecture of solids, architectural ambiance is based essentially on the modulations of time and energy of these sensory flows. Although it may seem as if there is no reality without temporality or movement, how can we grasp those phenomena that are by their very nature fleeting? We should think less in terms of static forms and more in terms of forms yet to be made, forms consisting of the capacity for variation and potential for action. Therefore, we must explore and create architectural objects in relation to how they transform and how they embody usage.

However, the considerable primacy of SENSE (i.e., meaning) over the senses, which are several and unstable, has tended to erase the subtle variations that are sources of both our imagination and of usages. These slight variations nevertheless produce the essential articulations on which our perceptions and actions rest. Thus, the role of a detail may be extremely significant for triggering a process of visualization. "I perceive an element of the whole, however slight it may be, and I find it determinant".[15] We may thus observe the role of individual elements in the perceptual organization of a whole; as a result, we can say that at times, ambiance resides in a detail when it is described.

Gaston Bachelard's observations and reflections provide very suggestive elements of this polarization, as a result of which a characteristic may bring about or orient an ambiance. To take one example from his investigation of our imagination, he describes the image *of a cell* as follows: "the being feels surrounded by the whiteness of a lime milk more than by massive walls. The secret's cell is white. One sole value is enough to coordinate dreams well".[16] The poetic image is thus subjected to a quality that is somehow enhanced. The whiteness of the walls is enough; "it is stronger than any geometry".[17] The idea that it "coordinates" our dreams suggests that a phenomenon can orient a whole. Whiteness is not just a quality like any other; in this specific instance, it is an essential phenomenon of the image of the cell that bolsters the idea of containing and keeping something within a cell. We may therefore distinguish between an image of an ambiance (here, that of a prison cell) and what can trigger and also form it, here the enveloping whiteness of the walls. This clearly shows that the image of a place can be formed by a sensory keynote that coordinates the other dimensions of the context and leads them towards the same meaning. As in music, this is a question of the dominant note, or rather, of tonality. By modulating factors such as light, sound, air, and heat, architecture shapes the tonalities of an ambiance that we can tune into or out of.

Inter-sensoriality: the details and conjugation of the senses

Can we imagine, feel, or create an ambiance without summoning all our senses at once? Multi-sensoriality seems a given when it comes to ordinary perception and usage, but we have fallen into the habit of thinking about the sensory by splitting up the senses. The ambiance-related trades (such as acousticians and lighting experts) work in specific and clearly separated sensory fields. So, if we observe each sensory domain separately, how can an ambiance take *shape*? The interactions between the different sensory registers should be better delineated and envisioned. Certain literary types of observations can also help us to understand the workings of our sensory imagination. In writing about the moonlight bathing the Milan Cathedral, Stendhal noted: "I found the most beautiful silence there", an experience that was visual and then also thermal.[18] In one sense, a phenomenon expresses what happens analogously in the others… an effective poetic process for a writer, undoubtedly riskier for an architect, but which nevertheless expresses a lived situation.

We may thus envisage a sensorial coordination that shows how several sensory registers at times follow what we might call a common scheme of variation. Literary description can once again help us here. In the previous excerpt from Balzac,[19] the (nocturnal) ambiance is expressed by the relationship between the objective narrowness of the street and the perceptible form of all the events relating to the usage of the place. The terms he uses illustrate sporadic phenomena that appear to slink in between everything (which corresponds to what we have called an organic urban ambiance). The whole that he expresses describes a unity of perceptible form: a manner of perceiving and being, the common thread of which, through what one sees and hears, is formed according to a common measure. The narrowness of the streets generates a certain evanescence to things, a thinning of events that borders on the fantastical. Everything happens as if, to signify this ambiance, one must lean towards the same type of conduct or manifestation of things, a common structure of sensory relations within which the discrepancies acquire even greater significance (a provocative heel, or the inversion of the climate).

We may suppose that these covariations of spatial, light, and acoustic structures are particularly interesting to observe, and that they definitely represent a major step towards "orienting an ambiance", rendering it as something that can be perceived or shared.

Sharing the sensory: presence and mediation instruments

What we perceive is not a strictly subjective, individual affair that is purely sensorial and receptive; it is also a matter of ways of feeling and acting together in a shared world. The social aspect is consubstantial with the concept of ambiance. It is thereby often a question of agreement, tact, and investment in an environment. Ensuring that an environment is experienced in a relatively similar manner by different users may be a significant factor of architectural ambiance. The sensorial framing of a shared imagination expresses itself in the way that a place is used. In this sense, ambiance is active, and not just something that is perceived. We can study how ambiances act by analysing and observing the relations between the constructed systems and the usages, identifying ways of doing[20] and perceiving things that involve architectural elements and ambient flows. We may thus conceive of built systems as *instruments* of sensory mediation that entail formalities of usage, as instruments of visibility, mobility, and audibility. Each one of these instruments has a way of *putting us in the presence of the world* and our ways of playing with them. This can give rise to a certain intimacy between inhabitants and these instruments, which expresses a style of inhabiting. Everyone, whether an inhabitant or a passer-by, intuitively uses strategic positions for looking and hearing based on distance or involvement, withdrawal or appropriation. Considering spaces as instruments clearly asks us to imagine all of the game's possibilities, what they allow us to do and the usage competencies they involve. We should recall that architecture grants specific propagation qualities in the space to these phenomena, and that the roles of these elements in social interactions and usages are not anodyne. Architecture acts using modalities of sensory *presence*, both in the way it filters light, sound, and thermal signals and therefore qualifies or transforms them, and in the way, it renders us accessible to one another. Our ways of doing, inhabiting, walking, and talking are of course connoted by social and cultural dimensions that lie beyond the architect's purview; however, when they colour a place, they seem to merge into the context and put the capacities of sensory presence into play. These principles lead us to consider the ways in which an architectural system initiates or predisposes certain forms of experience that characterize a lived ambiance and the memorization[21] of a place.

Sensory formants of experience

In light of these observations, conceiving architecture as an ambiance suggests that we apprehend built forms no longer as mere spatial arrangements, instead through the sensory *formants*[22] of perception and action that constitute them for use. Two challenges arise when we do this: better addressing the sensory conjugations that define ambiances and incorporating the idea that ambiances involve the human body and action. A built system has a formant dimension because it favours certain propagations of light, heat, and sound, certain perceptual potentialities, but also because it provides a style of appearing and executing certain actions and usages. In other words, the notion of the formant takes us from a definition of form, as it has been conceived of or given *objectively* on the basis of visual and geometric reasoning, to a notion of form that is in the process of formation, to a network of flows that make a form.

Thus, the goal is not to describe form as it may be or appear to be, rather to describe how it sensitizes a world, relationships, and thereby intensifies it, literally rendering it present, and rendering us present to one another. We must therefore explore and conceive of a form by involving the experience that drives it. This methodological perspective has us approach the built form in relation to the possibilities of experience and action to experience or define some of its ambient capacities. The question guiding this entire approach consists of asking oneself how architecture *makes* us perceive things, how it makes us perceive the world, and not how it itself is perceived.

The notion of formant is based on the potential for action and perception made possible by the form, and which causes the form to vary. What is definitively formed is action, a way of being and acting: our experience. The architectured forms and the built systems empower ambiance, in that they offer styles of being in the world, aesthetics that help shape how we see, hear, walk, or stop, how the body steps into action. It is in this close interrelation between built systems, usages, and the sensory environment that we may situate the question of formants. Cogeneration mechanisms between spatial form and perceptive action should thus be explored. It is a question of making things present so that a form or a system offers ways of doing (postures, positions, movements) that resemble an ambiance.

An old debate[23] on the issue of openings may serve as an illustration of this point. For the architect Auguste Perret, windows had to be vertical. He felt that this corresponded to the "upright human." This static vision suggests a human who may be more attentive to what he sees and hears from a fixed vantage point. The horizontal window favouring a panoramic view that Le Corbusier advocated for corresponds to a human being who walks by, who moves, and whose motion becomes the crucial criterion of modern aesthetics.

In my opinion, this debate serves as the perfect illustration of the idea that an opening, whether vertical or horizontal, forms or deforms relations to different sensory flows (especially light, visibility, and air), and that it shows how a specific architectural element articulates several phenomena according to a shared sensory structure. By advocating for the horizontal window, Le Corbusier generated an optical mechanism combining form, motion, and a panoramic vision of the landscape. In this sense, we could say that he saw it as an essential optical and kinetic formant of the landscape seen from the interior by creating an effect of optical and light-related continuity (as it relies on the distribution of natural light). The formant network of a horizontal opening system is characterized in our minds as a "continuity" of the perceptual style: a continuity produced by observation, which can follow the landscape in motion, a notion of uninterrupted panoramic vision, a continuity of natural lighting in the room, a continuity of openings out to the side walls so as not to cut the walls, and so on. The formant power of the vertical window, according to Perret, resides in its "depth": the sensorial and motion-related analogy evoked by Perret between the verticality of the opening and the posture of the "upright human" refers to a possible perception of a taller vertical depth. The opening allows one to see the landscape from top to bottom, thus accentuating its depth, and the lighting is also conceived of as penetrating the room more profoundly.

The method for thinking in terms of formants consists of varying architectural concepts and systems in terms of the sensory registers that they bring to life in our experience. Thus, lightness, in terms of a formant, would no longer be reduced to a question of mass. It would instead be developed in terms of how it is experienced through kinetic resistance, luminosity, thermal sensations, and sound. It is possible to explore the traits shared by several senses. Smoothness can be expressed as light (for example, reflections) and as sound (modes of reverberating propagation), which render the elements or signs that disturb its legibility less perceptible and at times less comprehensible. By looking at one quality in several sensory registers, smoothness can thus be characterized as a formant of a slippage that renders objects partially imperceptible. The sensory formant of a floor may be described in terms of the contact that specifies the existence and action of the step that adapts to it; for example, the step that slides on a smooth floor expresses a way of walking that doesn't exist if the conditions didn't allow for it.

Therefore, formants are not objects, systems, elements, or sensory indicators, but instead structures of sensory relations in which system and action cogenerate one another. The notion of formant does not refer to an external entity, an object that exists per se, instead to a coupling of action and environment. Consequently, our analysis does not focus so much on a perceived object (the building, the door, or the outside) as it does on the ways an object has us perceive the world and act, how it makes us apprehend this world with our bodies. We move from a notion of an objectified, reified form to a formant form, a form that breathes life into perception and is in turn brought to life by this very perception.

On this basis, many elements remain to be explored to develop this architectural approach to ambiance; it will be the task of research to specify the relevant means and resources.

In fact, from a methodological standpoint, on-site observation is a tool that makes our designing imaginations work and nourishes our inductive references. Sensory observation is an essential component of enlarging our understanding of the phenomena of ambiance and our thinking about types of formants as tools that help us describe and design. This is equivalent to saying that architectural ambiance exists only when we are able to express in different ways what renders a form or a system perceptible within the context of an experience that is at once ordinary and extraordinary.

Notes

1 Henri Maldiney, *Regard, parole, espace*, ["View, Word, Space"] and *L'age d'homme* ["The Age of Man"] (Lausanne: Collection Amers, 1973), 164.

2 The research and thoughts set out in this article are largely based on the following document: G. Chelkoff, *De l'espace à l'ambiance, Habilitation à diriger des recherches* ["From Space to Ambiance: Authorization to Direct Research"](Grenoble: Université Pierre Mendes France, 2005).

3 The notion of ambiance became central with the partnership between the CRESSON and CERMA laboratories in 1992.

4 Rayner Banham, *The Architecture of the Well-Tempered Environment* (Chicago: University of Chicago Press, 1984).

5 See the article by Jean François Augoyard: "Eléments pour une théorie des ambiances architecturales et urbaines" ["Elements for a theory of architectural and urban ambiances"], in *Les cahiers de la recherche architecturale et urbaine* (Marseilles: Editions Parenthèses, 1998).

6 J.F. Augoyard, *Pas à pas, essai sur le cheminement quotidien en milieu urbain* ["Step by step: an essay on the daily course of an urban milieu"] (Paris: Seuil, 1979) adopts this approach. For a definition of the modal approach in anthropology that is distinct from structural approaches, see François Laplantine, *Le sensible et le social – Anthropologie au coin de la rue* ["The sensorial and the social: Anthropology on the street corner"] (Paris: Tetraedre, 2005).

7 It developed as a term at the CRESSON laboratory, which, together with CERMA, established a master's program in "Architectural and urban ambiances". CRESSON's role was decisive for the question of interdisciplinarity and the on-site approach within the Ambiances UMR mixed research unit. To learn more about the internal debates within the laboratory, see the collective volume edited by P. Amphoux, G. Chelkoff, and J.P. Thibaud, *Ambiances en débat* ["The Debate over Ambiances"] (Paris: Edition A la Croisée, 2004).

8 At the CRESSON laboratory, the first interdisciplinary sensory description of urban ambiances that offered theoretical tools was formalized in: G. Chelkoff, *L'urbanité des sens* ["The Urbanity of the Senses"], doctoral thesis, Grenoble Urban Planning Institute, 1996, and in research on underground public spaces: G. Chelkoff and J.P. Thibaud, *Ambiances sous la ville – Une approche écologique des espaces publics souterrains* ["Ambiances under the City: An Ecological Approach to Underground Public Spaces"] (Paris: PUCA research report, 1997).

9 See P. Amphoux, *La notion d'ambiance, une mutation de la pensée urbaine et de la pratique architecturale* ["The Notion of Ambiance: A Change in Urban Thought

and Architectural Practice"] (Paris: Plan Urbanisme Construction Architecture (PUCA), Ministry of Public Facilities, Transportation, and Housing, 1998), 168.

10 For example, we tried to categorize urban ambiances on the basis of their underlying urban planning models. See *L'urbanité des sens.*

11 François Laplantine, Le sensible et le social – Anthropologie au coin de la rue ["The sensorial and the social: Anthropology on the street corner"] (Paris: Tetraedre, 2005).

12 Gaston Bachelard, *The Poetics of Space*, Translated by Maria Jolas, (Boston: Beacon Press, 1994), 194.

13 In terms of acoustics, a qualitative approach to the sound environment has developed that has resulted in the creation at CRESSON of a repertory of sound effects in the urban environment: J.F. Augoyard (ed.) & H. Torgue (Collective edition), *A l'écoute de l'environnement - Répertoire des effets sonores* ["Listening to the Environment – A Catalogue of Sound Effects"] (Marseille, Parenthèses, 1995).

14 Honoré Balzac, *The Splendors and Miseries of Courtesans*, Translated by Ellery Sedgwick, (Philadelphia: G. Barrie & Son, 1895).

15 Pierre Sansot, *La poétique de la ville* ["The Poetics of the City"] (Paris: Payot & Rivages, 2004), 35.

16 Bachelard, *The Poetics of Space.*

17 Bachelard, *The Poetics of Space.*

18 "I found the most beautiful silence there. So gothic and slender, jutting up into the air against a dark blue southern sky adorned with shimmering stars, these pyramids of white marble provide a spectacle unlike anywhere else in the world. And the sky was like velvet, in harmony with the soft beams of a beautiful moon. A warm breeze blew through the narrow passageways that run along certain sides of the massive cathedral – a magical moment" (our translation). Stendhal, *Rome, Florence, Naples*, (Paris: Gallimard Folio, 1987), 63.

19 Honoré Balzac, *The Splendors and Miseries of Courtesans.* This is the excerpt: "Perhaps it makes Parisians happy to leave these streets in their filthy state. Walking through during the day, it's impossible to imagine what all these streets become at night: frequented by strange beings from some other planet, half-naked, white forms adorning the walls, and the shadows buzzing with activity. Between the wall and a passer-by, a running public toilet walks and talks. Some doors left ajar laugh wildly. Words that Rabelais claimed were frozen melt, falling into one's ears. Melodies rise from between the cobblestones. The sound isn't vague; it means something. Raucous, it is a voice, but it resembles a song. There is nothing human about it anymore; it seems more like a whistling. There is often someone blowing on a whistle. The heels of boots have something provocative and mocking about them. All these things become dizzying. The atmospheric conditions have changed; it's warm in the winter and cold in summer, but regardless of what the weather is doing, this strange nature always puts on the same show. You can see the fantastical world of Hoffman the Berliner there".

20 Michel De Certeau called these "formalities". See *The Practice of Everyday Life*, Translated by Steven Rendall, (Berkeley: University of California Press, 2011).

21 Therefore, it is possible to work on either small or large-scale systems that would contribute to the formation of the sensory environment. We have developed this idea in several experimental prototypes: *Prototypes sonores architecturaux. Catalogue raisonné et expérimentations constructives* ["Architectural acoustic prototypes: a catalogue raisonné and construction experiments"] (PUCA research report, Cresson, 2003).

22 We broached this notion in "L'hypothèse des formants" ["The hypothesis of formants"], in *Ambiances en débat.*

23 This debate was analysed by B. Reichlin: "The pros and cons of the horizontal window: The Perret – Le Corbusier controversy", in *Daidalos*, issue no. 13 (1984).

17 Atmospheres in formation

Affective thresholds, immersive fields, and active boundaries

Izabela Wieczorek

In 1998, the architectural journal *Daidalos* raised the question of how to construct atmospheres. In response to this question, which still resonates today, the architectural theorist Mark Wigley presented atmosphere as one of the most enduring and challenging themes in architectural discourse and practice.[1] Thirty-five years earlier, elaborating on the notion of "mood", which is often considered a precursor of atmosphere, the philosopher Otto Friedrich Bollnow had also noted: "One can look at the whole of architectural history from the viewpoint of the different mood content of the interior spaces constructed in it".[2] By recognizing atmosphere as symptomatic of architecture, Bollnow and Wigley do not only present atmospheres as material formations. Their observations also imply that, although concerns for atmospheres have only recently crystallized into an influential transdisciplinary debate, the conceptual foundations and protocols for the production of atmospheres can be found beyond contemporary examples. Consequently, the recent *atmospheric turn* can be identified not as new, but rather as newly significant, that is, a *re-turn* to the principles that have over the centuries shaped what might be termed an *atmospheric awareness*.

Although atmospheres can emerge from unconsciously or spontaneously arranged relational configurations – and it is difficult to imagine any space without an atmosphere –, atmospheric awareness implies a particular mode of existence of atmospheres. Defining them as "staged",[3] the philosopher Gernot Böhme refers to atmospheres that are intentionally produced, conditioned, or manipulated. The idea of staging is, therefore, closely intertwined with the designer as "an active engineer of atmosphere".[4] This is how the sociologist and cultural theorist Jean Baudrillard described those who approach space and atmosphere as a "distributed system" in which it is possible to "determine responses".[5] Hence, considering space as such a system prompts reflection on a multiplicity of signs inscribed in the material fabric of the environment, presenting matter not as merely a tectonic medium, but also a catalyst of effects and appearances, that is, an active element of atmosphere that has performative, transformative and affective capacities, conjuring up a notion of involvement, beckoning our sensorium as well as inciting movements and behaviours.

DOI: 10.4324/9781032655307-25

Building on the above arguments, the title "Atmospheres in formation" carries a twofold meaning. It invites us to examine heterogeneous constituents, conditions, properties, and processes from which spatial atmospheres emanate. It also suggests a tracing of how the concept of atmosphere has been *formed* in architectural discourse, exploring atmosphere's challenging yet inspiring plurality as well as revealing the earlier strata of its history. To explore atmospheres in formation is, therefore, to adopt a quasi-archaeological approach and search for both material and discursive traces, illustrating the rise of an atmospheric sensibility and its translation into specific design procedures and processes, triggering new conceptions, perceptions, and experiences of space.

In search of atmospheric awareness

To begin with, then, we must recall the edifice recognized in its time as "perhaps the only building in the world in which *atmosphere* [was] perceptible".[6] The Crystal Palace was designed by the horticulture expert Sir Joseph Paxton and the civil engineer Sir Charles Fox for the Great Exhibition in London in 1851. Conceived as a vast, exteriorized interior abundant with artefacts and vegetation, in which, as the contemporary publicist Lothar Bucher pointed out: "all materiality [was] blended into the atmosphere",[7] this innovative ferro-vitreous construction became an apotheosis of both ephemerality and material exuberance. Associated with hothouses, the Crystal Palace was also the epitome of climatic manipulation. More recently, it has been identified with experience-oriented and artificially controlled environments, defining "a new aesthetic of immersion".[8] One could say that within its startling and multifaceted nature, and embedded heterogeneity of cultural and symbolic meanings, the Crystal Palace was a place in which all atmospheric categories coalesced into a single and *hybrid* atmosphere.

The anthropologist Tim Ingold states that for "architectural historians the atmosphere gets in the way of their apprehension of the forms of buildings".[9] But in 1941, the architectural historian Sigfried Giedion compared the Crystal Palace with the airy atmosphere of *Simplon Pass*, c.1848, by the Romantic painter J.M.W. Turner, who is widely known for his sentence "atmosphere is my style".[10] Giedion noted that the "insubstantial and hovering effect"[11] in Turner's painting also occurred in the Crystal Palace but "through the agency of the transparent glass surfaces and iron structural members".[12] In a similar vein, in 1907, the art historian Alfred Gotthold Meyer found "the sensuous perception"[13] of the Crystal Palace in earlier precedents such as the Versailles' Hall of Mirrors from 1684. Demarcated by glass surfaces dissolved in lustre, they both merged interior and exterior, confusing the orthodox experience of interiority.

Defined by Baudrillard as "the most effective conceivable material expression of the fundamental ambiguity of 'atmosphere'",[14] glass, used on such a vast scale in the Crystal Palace, created a spectral image described by contemporary commentators as "exceeding airiness", "fairy-like brilliancy", and "haze".[15] These tropes certainly represented a "glamour of the fogs and mists"[16] so often associated with atmosphere in architectural discourse, and ascribed by Steven Connor to the modern dream. This dream replaced static conceptions of the material universe with dynamic processes, leading to the dissolution of the architectural *object* and its fixity, and, consequently, offering a new understanding of space as a field of dynamic relationships. Such a dissolution of boundaries, seen not as physical, but rather as relational and processual, defines the design of immersions, according to the philosopher Peter Sloterdijk[17]. In that immersive field, the immaterial is not understood as a background condition, but as a productive force that shapes both architectural materializations and human behaviours.

Affective thresholds, immersive fields, and active boundaries

As if in response to Bollnow's claim, in 1948 Giedion, in *Mechanisation Takes Command,* traced a particular evolutionary timeline of the interior, guiding the reader from the medieval atmosphere of seclusion, through the erotic atmosphere of the eighteenth-century apartment, to the nineteenth and twentieth centuries, in which interiors embodied the power of mechanization[18]. Not only did Giedion point out that atmosphere "can arise from the free interplay of heterogenous elements",[19] he also identified comfort as "the atmosphere with which man surrounds himself and in which he lives".[20] In doing so, Giedion drew our attention to the fact that comfort, etymologically rooted in Latin *confortare* ("to strengthen"), has always been related to the ordering and control of the human body and its intimate surroundings[21]. To unfold his theory, Giedion used Albrecht Dürer's painting *Saint Jerome in his Study,* 1514, analysing the "dominant" atmosphere permeating the depicted interior and drawing Jerome's body into a state of stillness and contemplation[22]. Even though Giedion did not mention Baruch Spinoza, the "intensity of the room"[23] that he did refer to can be understood in a Spinozian sense. It can be seen as an intensity of affect "by which the body's power of activity is increased or diminished"[24] and turned into either action or inaction, revealing the affective dimension of atmosphere.

Awareness of such an affective dimension can be also found in the work of the architect Frank Lloyd Wright. In 1894, in his very first article, Wright defined "a set of golden rules" for the construction of a "vital atmosphere" in domestic interiors.[25] For him, atmosphere was a unifying factor between aesthetic pleasure and environmental management. It entailed a clever arrangement of material and immaterial components: spatial configuration, light, colours, furniture, and textures. Following a similar premise, in 1923, the opening of The Houston Majestic Theatre, designed by the Austrian-born and Chicago-based architect John Eberson, was seen as the emergence of a new movie palace typology defined as the "Atmospheric Theatre".[26] Deliberately chosen colour schemes, fabrics and finishes, statues and adornments, artificial plants, heating and cooling systems as well as lighting and sound effects constituted a wide pallet of atmospheric techniques. They were used to create an exteriorized interior with technologically simulated artificial weather, *transporting* the audiences to exotic realms, and revealing the atmospheric power of décor.

In parallel, in the Hannover Landesmuseum, "Atmosphere Rooms" (*Stimmungsräume*) defined an innovative curatorial strategy implemented by the museum director and art historian Alexander Dorner. Dorner's atmosphere-room principle entailed replacing the conception of a museum as a mere collection of objects with a collection of synaesthetic environments, in which spatial arrangements, visual and sound effects as well as a strategic display of artefacts created a dynamic assemblage. The culmination of this enterprise was his commission of the *Abstract Cabinet* (*Kabinett der Abstrakten*) designed in 1926 by the Constructivist El Lissitzky. The *Abstract Cabinet* was part of El Lissitzky's larger project known as "Demonstration Rooms" (*Demonstrationsräume*), which originated in the mid-1920s, blurring boundaries between spaces and artworks and translating his "optical dynamic"[27] into embodied, multisensory experiences. In his 1929 review, Giedion identified the *Abstract Cabinet* with a "fluid atmosphere", the emergence of which "depend[ed] on the hand whose touch [gave] life to the material".[28] In Constructivist imagery, materials were assigned transformative and affective agency. The conceptual premise of the *Abstract Cabinet* was to achieve choreographic effects through disrupting the wall surfaces through astigmatic effects produced by a modular lath system. The "shimmering"[29] walls defied both spatial logic and the logic of the eye, transforming space into a dynamic force-field charged with affective power and articulated through bodily engagement. Accordingly, Laszlò Moholy Nagy, also known for the creation of atmospheric machines such as the *Light-Space Modulator*, 1930, reflected on the relational nature of Constructivist space, describing the surface as "a part … of the atmospheric background".[30]

The *Abstract Cabinet* created a double immersion: the person in the space and the person in the artwork. Such an immersive loop was defined as "environment in the environment"[31] by the Italian architect and designer Franco Albini, whose museum and exhibition display designs, created from 1930s to 1960s, have been called "Atmospheric Spaces" (*Spazi Atmosferici*). Using surfaces, grids, and transparencies with Constructivist-Futurist resonances, Albini played with the illusory dissolution of limits, liberating both the space and the viewer from gravity. Like the Crystal Palace, Albini's designs were flooded with light and shadow, converting space into an almost imperceptible yet powerful medium (like the air itself), in which the visitor would "feel immersed and stimulated without noticing what [was] happening"[32] – to use Albini's words. Both Albini's slender grid constructions almost lacking in tangible substance and El Lissitzky's active surfaces with their material presence, orchestrated visitors' movements and reception of the artworks, revealing what the philosopher Tonino Griffero defines as "atmospheric authority".[33]

Awareness of this atmospheric authority was pivotal to a series of installations entitled "Immersions" (*Le Immersioni*) designed from 1967 onwards by one of the representatives of the Italian Radical Design movement Ugo La Pietra as a part of a larger project called "Spatial models of behaviour" (*modello spaziale di comportamento*). Conceived as both perceptual devices and immersive micro-environments defined by new materials, particularly methacrylate, acrylics, and resins, in combination with audio-visual technologies, La Pietra's installations provide further evidence for "the notion of atmosphere as a datum from which to pursue a recomposition of our relationships with things", [34] according to the critic Maurizio Vitta. As with many other devices developed at that time, such as Haus-Rucker Co's "Environment Transformers", La Pietra's devices served as a medium for challenging objective reality and for raising bodily awareness and sensitivity, promoting a co-production of new spatial, somatic, and collective realities.

Such an approach certainly brings to mind Situationist practices recognized by La Pietra himself as important references. It was through the Situationist International that terms such as "atmosphere", "ambiance", and "constructed situation" consolidated their position in architectural vocabulary. Building on the principles of psychogeography and the *dérive*, Situationists traced conceptual contours of "unitary urbanism",[35] which became a cornerstone for the project developed between 1956 and 1974 by the painter, sculptor and then visionary architect Constant Nieuwenhuys. *New Babylon* was conceived as a city lacking any permanent structure or form – transient, ephemeral, capable of re-arrangement and open to appropriation. It was a city dissolved into an endless network of constantly shifting atmospheres, "determined by an abundant manipulation of colour, sound, light, climate, by the use of the most varied kinds of apparatus, and by psychological procedures".[36] Similar to the project *Air Architecture*, on which the artist Yves Klein worked between 1957 and 1962 with the architects Werner Ruhnau and Claude Parent, *New Babylon* was a model for a totally new ludic society, building on the intrinsic relation between atmosphere and social and urban dynamics. While in *New Babylon*, materiality was one of the means to re-model these relationships, in *Air Architecture* it was the air and other immaterialities that were integral to spatial constitution and social organization, defining a "new sensitivity".[37] This new sensitivity, which bridged the environmental and the aesthetic, has been described as an "atmospheric revolution"[38] by the artist and architect Graham Stevens, who created a series of pneumatic installations in the 1960s called "Atmosfields". This revolution continues today.

Archaeology of the future

Despite the established assumption that the history of architecture has largely neglected its atmospheric component,[39] these and many other examples – forgotten or unexplored – reveal the existence of an entire atmospheric genealogy, indicating an alternative architectural history. Hence, they may today achieve a new relevance. With embedded discontinuities and conceptual thresholds, these works define what can be called an archaeology of the future. They become instruments for mapping the fleeting shadows of forgotten knowledge, illuminating what has remained overlooked or hidden and, at the same time, pointing towards what is yet to come.

Notes

1 Mark Wigley, "The Architecture of Atmosphere", *Constructing Atmospheres, Daidalos,* No. 68, June 1998, pp. 18–27.
2 Otto Friedrich Bollnow, *Human Space*, London, Hyphen Press, 2011, p. 220. First published in 1963.
3 Gernot Böhme, "Atmosphere as the subject matter of architecture", in P. Ursprung (ed.), *Herzog & de Meuron. Natural History*, Zurich, Canadian Center for Architecture and Lars Müller Publishers, 2005, pp. 388–406.
4 Jean Baudrillard, *The System of Objects*, London, New York, Verso, 1996, p. 26. First published in 1968.
5 Jean Baudrillard, *The System of Objects.*
6 Mary Philadelphia Merrifield, "Essay on the Harmony and Contrast of Colours as Exemplified in the Exhibition", *The Art Journal Illustrated Catalogue: The Industry of All Nations*, London, George Virtue, 1851, pp. I++-VIII++: II++, emphasis in original.
7 Quoted in Sigfried Giedion, *Space, Time and Architecture: The Growth of a New Tradition,* New York, Harvard University Press, 1982, p. 254. First published in 1941.
8 Peter Sloterdijk, "The Crystal Palace", *Public? Public. Art, Culture, Ideas, 37,* 2008, pp. 12–15: 13. First published in 2005.
9 Tim Ingold, "Lighting up the atmosphere", In M. Bille and T. Flohr Sørensen (eds.), *Elements of Architecture. Assembling Archaeology, Atmosphere and the Performance of Building Spaces*, Routledge, 2016, pp. 163–176: 163.
10 Quoted in Peter Zumthor, *Atmospheres. Architectural Environments. Surrounding Objects*, Basel, Birkhäuser Verlag, 2006, p. 5.
11 Giedion, *Space, Time and Architecture,* p. 254.
12 Giedion, *Space, Time and Architecture*, p. 254.
13 Quoted in Walter Benjamin, *The Arcades Project,* New York, Harvard University Pres, 2002, p. 541. First published in 1982.
14 Baudrillard, *The System of Objects*, p. 42.
15 These epithets appeared in *Chamber's Edinburgh Journal* from 1 March 1851 and *Ecclesiologist* from 1851. See: Armstrong Isobel, *Victorian Glassworlds. Glass Culture and the Imagination 1830–1880*, Oxford, New York, Oxford University Press, 2008, pp. 152–153.
16 Steven Connor, *The Matter of Air: Science and Art of the Ethereal,* London, Reaktion Books, 2010, p. 180.
17 Peter Sloterdijk, «Architecture As an Art of Immersion», *Interstices. Journal of Architecture and Related Arts. Unsettled Containers: Aspects of Interiority*, no. 12, 2011, pp. 105–109. First published in 2006.
18 Sigfried Giedion, *Mechanisation Takes Command: A Contribution to Anonymous History*, Minneapolis, London, University of Minnesota Press, 2013. First published in 1948.
19 Giedion, *Mechanisation Takes Command,* p. 499
20 Giedion, *Mechanisation Takes Command,* p. 301.

21 Giedion, *Mechanisation Takes Command,* p. 260.
22 Giedion, *Mechanisation Takes Command,* pp. 303–304.
23 Giedion, *Mechanisation Takes Command,* p. 303.
24 Baruch Spinoza, *Complete Works*, Cambridge, Indianapolis: Hackett Publishing Company, Inc., 2002, p. 278. First published in 1677.
25 Frank Lloyd Wright, "The Architect and The Machine", in B., B. Pfeiffer (ed.), *Collected Writings, 1984-1930*, Vol 1., New York, Rizzoli, Frank Lloyd Wright Foundation, 1992, pp. 20–26: 21. First published in 1894.
26 See, for instance: Jane Preddy, *Glamour, Glitz and Sparkle: The Deco Theatres of John Eberson*, New York, Theatre Historical Society of America, 1989.
27 El Lissitzky, «Exhibition Rooms», in S. Lissitzky-Kuppers (ed.), *El Lissitzky: Life, Letters, Texts*, New York, Thames & Hudson, 1980, pp. 365–367: 366. First published in 1923.
28 Sigfried Giedion, «Live museum», in S. Lissitzky-Kuppers (ed.), New York, Thames & Hudson, 1980, pp. 382–383:383. First published in 1929.
29 El Lissitzky, «Exhibition Rooms», p.366.
30 László Moholy-Nagy, *The New Vision and Abstract of an Artist*, New York, Wittenborn, Schultz, Inc, 1947, p. 39.
31 Federico Bucci, *Franco Albini*, Milano, Electa architettura, 2009, p. 26.
32 Federico Bucci, *Franco Albini*, Milano, Electa architettura, 2009, p. 8.
33 Tonino Griffero, "Architectural Affordances: The Atmospheric Authority of Spaces", in P. Tidwell and T. Wirkkala-Rut (eds.), *Architecture and Atmosphere*, Tapio Wirkkala-Rut Bryk Foundation, 2016, pp. 15–47.
34 Maurizio Vitta, "The point of encounter between art and design", in P. Restany and M. Vitta (eds.), *Ugo La Pietra. Monographs on Contemporary Design*, Barcelona, Editorial Gustavo Gili, 1991, pp. 9–13: 10.
35 See, for instance: Guy Debord, "Report on the Construction of Situations and on the International Situationist Tendency's Condition of Organisation and Action", in K. Knabb (ed.) *Situationist International. Anthology*, Berkeley, Bureau of Public Secrets, 1995, pp. 25–43. First published in 1957.
36 Constant, "New Babylon", in U. Conrads (ed.), *Programs and Manifestoes on 20th-Century Architecture*, Cambridge, MA, MIT Press, 1971, pp. 177–178: 178. First published in 1960.
37 Yves Klein and Werner Ruhnau, "Project of Air Architecture", in P. Noever and F. Perrin (eds.), *Yves Klein Air Architecture*, MAK Center, Hatje Cantz, 2004, p. 77. First published in 1961.
38 Stevens Graham, "Pneumatics and Atmosphere", *Architectural Design* 3, Vol XLIII, 1972, pp. 166–171.
39 See: Ingold, "Lighting up the atmosphere", p. 163.

18 Ambiance's affordances/ common architectures

Emmanuel Doutriaux

If we were to admit the failure of the notion of space and acknowledge the concomitant success of ambiance, this would yield a formidable experiential possibility for architecture. This was the refrain that led me in my doctoral research to examine "the (potential) gifts offered by ambiances, which, when sheltered by covers of a certain size, may lead to the exercise of a certain kind of commons".

Let us describe this proposition in greater detail. For starters, we must point out the vacillation between singular and plural that affects these notions – *ambiance/ambiances* and *common/commons* – because it expresses their ambivalence, their irreducible capacity for equivocation. And we will do so by describing both their conceptual fragility and their powerful heuristics.

By summoning such *covers of a certain size* (which are in reality rather "large") and such *commons of a certain kind* (which are in fact rather "diffuse"), we should also point out a dual indeterminacy, both in terms of programmes (in which ambiances signal their performance) and projects (which host them).

"Potential gifts of ambiances" means the "potential gifts through which ambiances are brought about and of which they are a vector". Thus, the dispositive structure of a certain type of open *programme and project* might actively lead users to arbitrate between these possibilities and to trigger the ambiances for which they feel a need at a given point in time. In an era of the most obvious architectural affordances, the most unexpected appropriations by users would be accepted.

By "covers of a certain size", I refer to the now recurrent emergence – in these times when what we call the *public space* is in crisis – of "sites for the public" that, even for being covered, are overdetermined because of their size and even by their form, with a nevertheless uncertain designation, an often complex legal status, and a no less ambivalent figural presence, through which they elude the traditional dichotomy of public/private and allow for the emergence of the *common*.

Ambiance's affordances and common architectures. To deal with the tension between the free expression of ambiances and the inventive socialization that such architectures allow, I focused on two "constructed situations"

DOI: 10.4324/9781032655307-26

characterized by the indeterminacy of higher education programmes: the Nantes School of Architecture and the Learning Centre at the Lausanne Polytechnic.

Poetics and politics of the architectures of ambiance

Let us define, point by point, the argument that leads from ambiances to commons.

What is ambiance?

Ambiance? "A notion with a wide embrace", exclaimed a sceptic in a paper[1] that denied ambiance any scientific classification due to its suspension between various fields, weaknesses in its theorization and practical application, and temporalities in its design and "reception".

This semantic, disciplinary, experiential, and productive openness is undoubtedly the reason for the success of this notion, and why it touches on fields that are becoming increasingly open. It embraces – the Latin word *ambire* shaped it – generously. Unless, that is, this difficulty of determination is consonant with epistemological inconsistency, if we are to consider the trial for being (un)scientific that has been brought against it. With the major nuance that ambiance probably forms part of our new age of knowledge and experience, which has rubbed up against the sciences of the imprecise [Abraham Moles] and non-linear epistemes, and escaping – if needs be, and if only through inductive approaches and trial-and-error protocols – from the sole law of hypothetical-deductive reasoning.

Grégoire Chelkoff coined the concise formula[2] that sums up operativity of the notion, the content of which was first sketched out by Jean-François Augoyard, and which continues to be developed by Jean-Paul Thibaud. Thus, ambiance can be *forms* that take built systems and the fabrication of their physical surroundings into account, and it can be distinguished by its measurability. It may consist of *formants*, pertaining to a sensorial perception of the environment, of which descriptive language can give an account. It can be *formalities*, or rather, those conducts, actions, relations that can be observed in the context shared by the very spectators who produce them.

This definition has been criticized as reifying, and its primary assignment to form is indeed questionable. But beyond this critical point, let us consider the idea of a weighing scale with three modes that are rigorously articulated on the basis of these "prescriptive areas".

The thesis has given shape to my own tabular inscriptions by dealing (instead of "forms") with "conscious formants" (for the designing actors) as a configuration and modulation of physical data that places the various ambiances, one by one, on the table of *poïesis*; by dealing with "semi-conscious" or "perceptive formants" (for the receiving subject), which is the place or moment of the "Ambiance's" recollection, as part of an *aisthesis* that engages with a comprehensive, sensory lived experience; and, lastly, by dealing with "relational formants" (as "formalities" engender too much confusion), which are the mechanisms of the *capture* and effect of their modulation by the users/actors who are partners in the ambiance's transformation. I see in it a *politikè* in the making, and I have seen in it the promise of the *common*.

But this triptych also expresses the intertwining of three temporal, scientific, and experiential infrastructures: time – perception/design/reception – a spiralling, looped, or untimely time in which lived experience and designed experience constantly call out to one another; disciplines – architecture, arts, sociology, anthropology, psychology, aesthetics – and "political art", in which the respective skills and competencies of the actors come into play; and performances – shared between the experimentations of seasoned experts/designers and the experiences authenticating their users.

What does ambiance do to architecture?

The inclusion of this notion within the research circles of the "soft" sciences oddly still seems to be confined largely to urban territories and scales. And yet, aren't the closed spaces of a building, or rather, the semi-closed spaces of a built situation (more) able to *create ambiance*? If we accept that ambiance is an artefact, this is the result of an artificialization.

Ambiance exists when there is a project – and a milieu, insofar as it concerns life. That is the difference I draw between milieu and ambiance, which some may dispute as being perhaps too anthropocentric. If all living things *result from* and *contribute to* this milieu which is their own (even if unconsciously), we need *a disposition to* (feel and perceive), *resolution to* (react), and, if applicable, the *will to act* to name this ambiance that acts on us.

Clearly, it forms part of a flow of consciousness that varies in its degrees. This allows us to distinguish between the milieu of the tick in Uexküll's underbrush and the ambiance that we contribute to individually and together, which forms in a given room where we find ourselves gathered on a given day. Architecture is especially able to summon and host these degrees of intent and flows of consciousness.

That having been said, I will now measure the risk: in the folds of this notion to which the acts and time of the designer and those of the "perceiver/user/interpreter" wish to attach themselves at any cost, a risk insinuates itself of obscuring the difference between a position of decision and responsibility, and one succumbing to indecision, or rather, to "irresponsibility". But isn't the user's relationship to, and actions on, the ambiance just as crucial for qualifying it? There is no shortage of real-life experiences that prove this.

Ambiance is not space

Ambiance provides a remarkable heuristic wealth of opportunities for discussing architecture, starting with its capacity to thwart the modernist doxy of space. If it is true that *we have never been modern…* it would appear that architecture could never have been space. A reading of Bruno Latour would have been seminal in this instance.[3]

Several intertwining branches were bundled together in this quest to view architecture as being caged within space. I will limit myself to listing them due to a lack of "space" to describe their sequence in any detail.

Let us begin with the dominance of Brunelleschi's mirrors; the optical codification of the Renaissance needed space to occur.

Let us then refer to the Newtonian faith in an "ether-space": an atomistic conception of an ether as a granular structure of extremely small particles between which lies a void.

We should recall the structural codification of ancient times (as reread in the eighteenth century by Durand) and the Gothic age (as envisaged in the nineteenth century by Labrouste and then Perret); what was there to envisage aside from a structure (and its envelope), if not a pure spacing?

And we have the prism of Hegelian idealism: the philosopher described the Gothic church as "the concentration of the essential soul-life enclosed within a spatial relationship".

"Modern" thinking supposedly became purer the more that space was described in a limpid manner. Hegel invented this in the bays of cathedrals, as Peter Collins has shown.[4] He was the first to claim this in architectural terms. This was the approach that Wölfflin then took to describe Wright's spatialism. Collins also shows how Giedion dismissed Einsteinian relativity to furnish the appearance of a scientific basis to his conception of architectural space and time.

Finally, let us consider the Foucauldian empire of *societies of discipline*: Hygienism expressed the new power of "making things live" and "letting things die" as opposed to "making things die or letting them live".

In the stratification that gradually formed over the centuries for all these intersecting reasons, no [Foucauldian] panopticon appeared possible without optical effectiveness, without a visual exclusivity. No modernity was conceivable in all its mad abstraction without this spatial mechanism.

I want to point out how much this space *had to be* written – under the empire of modern doxy – to combine visibility and distance, reason and clarity, and abstraction and power.

I have described how long it took architecture to bring forth the notion (from the late nineteenth to the mid-twentieth centuries) from within these consecutive and concurrent histories. And this was the case for the long maturation of the discipline's own pedagogy, which took until 1893 to formulate space as "the essence of architectural creation",[5] despite the fact that modernity had slowly been building itself up for the last four centuries. And it took all that time for the notion, once edified, to reveal its own limits.

When the time of this modern empire had passed, if the space was obsolete, if it was ineffective, or rigorously empty, and as such uninhabitable, ambiance seemed "clearly" desirable. Not to substitute one term for the other – this would have recreated a *tabula rasa* and renewed an impetus for purification – instead to breathe life into a notion in its transience, its capacity to "embrace widely", and to understand itself in terms of its own *lack of definition*, *purity*, and *discipline*.

The choice of air

Air thus presented itself as the lowest common denominator for ambiances that elude space. Air is *invisible* (but essential) *by nature*... Let this be a methodological and epistemological argument.

This choice was probably also construed as the primordial (primary) element for our existence, as the atmosphere is the space in which living things breathe. I espy a phenomenological argument in that.

Lastly, this choice was probably envisioned because air is a central agent in the issue of climate. This takes the ecological argument into account, which is having a profound effect on contemporary production.

In reality, air is not just a matter of technical conditions. Aside from all technical considerations, breathing is indeed an action that involves "being in the world", to use a term of Merleau-Ponty's. Air is anything but neutral. On the contrary, it is loaded, variable, adjustable, "mouldable", as ambiance itself should be. Air is a project for inhabiting. This counteracts, following Irigaray, Heidegger's earth-based, telluric, tectonic position:

> But is not [man's] being a kind of living insensibly that is passed from one air to another? Fluid matter, voice, appearance. The possibility of breathing-living, the possibility of calling-naming, the possibility of appearing-becoming present. Heidegger does not recall this passage. He forgot about the difference in air(s). What stands in for this oversight? A certain void.[6]

But we must also agree that choosing air, and sticking with it, expresses a reservation with regard to the invasive, over-determining, and even demiurgic potential of ambiance. Instead of architectures *of ambiance,* in which ambiance appears expertly orchestrated in a score that has already been written, prescribed ahead of time, and staged (as is true for Peter Zumthor's work, which represents a reinvention of the experience of the *promenade architecturale* that is as precise as its model was deterministic), I believe it is more fruitful to think of architectures *by ambiance*, in which factors are placed at the disposal of the public (the users), who are responsible for appropriating and activating them (as in Lacaton & Vassal's proposals, which made inhabitants the operators of their own climates).

This ambiance *of discretion/by discretion* (to echo the subtle and potentially disturbing characterizations made by Thibaud in Volos[7]) – in which ambiance, as long as it remains discreet, anisotropic, and adjustable, plays out primarily in the notion of "creating climate" on the basis of the "air conditions" that it mobilizes – seems to be in harmony with a "democratizing" and democratized approach to architecture.

In its "before" phase, it cancels out what we had in mind: typology – composition – staging. From the "after" phase, it brings out questions concerning usage: upkeep – maintenance – (the possibility of) transformation. Such an ambiance is in fact characterized by its "adjustability". It is available. Or rather, architecture declares itself to be a mechanism for mobilising and adjusting ambiances. This leads to the emergence of the hypothesis, wager, choice of ambiances that are "weak" (that can be activated, modulated, and reversed) and which do not impose themselves in terms of usage.

In this sense of empowering ambiances for action, I thought it would be fruitful to look at the relationship between affordances and appropriations: at architectural affordances, starting with their capacity *to give an aeraulic foothold* for user skills and actions, and at the appropriations caused by such affordances, which can be evaluated in terms of their use. We will return to this in "the test".

Ambiance of the commons/of the common

Ambiance's *paradigm of discretion* – an ambiance at one's discretion; the *paradigm of air conditions* – conditions made for air, at least in the form of conditions that this air would impose on us; the *interplay of affordances and appropriations* – contributes to a political qualification of ambiance. *Techniké, poïesis,* and *politikè* are all connected.

Such a general context would have led me to "happen upon" two very particular "constructed situations" emanating from very different *topoi* in terms of the design process and programmes, and on a technical and aesthetic level. And both of them would appear as functioning more or less explicitly as a palliative, a refuge, a recourse, an alternative, or even an antidote to the public space in all its redeployments [Dacheux], inasmuch as this would be a centralized, authoritarian government *delegating* the collective.

As a theatre play about the indeterminacy of "living together", this would form an unprecedented and fruitful articulation between ambiances (above all, these air conditions), affordances (and with them, their appropriations), and practices of the common, because this common is more diffuse (as humble relational practices, like those that happen spontaneously in the street) than a formal or declarative qualification (as an explicit, established principle [Dardot & Laval]), or an insurrectionary/oppositional one (as militant actions arising from practice [Nicolas-Le Strat]).

The common is thus considered here as something that questions, shifts, and transforms the disputed reality of the public space and subsumes the chiasmus of public and private. To simplify things, we could say that the public is under everyone's control (whose, though, ultimately?), the private, under the control of one sole person, and that the common transcends this dichotomy by distinguishing between ownership and use (and the right to use) and by re-assessing the capacity of the relevant users to govern its functioning [Coriat].

***

Ambiance *vs* space.

Let an architecture be freed of its physician's complex, of the typological and tectonic prism within which it was imprisoned, which evolves according to a regime of flows within the topological. Air is included as the lowest common denominator for ambiance (in terms of its phenomenology and its production) and as the relationship between ambiance and climate, in the full technical and aesthetic, or techno-aesthetic sense [Simondon]. In this qualification that is both weak and rich, envisaged as the release for the common, ambiance *belonging to/subject to* becomes an expression of the political.

The test

Let us come to the test. Among these constructed situations we have the Nantes School of Architecture and the Learning Centre in Lausanne, so different in aspect, form, figure, and status. Yet they bathe in the same programme waters, reinventing, in the spirit of "learning", the conditions of the existence of higher education. Each one assesses the results of these modes of learning on the design of common spaces, their democracy of usage, and their air conditions.

Let there be no mistake. This is not a question of pitting productions and doctrines that appear so dissimilar against one another, nor of overusing dialectical argument. Rather, it is a question of suggesting how a broad spectrum of experimentation can be opened up between such centres as embodied by these two experiences that resemble manifestos.

Please refer to the visual notebook on these experiences that is attached to this essay.

Conclusion

While it is clear that ordinary architectural experiences emanate outward towards exceptional situations, this influence also occurs in the reverse direction, through the folly of their lack of limits (the "excess" in their size, the indeterminacy of their programmes, and the informality of their projects) and their hybridizations (above and beyond the dichotomies of public/private and interior/exterior) – and which their liberalities sanction in the totally different registers of these two operations.

The agricultural shed abuts the farm, the market hall adjoins the town hall, and the greenhouse is built next to a house: "architecture as a factory" is made up of this history, but also of previous cultural factories. This is a form of shared domesticity at work, the potential of which the Nantes School of Architecture is exploring and developing.

Whether it's the bushy identities of familiar-looking parks, the disquieting indoor campsites of No Stop City, or the spectral figures of Ukiyo-e, a number of evocations are leading the notion of "architecture as park" in Lausanne to lend itself to a new topological experience – by inventing the climate suited to the undertaking.

An architectural qualification confined, more or less, to the treatment of a floor, seen as the ultimate surface for resolution.[8] A floor, a sky. An air mattress. The common?

Visual workbook

A metaphor of praxis

Nantes Architecture School (Lacaton & Vassal Architects)

In describing the rhetorical, semantic, and hermeneutic levels of metaphor, Paul Ricœur showed how metaphor contains a discursive strategy that preserves and develops the heuristic power of fiction. This quite literally involves "allowing for a sense of play" (Figures 18.1–18.3).

Figure 18.1 The "factory", here Summerlab 2014.

Source: Ping – Creative Commons BY-NC-SA licences.

Figure 18.2 The "forum": A workshop at the start of the schoolyear for first-year students, 2016, directed by Emmanuelle Huynh.

Source: Photo by Aanor Rodrigues.

Figure 18.3 In Nantes, large sliding doors and tall windows provide clear views of the sky.

Source: Photo by Emmanuel Doutriaux.

An allegory of the relational

Learning Centre at the École Polytechnique Fédérale de Lausanne (Sanaa Architects)

Kazuo Sejima said: "A park lets different types of people stay in a space at the same time. Moreover, in a park, you can join a large group, but at the same time, someone can stand alone next to you, reading a book or just drinking a fruit juice. I love this feeling, this character for [defining] public buildings".

The allegory of the relational, used by Sanaa, based on the model of the park, is based on a fictional spirit comparable to that of the metaphor used in Nantes (Figures 18.4, 18.5).

In Lausanne, new air under the carpeting (Figure 18.6, *Photo by Sorane*), while exhaust systems pull out the bad air (Figure 18.7, *Photo by Emmanuel Doutriaux*)

Figure 18.4 Example of the experience of "being alone together" in this "intimate public space".

Source: Photo by Yafiza Zorro.

Figure 18.5 Random relationships, working/not working, outside/beneath, outside/outside, inside/outside, on sliding surfaces.

Source: Photo by Yafiza Zorro.

Figure 18.6 New air under the carpeting.

Source: Photo by Sorane.

Figure 18.7 Exhaust systems pull out the bad air.

Source: Photo by Emmanuel Doutriaux.

Notes

1 Olivier Chadoin, "La notion d'ambiance" ["The Notion of Ambiance"], in *Les annales de la recherche urbaine* ["Annals of Urban Research"], No. 106, PUCA, (July 2010): 153–159.

2 Grégoire Chelkoff, "Formes, formants et formalités" ["Forms, formants, and formalities"], in *L'espace urbain en méthodes* ["The Methods of Urban Space"], Jean-Paul Thibaud and Michèle Grosjean (ed.), (Marseille: Parenthèses, 1997), 103–108.

3 Latour thus refers to Vogt's brilliant analysis of Le Corbusier to show the obsessive recurrence in the architect's works of the model of "the neolithic fishermen's houses on stilts on Lake Neuchâtel whose ruins he had visited as a child". This showed a modern schizophrenia unable to reconcile practice and theoretical positions. Bruno Latour: "Qu'est-ce qu'un style non-moderne?" [What is a Non-Modern Style"], in *La parenthèse du moderne* ["The parenthesis of the modern"],

Records of the conference held on 21–22 May 2004, Catherine Grenier (ed.), (Paris: Editions du Centre Pompidou, 2005), 31–46.

4 Peter Collins, *Changing Ideals in Modern Architecture* (London: Faber and Faber, 1965).

5 August Schmarsow, "L'essence de la création architecturale" ["The Essence of Architectural Creation"], in *L'espace du jeu architectural, mélanges offerts à Jean Castex* ["The Space of Architectural Play: mixtures in honour of Jean Castex"] (Versailles: Ed recherches, ENSA Versailles, 2007).

6 Luce Irigaray, *The Forgetting of Air in Martin Heidegger*, translated by Mary Beth Mader (Austin: University of Texas Press, 1999 (originally published in 1983)).

7 Jean-Paul Thibaud, "Les puissances de l'ambiance" ["The Powers of Ambiance"], in Nicolas Rémy et Nicolas Tixier (ed.), *Ambiances demain. Actes 3è congrès international sur les ambiances* ['Ambiances Tomorrow: Acts of the 3rd International Conference on Ambiances, International network of ambiances & University of Thessaly, Volos (2016): 689–694. Unsettling, discreet atmospheres when they are the result of an invisible hold; "liberal", discreet atmospheres when there is a *hold* for the user activating them. I will allow myself this distinction, thus challenging Jean-Paul Thibaud.

8 Hugues Fontenas, "Le temps improbable de la construction" ["Improbable Construction Time"], in Finn Geipel, Jac Fol, and Nicolas Michelin (ed.), *Labfac, Finn Geipel, Nicolas Michelin Laboratory for Architecture*, (Paris: Centre Pompidou, 1998), 97–108.

Bibliography

Chadoin, Olivier. "La notion d'ambiance". In *Les annales de la recherche urbaine*, Vol. No. 106, (July 2010).

Chelkoff, Grégoire. "Formes, formants et formalités". In *L'espace urbain en méthodes*, edited by Jean-Paul Thibaud and Michèle Grosjean. Marseille: Parenthèses, 1997.

Collins, Peter. *Changing Ideals in Modern Architecture*. London: Faber and Faber, 1965.

Coriat, Benjamin, ed. *Le retour des communs*. Paris: Les Liens qui libèrent, 2015.

Dacheux, Eric. "Un nouveau regard sur l'espace public et la crise démocratique". *Hermès, La Revue* 2003/2 (No. 36) (2003): 195–204.

Dardot, Pierre and Christian Laval. *Commun, essai sur la révolution au XXIème siècle*. Paris: La Découverte, 2014.

Deleuze, Gilles. *Negotiations* 1972-1990, translated by Martin Joughin. New York: Columbia University Press, 1997.

Doutriaux, Emmanuel. *Conditions d'air. Politique des architectures par l'ambiance*. Genève: Métis Presses, 2020.

Fontenas, Hugues. "Le temps improbable de la construction". In *Labfac, Finn Geipel,* Nicolas Michelin Laboratory for Architecture, edited by Finn Geipel, Jac Fol, and Nicolas Michelin. Paris, Centre Pompidou, 1998.

Foucault, Michel. *Discipline and Punish: The Birth of the Prison*, translated by Alan Sheridan. London: Penguin, 2020.

Gibson, James J. *The Ecological Approach to Visual Perception*. Boston: Houghton Mifflin, 1979.

Hegel, Georg Wilhelm Friedrich. *Introductory Lectures on Aesthetics*, translated by Bernard Bosanquet. London: Penguin Classics, 2004 (originally published in 1832).
Irigaray, Luce. *The Forgetting of Air in Martin Heidegger*, translated by Mary Beth Mader. Austin: University of Texas Press, 1999 (originally published in 1983).
Latour, Bruno. *Nous n'avons jamais été modernes, Essai d'anthropologie symétrique.* Paris: La découverte, 1991.
Latour, Bruno. "Qu'est-ce qu'un style non-moderne?" In *La parenthèse du modern*, Records of the conference held on 21-22 May 2004, edited by Catherine Grenier. Paris: Editions du Centre Pompidou, 2005.
Moles, Abraham. *Les sciences de l'imprécis.* Paris: Seuil, 1990.
Newton, Isaac. *The Mathematical Principles of Natural Philosophy,* translated by Andew Motte. Oxford: Oxford, 1729.
Newton, Isaac. *Opticks.* Amherst, NY: Prometheus, 2003 (originally published in 1704).
Nicolas-Le Strat, Pascal. *Le travail du commun.* Paris: Éditions du commun, 2016.
Schmarsow. August. "L'essence de la création architecturale". In *L'espace du jeu architectural, mélanges offerts à Jean Castex.* Versailles: Editions recherches, ENSA Versailles, 2007.
Simondon, Georges. *Du mode d'existence des objets techniques.* Paris: Aubier, 1958 (republished in 1989).
Thibaud, Jean-Paul. "Les puissances de l'ambiance". In *Ambiances demain. Actes 3è congrès international sur les ambiances,* edited by Nicolas Rémy et Nicolas Tixier. Volos: International network of ambiances & University of Thessaly, 2016.
Vassal, Jean-Philippe. "Ambiances et climats". In *Climats.* Paris: Infolio, 2012.
Zumthor, Peter. *Atmospheres: Architectural Environments. Surrounding Objects.* Basel: Birkhäuser, 2006.

19 In pursuit of the aesthetics of atmospheres

With and beyond Böhme

Céline Bonicco-Donato

The notion of "ambiance" that arose with the pioneering work of Jean-François Augoyard[1] exhibits numerous similarities with the concept of "atmosphere" elaborated jointly by Gernot Böhme.[2] At the same time, atmosphere forms an explicit part of a certain tradition in German philosophy that helps explain a number of its facets in a fairly original manner. In fact, Böhme developed his concept by intertwining two traditions often posited as antagonistic to each other: phenomenology and critical theory. Perusing these works shows the capacity of ambiances, understood in the broadest sense, to test the philosophy of aesthetics and how it envisions architecture. However, architecture is not just a privileged object for the aesthetics of atmospheres; it also serves as a point of resistance that should incite the further development of such analyses. In fact, spatial design addresses the issue of atmospheres on a daily basis, less as a concept for mapping out part of the world than as a phenomenological reality. But when it comes time to create projects, Böhme's *poeitic* use of the concept runs into a series of difficulties that forces it to shift, thereby revealing several insufficiencies to his formulations.

This contribution does not place itself up above and looking downward, from a perspective where philosophy would exhaust architecture. It instead wishes to create a dialogue between them, insofar as the singularities of architectural design ask us to continue Böhme's theoretical work. Even though the aesthetics of atmospheres is formalized to its fullest extent in his work, it is not a set archive, rather a space for diffraction that calls for further exploration and therefore applies more broadly to thinking about ambiances. Hence, we will go with Böhme and then beyond Böhme.

The archaeology of the notion of atmosphere for Gernot Böhme[3]

Böhme is responsible for enshrining the concept of atmosphere in philosophy by establishing an aesthetics for it. This had begun in the 1960s at the initiative of Hermann Schmitz, who looked at inhabited space and atmospheres in his ten-volume work *System der Philosophie*,[4] but it wasn't until the publication of Böhme's decisive text in 1995, *Atmosphäre: Essays zur neuen Esthetik,* that the concept became the main focus of a work instead of merely a peripheral

DOI: 10.4324/9781032655307-27

notion invoked in relation to other arguments. It contributed to the thinking and production of various architects, including Peter Zumthor and Juhani Pallasmaa,[5] and it also fostered a theoretical discussion that has extended past the realm of German philosophy, as shown by the works of Tonino Griffero[6] and Richard Shusterman.[7] Its archaeology is fascinating, as it is the heir to two very different traditions: phenomenology on one side and Benjamin's critical theory on the other, both of which shed light on its main attributes.

An unclear ontology inherited from Heideggerian phenomenology

For Böhme, atmosphere signifies the original union of a space and its experience.[8] It is the original intertwining of an objective physical reality whose various parameters and affective tonality one may be able to measure. So, when we speak of the melancholy atmosphere of the Galerie Vivienne or the euphoric one of the Centre Beaubourg, we are talking about spaces not just in terms of how they are perceived, but also how they are felt. Thus, atmosphere can be defined as either an affective space or a spatialized affect, it is understood that the two terms that make up this definition are first intertwined in the act of perception; therefore, their linguistic separation is phenomenologically arbitrary. We may thus speak of a dual synthesis, as long as we specify that this is not formed of separate elements, instead primary ones; we have, on the one hand, a fusion of the different sensations of space that are initially unified into a comprehensive experience of the place, and, on the other, of this sensation and a way of feeling oneself to be oneself.

With regard to the first synthesis, the immediate experience of a space appears to be *one* general experience in which one does not perceive isolated "sense data", instead a union of visual, acoustic, thermal, kinaesthetic, and other sensations. We can use the term "synaesthesia",[9] and the term "diffuse" (*unbestimmt*), often invoked by Böhme to describe this background tonality from which precise sensations then spring forth, seems particularly appropriate.

The second synthesis describes the fact that, when we are taken with the atmosphere of a space we have entered, we are neither neutral nor distant, instead "intoned" by it. In this phenomenon of affective impregnation, the pathic charge of a place resonates within us and attunes us to it. In other words, we cannot find ourselves within a space without finding ourselves in this or that way, as even the polysemy of the German *sich befinden* indicates.[10]

In relation to these two syntheses, atmosphere lies at the intersection of the objective and the subjective. It emanates from an objective space for being perceived *as a sensory whole* and for being felt immediately. This inspired Böhme to propose a new ontological category, the in-between (*zwischen*),[11] the existential implications of which are particularly fascinating. In effect, the experience of an atmosphere includes that of our inclusion in the world; our subjective sphere, far from being hermetic, is ultimately porous to its surroundings.

This initial characteristic of atmosphere is the heir of a previous notion developed by Martin Heidegger in *Being and Time* (1927), *Stimmung*,[12], which is translated into English as "attunement" or "mood". Böhme emphasizes that "Heidegger's masterful analysis highlights the reciprocal involvement of perception, the experience of existence, and the experience of the world".[13] In German, *Stimmung* means someone's mood, the feeling of a place, and being musically in tune. Its examination feeds the Heideggerian critique of substantive understandings of subjectivity. It enables the philosopher to show that, far from being a self-referential monad, *Dasein* communes with its surroundings. Thus, the famous "Being-in-the-world" is a marker of its dependence on, and participation in, a milieu, thus nullifying any overly rigid distinctions between interiority and exteriority.

An epistemology of the flesh to extend the theories of empathy and the new phenomenology

Böhme believes that one accesses an atmosphere through *an experience of a sensory copresence* with the space,[14] and not through a judgement. Therefore, we must consider the field of aesthetics from the vantage point of sensory experience and one of its very particular modes: immersion. This resuscitates the opposition between Baumgarten[15] and Kant, by thinking about the aesthetic subject starting with one's own body, not one's intellectual faculties. Atmosphere is a somatic experience, such that we can identify the "space of our bodily presence" and "atmosphere".[16] Thus, atmosphere is the way in which space, present to one's body, renders one present to oneself as an affected body.

This epistemology of the flesh extends the theories of empathy[17] from the early twentieth century and Schmitz's new phenomenology.[18] Böhme relies on the organic metaphor applied by Heinrich Wölfflin to architecture in his *Prolegomena to a Psychology of Architecture* (1898)[19] to describe the union of spatial perception and spatial experience. According to Wölfflin, because architecture is itself a body, we can say that its forms are corporal. Precisely for the qualities that architecture shares with our bodies, these forms resonate affectively within our bodies. In fact, Wölfflin believed that we are our bodies more than their being something we possess, such that the dynamics that work their way through our bodily spatiality (tightening, dilation, etc.) are also affective motions. When we perceive architectural dynamics, we experience emotions tied to body movements that resemble them. Thus, spatial perception is not "a matter of the eye": it instead "expresses itself as a direct bodily sensation".[20]

Schmitz's perspective inspired Böhme to delve further into this epistemology of the flesh to describe the ways in which we experience atmosphere. By developing a new phenomenology that "seeks to enable us to conceptualize involuntarily felt experience",[21] Schmitz accorded a central role to the experience of the body, to everything that "we can feel ourselves to be within the realm of our own body, without using its five senses or perceptive corporal framework".[22]

A critical horizon in Walter Benjamin's wake

For Böhme, atmosphere as a notion for describing our experience of space is both descriptive and critical. It allows us to assess built spaces on the basis of how they are experienced, and not just in terms of their intrinsic, formal qualities. This aspect is present already in Böhme's first writings. In 1995, he asserted that "the aesthetics of atmospheres entails a crucial critical duty"[23]; in fact, atmospheres possess an undeniable power to rattle the subjects who perceive them, affecting them and revealing the subject to him- or herself according to a certain mode.[24] Atmospheres' power of subjectivation may thus appear as a subjection, which allowed Böhme to emphasize the political aspects of his aesthetic philosophy; an examination of atmospheres reveals insidious power mechanisms, as certain spatial designs, in affecting the body's space, ultimately affect its conduct. By articulating *poïesis* and *aisthesis*, he thereby looked at the aesthetic workings that drape various spaces – even virtual ones – with an atmosphere that increases their exchange value as a scenic value. In so doing, Böhme contributed to the debate over "the aestheticization of the world of life".[25]

This third characteristic of the notion of atmosphere relies on two concepts developed by Walter Benjamin to which Böhme explicitly refers: the definition of aura in *The Work of Art in the Age of Mechanical Reproduction* as the unique apparition of the faraway, "however close it may be",[26] and at the very end of this essay, where Benjamin examines as the "aestheticization of politics",[27] how totalitarian regimes rely paradoxically on this technique to exacerbate the aura of certain artistic forms that it nevertheless desecrates. The emphasis given to the archaic and irrational elements of aura and their potential instrumentalization underlies Böhme's condemnation of the aesthetic workings of contemporary capitalism.

Atmosphere as the horizon of a dialogue between aesthetic philosophy, architectural theory, and architectural design

Böhme's aesthetics of atmospheres describes an original understanding of the aesthetic experience and rehabilitates architecture within the classification of the arts. It signals an experience that is not distant, in which the subjective component is involved through his/her emotions, which constitute a way of feeling oneself to be oneself. In a break from the Kantian tradition of disinterested pleasure,[28] it accepts utility on the condition that it can be understood in existential terms. Architecture allows us to feel a presence of the self and the world, a presence of the self through its presence in the world, and through the presence of the world to the self. In this sense, we could talk of an infra-subjectivation to designate this mode of the self's affirmation as emanating from an initial gap. Furthermore, as the body experiences the nakedness of its own existence, it is in motion; it is not immobile. Far from settling into a face-to-face relationship, it moves in an immersive manner. Lastly, its attention span is distracted; although its fluttering attention may at times settle on certain details, it is primarily characterized by its receptivity to weak signals and sensory intervals, insofar as the process of perception gathers the grasped space as a whole. From this perspective, architecture is rehabilitated in aesthetic terms, both for its utilitarian and material aspects. But for however inspiring Böhme's thinking is, it calls for further development.

Expanding architectural atmosphere past the limits of the building

In his attempt to differentiate architecture from the visual arts, Böhme forcefully and meticulously examines architectural atmospheres within buildings. In so doing, he insists on the fact that architecture configures spaces that one must penetrate[29] to experience them in atmospheric terms. However, such an identification of architecture with the building may be reductive. In fact, as shown by Frank Lloyd Wright's famous criticism of the closed box,[30] a well-designed architectural space is a *relational* space that exceeds what the building's facades contain. It reconfigures the space of the site in the broadest sense, in both tangible and intangible terms. The articulation of form, interior layout, and natural environment in his Prairie Houses serve as an example of this. We should remember that an architecture's atmosphere is certainly the result of the built space *insofar* as it spatializes its surroundings. The atmospheric experience of the Church of Saint-Pierre d'Extravache in the Haute-Maurienne Valley in Savoy arises from one's perception of the peaks of La Vanoise and their slopes, the scent of the lady's slippers, and the coolness of the breeze on one's skin, among many other sensations.

This point enriches the aesthetics of atmospheres by not limiting it to a simple analysis of sensory reception, as an etymology of the term on which Böhme mainly insists would appear to dictate. It can incorporate *the study of the pleasure of perception*, an essential aspect of aesthetic experience understood properly,[31] which would, thus, enable an assessment of different architectural atmospheres. I believe that the perception of an architecture is experienced not just affectively but also happily when the building's atmosphere, which integrates with that of the site that situates it, *suits* it. The aesthetics of atmospheres calls for a reactivation of the traditional notion of conformity as it appeared in Stoicism.[32] In addition to the alphabet of atmospheres proposed by Böhme in line with Schmitz,[33] we should accordingly develop a typology of potential modalizations for this conformity, from a mimetic reprise to a rupture, by way of a reinterpretation and its many facets.

Such an effort involves architectural design as well as theory, as it asks us to reassess the notion of ornamentation. Unlike the arbitrary decoration denounced by Loos,[34] we may in fact distinguish it by referring to the Greek etymology of the term referred to by Semper,[35] as supplements expressing the relationship between a built form and its surroundings, thereby contributing to the emergence of *one* architectural atmosphere that emanates from the building and its way of spatializing the site.

Envisaging bodily copresence as a rhytmicization of the body

While Schmitz pays close attention to architecture as an incitement to movement,[36] Böhme does not specify the exact nature of the correspondence between the architectural corpus and our own bodies. Can we go past an observation of this relationship to describe the fact that the simple perception of architecture's spatial dynamics echoes those of our own bodies and elicits the related affects? His notion of bodily copresence should be specified by paying closer attention to the *dynamic* of this understanding. Here as well, architectural theory may provide an incentive to this end.

If "perception entails being fully, personally committed in the world and being in touch with what one perceives",[37] I believe that the singularity of architectural perception resides in its *duration*, which implies a measure of moving about. It is neither instantaneous nor static, instead forming over time through the person's movements. In this regard, I disagree with Böhme[38] when he considers it merely to be "an art of space". Its points of convergence with other time-based arts can be understood because of its dimensions and because of our utilitarian relationship to it. Le Corbusier wrote, "Architecture is a not a synchronous phenomenon, rather a successive one consisting of performances that are added to one another".[39] We should follow Claude Parent when he exclaimed:

> How many times have we said that mankind is vertical (gravity) and frontal (aesthetics)? And how many times must we answer that mankind only acts in a state of permanent imbalance, in projection of his/her own weight, in a condition of instability? [...] Always in motion, mankind breaks the state of stability at every turn.[40]

This allows us to characterize the perception of architectural space as rhythmic. As we grasp its dynamics, it takes hold of our bodies, through a process in which we cannot feel a rhythm without ourselves being rhythmic. And it is moving because of the link between the rhythmic component of the movements that run through us and their affective tonality. This point allows us to add a nomenclature of various architectural processes to the alphabet of atmospheres that carefully considers how a building's rhythms exist in sync with those of its surroundings, how they become imprinted in our bodies' movements, and how they elicit our emotions. I hypothesize that this yields a perspective on architectural styles (baroque, classical, modern, etc.) based on how our bodies *receive* them.

From the aesthetics of atmospheres to their poïesis

Thus enriched by architectural theory, the aesthetics of atmospheres can help nurture the design process. These possibilities point to a set of exercises that can be developed throughout a project. To make just a few examples: spending a lot of time walking around the site so that it sinks into you; observing how people move across it to understand its rhythms, representing it in the form of sketches to capture its atmosphere, on the assumption that the unfinished nature of a quick drawing cements an overall perception that remains in movement, that is not denatured, and, thus, thinks about it without betraying its infra-discursive status; building sensory models to explore the atmospheres to be brought forth, as well as their potential objective substrata; imagining different potential pathways by viewing the building as a generator of its own space-time continuum, according to the rhythmic analysis proposed by Maldiney[41]; and thinking about the affective tonalities in terms of music and poetry – these among many.

Emerging at the intersection of phenomenology and critical theory, the notion of atmosphere is very fertile in both aesthetic and poietic terms. It is, thus, of interest to both philosophers and architects. To philosophers because it gives them the means to think about how space offers itself to us; it sheds light on its own diffuse hold at the edge of consciousness, which comes to reveal our way of being present in the world as an immersed, sentient, and receptive body that cannot feel without feeling itself. And for architects, by shifting the process of the project past the building's meaning or form in the strictest sense; it encourages them to consider the affective and somatic aspects of a space, which in turn impels them to marry a sense of finesse and geometry in their determination and selection of a certain number of physical parameters. And aside from members of these two professions, the notion of atmosphere is relevant for ordinary citizens, as it generates an original perspective on a number of contemporary socio-economic realities from a micro, situated, and sensory vantage point. But it should not be considered as a closed concept that has been set in stone by various texts, instead as an open concept that welcomes theoretical and practical developments. The aesthetics of atmospheres should be developed by thinking about the relational space of architecture and how this suits its environs (or not), and about the rhythm that sustains it and that takes hold of users, aspects that architectural theory has considered at length. The former enriches the aesthetics of atmospheres by integrating the question of pleasure and not just emotion; the latter demands that we take time into account. These two aspects directly concern architectural design, especially its process-like nature, both in the space created and the space perceived. The extension that we have sketched out thus forms part of a vaster horizon of thinking about ambiances in practical terms.

Notes

1 See J.F. Augoyard, "Éléments pour une théorie des ambiances architecturales et urbaines" ["Elements for a Theory of Architectural and Urban Ambiances"], in *Les Cahiers de la recherche architecturale*, 42/43, (1998): 13–23.

2 See G. Böhme, *The Aesthetics of Atmospheres* (Abingdon-on-Thames: Routledge, 2018).

3 See J.P. Thibaud, "Petite archéologie de la notion d'ambiance" ["A Brief Archaeology of the Notion of Ambiance"], in *Communications,* Vol. 90 (2012): 155–174.
4 See H. Schmitz, *Der Gefühlsraum* ["The Space of Feelings"], (Bonn: Bouvier, 1969), Vol. III/2 of *System der Philosophie,* 10 volumes (Bonn: Bouvier, 1964–1980).
5 See issue no. 91 of the journal *OASE.*
6 See in particular, T. Griffero, *Atmospheres: Aesthetics of Emotional Spaces* (Abingdon-on-Thames: Routledge, 2016).
7 See R. Shusterman, *Architecture et soma-esthétique* ["Architecture and Soma-Aesthetics"] (Geneva: Haute école d'art et de design (HEAD), 2010).
8 See G. Böhme, *Aisthetique. Pour une esthétique de l'expérience sensible* ["*Aisthetique*: Towards an Aesthetics of Sensory Experience"] (Dijon: Les Presses du réel, 2021), 55.
9 G. Böhme, *Aisthetique. Pour une esthétique de l'expérience sensible* ["*Aisthetique*: Towards an Aesthetics of Sensory Experience"] (Dijon: Les Presses du réel, 2021), p. 109 *et seq.*
10 G. Böhme, *Aisthetique. Pour une esthétique de l'expérience sensible* ["*Aisthetique*: Towards an Aesthetics of Sensory Experience"] (Dijon: Les Presses du réel, 2021), p. 97 *et seq.*
11 G. Böhme, "L'atmosphère, fondement d'une nouvelle esthétique" ["Atmosphere as the Basis of a New Aesthetics"], the first chapter in Atmosphäre: Essays zur neuen Ästhetik, (Berlin: suhrkamp Verlag, 2013.), translated into French in Communications, no. 102, (2018): 26.
12 M. Heidegger, *Being and Time,* translated by Joan Stambaugh, (Albany: SUNY Press, 2010), 130.
13 G. Böhme, *Aisthetique,* 101.
14 See G. Böhme, "Les atmosphères comme objets de l'architecture", 33.
15 A. G. Baumgarten, *Aesthetica* [1750–1758], (Hildesheim: Georg Oms Verlag, 1961), preceded by *Metaphysics. A Critical Translation with Kant's Elucidations, Selected Notes, and Related Materials,* translated and edited by Courtney D. Fugate and John Hymers (London, New York: Bloomsbury Publishing, 2013).
16 G. Böhme, "Les atmosphères comme objets de l'architecture", 188.
17 *Einfühlung* (empathy) literally means "feeling in". The *ein* expresses the unity of the form, of its perception, and of its experience as a manner of "feeling oneself to be oneself". See M. Galland-Szymkowiak, "L'engendrement de l'espace architectural: trois modèles (Wölfflin, Schmarsow, Lipps)" ["The Engendering of Architectural Space: *Three* Models (Wölfflin, Schmarsow, Lipps)"], in *Construire et éprouver, dans l'espace et dans la pensée. Points de rencontre entre l'architecture et la philosophie* ["Building and Experiencing in Space and in Thought: Points of Convergence between Architecture and Philosophy"], ed. M. Galland-Szymkowiak, P. Lohmann, 2017, 83–100.
18 The Heideggerian analysis to which Böhme assigns the notion of atmosphere is characterized by a forgetting of the body, which explains his recourse to other authors. See *Aisthetique, Aisthetique. Pour une estheÅLtique de l'expeÅLrience sensible* (Dijon: Les Presses du réel, 2021), 101.
19 G. Böhme, "Les atmosphères comme objets de l'architecture", 179.
20 H. Wölfflin, *Prolegomena to a Psychology of Architecture,* translated by Michael Selzer (Scotts Valley: CreateSpace Publishing, 2016).
21 H. Schmitz, "Les sentiments comme atmosphères" ["Feelings as Atmospheres"], translation into French of a chapter of *Atmosphären* (Freiburg-im-Breisgau: Karl Alber, 2014), in *Communications,* 102, (2018): 51.
22 H. Schmitz, "Les sentiments comme atmosphères" ["Feelings as Atmospheres"], translation into French of a chapter of *Atmosphären* (Freiburg-im-Breisgau: Karl Alber, 2014), in *Communications,* 102, (2018): 52.
23 G. Böhme, "L'atmosphère, fondement d'une nouvelle esthétique", 40.

24 See also the following works: G. Böhme, *Aisthetique,* p. 224 *et seq.*; G. Böhme, O. Eliasson, and J. Pallasmaa, *Architectural Atmospheres. On the Experience and Politics of Architecture* (New York/Berlin, De Gruyter, 2014); G. Böhme, *Critique of Aesthetic Capitalism,* Mimesis International, 2017.
25 See in particular Rüdiger Bubner, *Ästhetische Erfahrung* ["Aesthetic Experience"] (Frankfurt-am-Main: Suhrkamp, 1989), 143–156.
26 Walter Benjamin, *The Work of Art in the Age of Mechanical Reproduction*, translated by J.A. Underwood (London: Penguin Classics, 2008).
27 Walter Benjamin, *The Work of Art in the Age of Mechanical Reproduction*, translated by J.A. Underwood (London: Penguin Classics, 2008).
28 See I. Kant, *Critique of Judgment,* translated by Nicholas Walker (Oxford: Oxford World Classics, 2007).
29 G. Böhme, "Les atmosphères comme objets de l'architecture", 174.
30 F. L. Wright, *An Autobiography* (New York: Duell, Sloan and Pearce, 1943), 128.
31 See J.M. Schaeffer, *L'expérience esthétique* ["The Aesthetic Experience"] (Paris: Folio Essais, 2015), p. 182.
32 See, in particular, Cicero, *De Finibus Bonorum et Malorum* (Arma Virumque Editions, 2021); *On Duties,* translated by Walter Miller, independently published, 2017.
33 See, in particular, G. Böhme, "Les atmosphères comme objet de l'architecture", p. 189 and H. Schmitz, *Der Leib, der Raum und die Gefühle* ["The Body, Space, and Emotions"] (Bielefeld/Basel: Sirius, 2009), 57–64, and *New Phenomenology: A Brief Introduction* (Sesto San Giovanni: Mimesis International, 2019).
34 See A. Loos, *Ornament and Crime* (London: Penguin Classics, 2019).
35 G. Semper, *Du style et de l'architecture* ["On Style and Architecture"] (Marseille: Parenthèses, 2007), 235. *Kosmein* means both "to adorn" and "to arrange according to the order of the world".
36 See H. Schmitz, *Atmosphären,* 94.
37 G. Böhme, *Aisthetique,* 47.
38 See G. Böhme, "Atmosphären: die Beziehung von Musik und Architektur jenseits physicalischer Vorstellungen" ["Atmospheres: The Relationship between Music and Architecture beyond Physical Ideas"] , in *Musik und Architektur,* ed. C. Metzger (Saarbrücken: Pfau-Verlag, 2003), 98–108.
39 Le Corbusier, *Le Modulor* (Paris: Denoël, Gonthier, 1977), 75.
40 C. Parent, *Entrelacs de l'oblique* ["Intertwining the Oblique"] (Paris: éd. du Moniteur, 1981), 107.
41 H. Maldiney, "Notes sur le rythme" ["Notes on Rhythm"], in *Henri Maldiney: penser plus avant* ["Henri Maldiney: Thinking Farther Ahead"], ed. J.P. Charcosset (Chatou: éd. de la Transparence, 2012), 20.

Part 6

Ambiances, city, and urban space

20 The ambiances are the message

Transects, projects and urban situations

Nicolas Tixier

For geographers, the term "transect" means "a tool for observing a terrain or representation of a space along a straight line and according to its vertical dimension to highlight a superposition, spatial succession, or relationship between phenomena" (Marie-Claire Robic[1]). Applied to the development of a territory, a transect constitutes a practice whose elements were first proposed in the early twentieth century, specifically by the Scottish urban planner and botanist Patrick Geddes. Geddes insisted on the "synoptic" potential of a section, i.e., on its capacity to consider long-standing, historical relationships that connect forms of collective life to notions of physical geography and to render them visible in the present.[2] His use of this visual projection sought to bring together various perspectives and disciplines within the same, single visual representation. This undoubtedly explains why, in these same years, he outlined a convergence between the visual and cultural conventions underlying the architectural cross-section and the transect as practiced in the domains of human geography and the life sciences. Team X[3]'s reprise of the famous *Valley Section*, conceived by Patrick Geddes 50 years ago, which was in turn inspired by Humboldt's geographical work, allows for a hybridization of the two genres that can be placed at the service of architecture and urban planning, one that remains underdeveloped to date.

The transect has since been revisited, both as a technique of representation and a field practice. To us, it is a hybrid device between a technical cross-section and a sensorial path; it is constructed with the use of drawings, photographs, measurements, text, and even video, and when practised on site, with one's perceptions, words, perambulations, and in general, by walking and by moving across the terrain.[4] By rehabilitating the atmospheric dimension of urban representations and enabling the inclusion of inhabitant narratives in the specialized debates between disciplines, the transect serves as a tool of inquiry and expression of sensory space and lived practices.

Performing and describing a transect uses a method of sampling, selecting, and then editing. As a projective mechanism that reveals existing elements in their repetitiveness and differences, the question arises of what these existing factors will become through the demonstration (and suggestion) of potential

DOI: 10.4324/9781032655307-29

gestures, experiences, and transformations for inhabiting and for flows (of people, activities, the natural landscape, the atmosphere, and more). A transect is not a mono-disciplinary tool; it instead functions as a medium open to the knowledge and representations of all the disciplines that may engage in a multi-polar dialogue. From the practice of *inventory,* it derives the intent to locate and collect situations that are as unique as they are paradigmatic. And from the *album,* it takes the capacity to visually arrange things and situations. But it remains open, much like Aby Warburg's Mnemosyne Atlas.[5] It does not require formatted frameworks or a unity of modes of representation. One can go from one scale to another, from one document to another. One can add and remove elements as needed. It is quite literally a working surface (a table or a wall, physical or digital) that can be shared or amended, where documents placed side by side generate meaning in their relationship to a context by means of the section.

The transect opens up a unique space and a non-hierarchical timeframe for hatching different projects or experiences to be debated and carried out. The whole is no longer primarily a project defined by a traditional sequence of scale (from a reading of the territory to a definition of an overall plan, an architectural design, and the creation of all the details). Nor is it limited to the application of a technical solution. It is neither a project imposed by its designers alone, nor a project that, conversely, arises only from user practices. It is instead an intermediate space for dialogue and negotiation where all actors within a geographic area (from inhabitant to designer, and from politician to operator) can find project materials at their own scale, within their domains, and in relation to their practices. Constructing these spaces for interaction (between disciplines and actors, whom we will consider here as each having their own language in some way) also requires considering the role of translation, or more precisely, of the translator as a third party between two foreign languages, specifically as a third party who does not disappear, and instead remains present. In our work, because things are not translated from one language to another, we must create the conditions for reading and comprehension between the languages and modes of representation adopted by the various actors. We must create mechanisms that, as Paul Ricœur said,[6] welcome the language of the other. We must provide "linguistic hospitality", arranging the subject for its transmission, so that it can engage in a dialogue with other subjects. This third-party role should be a desire and not an obligation, as Ricœur also said, so that we make every effort to "expand our own language".

The transect offers an opportunity to remove things from their usual frame because it does not belong to predefined categories for the elements it mobilizes and interactions that it requires. It may even appear as a tacit critique of urban zoning regulations that refocuses the conversation on local singularities and the practices of inhabitants as a way of working on what things will become through and with their milieu in some way.

An approach to urban ordinariness through ambiances

The theoretical interest of the transect resides mainly in the fact that it is a way to generate a pluralist formal representation of what we find at the heart of any urban situation: the unity and uniqueness of a place in terms of its everyday life and its ambiances.

There is no obvious description of what constitutes urban ordinariness or an inhabitant's daily routine, and it is even more difficult to take this into account within the logic of a project. And yet, we all pay close attention and are at times even "captivated" by the ambiance we perceive upon arriving in a given place. We distinctly notice remarkable lighting or a particular quality of sound; we are inspired by a public space that is lively or calm, and we are prone to contemplation in a place where a sense of quiet predominates. Often unique and irreducible, the ambiance of a place varies depending on the day, time, weather, people present, and our own actions. And yet, despite these variations, it possesses certain general characteristics that grant it an identity and allow us to recognize it.

On a day-to-day basis, we experience ambiances, and they experience us. While we readily feel and share such sensory experiences, paradoxically the notion of ambiance eludes any overly strict formal definition. It is experienced in the singular, as a whole that does not separate our sensory channels or our acts of perception and representation. But we are only able to analyse it in the plural, sense by sense, and discipline by discipline. Thus, the more our knowledge of ambiances becomes precise in terms of their composition and means of formation – as they are enriched by acoustic, lighting, thermal, tactile, spatial, and social know-how –, the more we risk losing sight of what renders them a unified whole, this lived relationship to a place that is always unique. They constitute "a fleeting singularity, and a scattered plurality," as summed up by Jean-François Augoyard.[7]

The transect as the interaction and contextualization of the elements collected

The study of ambiances thus proceeds by sectioning different data (spatial typologies, measurements of sound, heat, water, air, sunlight, and lighting), as well as by directly observing usages, interviewing a place's users, inhabitants, and actors, and understanding the history of the built and natural environments – these factors among many. Accounting for this data and interpolating it to any significant extent is not readily obvious, as the data can be visual, metric, temporal, textual, photographic, videographic, or acoustic. And each of these registers contains considerable variations within itself in terms of the nature and status of the data. The transect furnishes the possibility of placing data of different registers literally side by side and getting these data sets talk to one another, depending on their placement and distance from each other. They answer, complete, nuance, contradict, ignore, and resonate with one another. They are not arranged solely in relation to each other. They have a very specific initial hook, namely that they always exist in relation to the section, the line of the transect. This represents a major difference from Aby Warburg's Mnemosyne Atlas, which does not presuppose any context on the worktable before the elements are laid down on it. This relationship between the data and the section can be localized very precisely (through a direct connection to a related place) or more generally, depending on the potential distance employed to place the element above (towards the sky) or below (on the ground).

The set of arranged elements generates a triply open urban narrative, in that:

- the person reading the transect decides what they will read and in what order, what they will connect, and what they will retain as important;
- new data may always be added to the transect and thus uphold or inflect the narratives and individual components included;
- what the place will become is not set in stone, but this can instead be inferred from the profile of the place, the data, and the narratives, thereby initiating a connection between past, present, and future.

We should specify an important point at this juncture, namely why each transect has to be constructed, i.e., the reason for the inquiry and the why of a potential project. Let it not be a surprise that this initial question heavily orients the line of the transect and the collection of data. To answer these questions, we must "investigate," but more importantly, we have to interpret, correlate, and then plan, rather than a solution, one or more possible evolutions based on the context's specificities, all the while retaining a sense of creative freedom. As in any investigation, elements are sampled in a context that will also serve as potential clues for future interpretations and *plans*. These interpretations and *plans* together form narratives that describe the possible futures of each situation.[8]

Therefore, there is an orientation to the type of data collected during the investigation, and there is also a selection of data that is retained and included in the transect. This requires that one find the right amount of what will be selected (neither too little, under penalty of having a transect that is poor in narratives and potential developments, nor too much, under penalty of not managing to establish visual connections between the sampled elements and their context). The fact of working collaboratively (and across different disciplines) both during the data collection and the formation of the transect helps prevent too much pre-selection beforehand, and it also helps people accept what each person has deemed pertinent for collection and collation. The size of the worktable and the scale chosen for the transect matter hugely in this process. Once the elements are placed on the transect, it can be proposed for debate and *planning* among colleagues (researchers and urban planners), as well as – and especially with – actors within the geographic area.

Thus, understanding a situation begins with, for example, quotes from inhabitants, a set of situated measurements, the form of the built environment, or the nature of the landscape, photographs, and historical elements. But this comprehension is not explicit; it is formed by each person as they examine the transect. It depends on each person's knowledge, both in terms of his or her professional discipline and personal experiences. It must in turn be recited and shared with everyone else present. A critical and planning discourse emerges fairly naturally that may itself come to form part of the transect, whose role of representation thereby becomes the matrix for the project.

Using the transect: from the roundtable to the long table

The "long table system"[9] consists of literally using a long table on which one places the urban cross-section one has created, which must contain a minimum number of elements that express the issues to be addressed in the project (such as quotes from inhabitants and from experts, photographs, an expression of the usages, quantitative data, a focus on specific points, elements for diagnosing issues, and project sketches). The actors concerned by the site or the question to be addressed must then be summoned, and the section must be used to initiate a debate by eliciting reactions to what has been included or left out, by adding comments, information, or new narratives, by listening to the discussions that take place at different areas along the table, and so forth. This debate of the issues at stake among the various actors has the advantage of keeping the context at the core of the conversation, because the section is always in front of and between people. Working at a roundtable is instead necessary for being able to step back, but this often frames the debate in terms of a specification of positions that may wander away from the site's characteristics. Soliciting comments is in itself a process of working with others, listening, and getting actors to debate with themselves and one another. This kind of method can provide a prospective diagnosis in three ways: as a means of generating words, a means of collecting notes, and as a means of revealing lived realities. The multiplicity of experiments conducted in different contexts also quickly leads to the identification of several principles that render the process more effective:

- After conducting the first long tables, it quickly became clear that it was important to have "thick description" elements, as Clifford Geertz termed them, ones that do not result in too hasty a conclusion that in turn prevents people from grasping the density of the situation (through too technical a representation) and of the lived experiences (through overly simplified narratives).
- Positioning data all along the transect that interact with one another ("two images make a story", as Jean-Luc Godard liked to say) and whose visual layout supports the context allows people to examine each contributed element according to its own mode of enunciation; a text remains a text,

an image stays an image, and a measurement is still a measurement. This deviates quite strongly from the logic of thematic layers to be placed one on top of the other that is typical of geographic information systems that generally require codification (colours, hachures, values, markers, etc.).

- To optimize the application of the above two points, one should avoid using codes as much as possible that require a reference to captions to be interpreted. If data requires a caption to contextualize or specify it, it is much clearer to indicate this simply as information within the transect under that every piece of data.

The analysis conducted during the long table becomes more precise as a result of the interactions between the actors present. It reveals consensuses, or, to the contrary, it expresses vehement local disagreements that often find their way into the project drawings made at all scales (from the inhabitant intervention to urban orientation) that support or help the situation evolve (as a built, social, and sensory reality). It constitutes a pluralistic means of putting issues of daily life up for debate and including them in the project.

Transect and project(s)

If the transect allows us to move through the project's different construction phases or to mediate between the various actors, is it possible to build projects the way we assemble transects?

Formally speaking, any project to create or renovate a line that crosses an existing context may be viewed as a transect. However, the transect's capacity to help people read the context traversed and to play on a reciprocal alteration is what gives it authority. Thus, projects that use an old railway line like the High Line in New York City, which opened in 2009, or the Promenade Plantée in Paris, which opened 20 years earlier in 1988, are well-known examples of construction systems that literally work like a transect. But these lines and sections in a territory sometimes acquire more ordinary aspects that are no less powerful in the practice and perception of the city they enable through the spatial, social, and historical densities they reveal. In the spirit of the nineteenth-century meanderings near and dear to Walter Benjamin and Charles Baudelaire (where every passer-by is a *flaneur*), might "transect" lines represent the passageways of the twenty-first century (in which every passer-by is a photographer without a camera[10])?

In a more symbolic vein, there are also projects that, due to their expert construction, can be compared to transect projects. A recent example of such a project that we find particularly exciting is the *Newtown Creek Nature Walk* in Greenpoint, Brooklyn. It was created in 2007 by the artist George Trakas and the landscape architects Quennell Rothschild & Partners. We see it as emblematic of what we could call a "transect" project. It was presented on the website as: "It is a linear waterfront walkway and an entry garden at the street access point. It's designed to evoke the rich, continually evolving environmental, industrial, and cultural histories surrounding Newtown Creek". Far from being just a place for the body and its sensations (as Catherine Grout demonstrated very clearly in one of the few studies conducted on this project[11]), we see it as a leitmotif through which everything can be expressed: history, geography, and society. We could use the context of this completed project to mobilize tools and concepts such as sections, clues, layers, atlases, miniature metropolises, and "taskscapes", because to us it represents such a clear paradigm of a transect project at the scale of the public space.

It is mainly through a structural approach to comprehending territory that the transect expresses its potential, because it reveals unique spaces and non-hierarchically ranked timeframes for a whole set of projects. The whole is thereby no longer a project defined by the traditional interplay of scales, or a project imposed solely by its designer or, conversely, through the practices of its users. It is instead a space for negotiation where all actors within a geographic area can find project materials at their scale and in relation to their practices.

Using the transect

What happens when we use these elements as indications? First, there are associations between the data, and then, a comparison of situations. There is, of course, deduction and induction when one attempts to generalize, define rules, or verify consequences; however, contexts that remain unique quickly exhibit their resistance to any kind of explanatory rule. Each place is a singular composition that needs to be understood in terms of its arrangement. Thus, abduction – which here instead consists of formulating hypotheses of narratives that supplement the indications – frequently takes place around a long table when a new connection is formed between two previously unrelated elements, when the actors present introduce new elements, when a new arrangement is formed based on a logic that applies to other kinds of elements (i.e., by analogy), and with the introduction of a new development that catalyses a situation and overturns the initial interpretations. The long table is a fertile, abductive moment for reciting narratives, first of all as a way to test one's own account of the place in response to what has literally been placed on the table, and then, as a way to elaborate, amend, or strengthen one's narrative, because we test its validity when we tell it to others. *The interpretation of a situation always entails sketching what it will become.* The novelty of a narrative or a proposal comes first from the creation of new connections between elements that are mostly already present, from a way of interweaving relations between elements arranged in an original manner that reflects each place's singularity. Said another way, innovation no longer resides solely within the domain of the new, in either technical or aesthetic terms, rather a consistently unique hybridization of various issues that are specific to each existing situation and which need to be brought to light.[12]

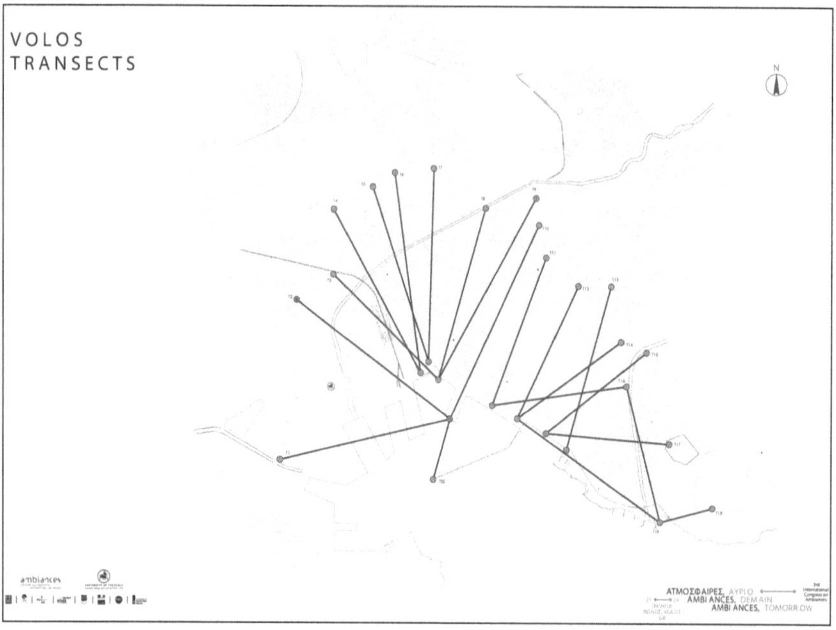

Figure 20.1 Volos Transects, 2016.

Source: Nicolas Rémy and Nicolas Tixier.

Figure 20.2 Long table – Amiens 2030, Metropolitan Project.

Source: By BazarUrbain, 2012.

This notion is fairly pragmatist, namely that every urban situation is a type of "public thing" that is a permanent construct. Once the elements have been placed in the transect, they can begin to act as clues for developing potentialities for the future or potentialities one imagines to give a history to the past and meaning to the present. This is why we feel that any work on the future of a place built on knowledge of the existing elements necessarily undergoes an abductive phase. The relevant clues for articulating a situation often lead to a collective of potentialities and help reveal a public, as Dewey might say. Thus, a local validity is constructed on which a collective of actors or a public policy (even top-down) must be able to rely.

More online iconography

http://cressound.grenoble.archi.fr/NT_TRANSECT.pdf

Notes

1 Marie-Claire Robic, "Coupe (Transect)" ["Section (Transect)"], *Hypergéo* (10 May 2004). Available at: www.hypergeo.eu/spip.php?article60#.
2 Patrick Geddes' first published his famous sections in 1925 (Geddes, 1925 a and b). For an archaeology of this concept, please see Federico Ferretti (2012).
3 See Peter Smithson's famous *Doorn Manifesto* (1968).
4 This article refers to and builds on two previously published articles:
Nicolas Tixier, "Le transect urbain. Pour une écriture corrélée des ambiances et de l'environnement" ["The Urban Transect: Towards a Correlated Design of Ambiances and Environment"], in Sabine Barles, Nathalie Blanc (ed.), *Écologies urbaines – Sur le terrain*" (Paris: Éd. Economica-Anthropos / PIR Ville et Environnement, 2016): 130–148.
Nicolas Tixier, "Le transect : un opérateur abductif" ["The Transect: An Abductive Operator"], in Remi Clot-Goudard, Viviane Huys, Denis Vernant (ed.) *Abduction*, in the journal *Recherches sur la philosophie et le langage*, No. 34, Paris: Éd. Vrin (November 2018).
5 Aby Warburg's Mnemosyne Atlas consisted of a large montage (on a flat surface) of photographic reproductions "substituting the question of transmission for that of its exhibition [thus organizing] a network of tensions and anachronisms between the images". Through a vertical juxtaposition of synoptic panels without the intervention of the verbal register, "Warburg's atlas presents multiple images on each panel to generate multi-polar comparisons"... "Warburg's atlas multiplies the frames and scales". Warburg did not conceive of an atlas as a book, instead as a worktable that was open to interpretation, manipulation, and to new images. "Atlases create *heterotopias* by working with divisions and montages of domain that have traditionally remained separate from one another". Taken from the work of Philippe-Alain Michaud and Georges Didi-Huberman in Philippe-Alain Michaud. *Aby Warburg et l'image en mouvement* ["Aby Warburg and the Image in Movement"] (Paris: Éd. Macula, 2012 [1st ed. 1998]), 298, and Georges Didi-Huberman. *L'album de l'art à l'époque du "Musée imaginaire"* ["The Art Album at the Time of the 'Imaginary Museum'"] (Paris: Éd. Hazan / Musée du Louvre, 2013), 208.
6 Paul Ricœur, *On Translation*, trans. Eileen Brennan, with an Introduction by Richard Kearney (Abingdon, UK and New York: Routledge, 2006).

7 Jean-François Augoyard, "Ambiance(s)", in the dossier "L'espace anthropologique" in *Les cahiers de la recherche architecturale et urbaine*, No. 20/21 (Paris: Éditions Monum, 2007): 33–37.
For the topics of debate over the definition and effectiveness of the notion of ambiances, we refer in chronological order to the following sources, beginning with the issue edited by Luc Adolphe, "Ambiances architecturales et urbaines" ["Architectural and urban ambiances"], *Les cahiers de la recherche architecturale et urbaine*, No. 42–43, 3rd quarter, 1998 (Marseille: Éditions Parenthèses, 1998), and then the collective work edited by Pascal Amphoux, Jean-Paul Thibaud, and Grégoire Chelkoff, *Ambiances en débat* ["The Debate Over Ambiances"] (Bernin: Éd. A la Croisée, 2004), and the work of Jean-Paul Thibaud, *En quête d'ambiances. Éprouver la ville en passant* ["In Search of Ambiances: Experiencing the City in Passing"] (Geneva: Éd. MetisPresses, 2015). And more recently, the works of Gernot Böhme, Tonino Griffero, Jean-Paul Thibaud, and Juhani Pallasmaa in *Architecture and Atmosphere*, P. Tidwell (ed.), (Espoo: Tapio Wirkkala-Rut Bryk Foundation, 2014), Gernot Böhme, *The Aesthetics of Atmospheres*, Jean-Paul Thibaud (ed.), for the collection *Ambiances, Atmospheres and Sensory Experiences of Spaces* (Abingdon: Routledge, 2016), and Tonino Griffero, *Quasi-Things. The Paradigm of Atmospheres*, trans. Sarah De Sanctis (Albany, SUNY Press, 2017).

8 For specific examples of a tZransect's application, please refer to the following studies available online:
Nicolas Tixier (ed.) et al. *L'ambiance est dans l'air. La dimension atmosphérique des ambiances architecturales et urbaines dans les approches environnementalistes* ["Ambiance is in the Air: The Atmospheric Dimension of Architetural and Urban Ambiances in Environmentalist Approaches"], a PIRVE research report (CNRS & PUCA), 250 p. + annexes (Grenoble: CRESSON, 2011).
Nicolas Tixier (ed.) et al. *Amiens 2030. Le quotidien en projets* ["Amiens 2030: Daily Life in Projects"], 488 p. + video (Grenoble: Éd. BazarUrbain, 2013).
Frédéric Pousin (ed.), Nicolas Tixier, et al., *DTRANSECT. Les délaissés des traversées de la vallée de l'Huveaune: dispersion des espèces, pratiques vernaculaires, médiations paysagères* ["The Forgotten along the Huveaune Valley: The Dispersing of Species, Vernacular Practices, and landscape Mediation"], an ITTECOP research program, French Ministry of the Environment, a LAREP, CRESSON, LPED, and COLOCO collaboration (Grenoble: BazarUrbain, 2012–2014). The research report, 152 p. + video + transects are available at http://dtransect.jeb-project.net, January 2015.

9 A terminology coined together with Pascal Amphoux.

10 See the very interesting article by Viva Paci and Martin Bonnard, "Le flâneur de la High Line de New York: un opérateur sans caméra et une vue urbaine sans film" ["The Flaneur on New York City's High Line: A Photographer without a Camera and a View of the City without any Film"], *Annales de géographie*, No. 695–696 (2014).

11 Catherine Grout, "Paysage et Art Contemporain (Erwin Straus et George Trakas)" ["Landscape and Contemporary Art (Erwin Straus and George Trakas)"], *DEMéter* (2012), and the presentation of the project at https://www.nyc.gov/site/dep/environment/newtown-creek-nature-walk.page.

12 Pascal Amphoux, « L'innovation architecturale n'est pas toujours là où on l'attend », in *Culture et recherche*, Ministère de la Culture et de la Communication, n°121, 2009, p. 22–23.

Bibliography

Amphoux, Pascal. "L'innovation architecturale n'est pas toujours là où on l'attend". *Culture & recherche*, No. 121 (2009):22–23.

Augoyard, Jean-François. "Ambiance(s)", in the dossier "L'espace anthropologique". In *Les cahiers de la recherche architecturale et urbaine*, No. 20/21. Paris, Ed. Monum (March 2007): 33–37.

Didi-Huberman, Georges. *L'album de l'art à l'époque du "Musée imaginaire"*. Paris: Éd. Hazan / Musée du Louvre, 2013.

Ferretti, Federico. "Aux origines de l'aménagement régional: le schéma de la Valley Section de Patrick Geddes (1925)". *M@ppemonde*, No. 108 (2012). Consulted 10 August 2015, https://mappemonde.mgm.fr/num36/articles/art12405.html.

Geddes, Patrick. "The Valley Plan of Civilization" & "The Valley in the Town". *Survey Graphic* (1925).

Geertz, Clifford. *The Interpretation of Cultures*. New York: Basic Books, 1973.

Grout, Catherine. "Paysage et Art Contemporain (Erwin Straus et George Trakas)". *DEMéter* (2012).

Michaud, Philippe-Alain. *Aby Warburg et l'image en movement*. Paris: Éd. Macula, 2012.

Paci, Viva and Martin Bonnard: "Le flâneur de la High Line de New York: un opérateur sans caméra et une vue urbaine sans film". *Annales de géographie*, No. 695–696 (2014).

Pousin, Frédéric (ed.), Nicolas Tixier, et al., *D TRANSECT. Les délaissés des traversées de la vallée de l'Huveaune: dispersion des espèces, pratiques vernaculaires, médiations paysagères*, an ITTECOP research program, French Ministry of the Environment, a LAREP, CRESSON, LPED, and COLOCO collaboration. Grenoble: BazarUrbain, 2012–2014. Research report, 152 p. + video + transects also available at https://dtransect.jeb-project.net (January 2015).

Ricœur, Paul. *On Translation*. Translated by Eileen Brennan. With an Introduction by Richard Kearney. Abingdon and New York: Routledge, 2006.

Thibaud, Jean-Paul. *En quête d'ambiances. Éprouver la ville en passant*. Geneva: Éd. MetisPresses, 2015.

Tixier, Nicolas (ed.) et al. *L'ambiance est dans l'air. La dimension atmosphérique des ambiances architecturales et urbaines dans les approches environnementalistes*, a PIRVE research report (CNRS & PUCA). Grenoble: CRESSON, 2011).

Tixier, Nicolas (ed.) et al. *Amiens 2030. Le quotidien en projets*. Grenoble: Éd. BazarUrbain, 2013.

Tixier, Nicolas. "Le transect urbain. Pour une écriture corrélée des ambiances et de l'environnement". In *Écologies urbaines – Sur le terrain*, edited by Sabine Barles and Nathalie Blanc. Paris: Éd. Economica-Anthropos / PIR Ville et Environnement, 2016.

Tixier, Nicolas. "Le transect: un opérateur abductif". In *Abduction*, edited by Remi Clot-Goudard, Viviane Huys, and Denis Vernant. In the journal *Recherches sur la philosophie et le langage*. Paris: Éd. Vrin, No. 34 (November 2018).

21 Atmospheres in the Metapolitan Anthropocene

Niels Albertsen

Introduction

The years after the Second World War saw the rise of the Metapolis (Ascher 1995) as a new type of urbanity assembling – through mobility infrastructures – separate urban areas and landscapes into vast conurbations of dispersed, discontinuous, heterogeneous, and multi-polarized territories. Urbanization including the Metapolis was a prime mover in the "Great Acceleration", which from the 1950s onward generated the Anthropocene as a new geological epoch of hitherto unseen human impact on the planet Earth, its climate, ecosystems, and geology. Urban aesthetic (Böhme 2001) atmospheres changed as well, diversifying and intensifying differently in various parts of the Metapolis. This chapter explores how Metapolitan diversification, anthropogenic processes, and atmospheric change link together. Special attention is given to the anthropogenic "call" for understanding the relation between meteorological and aesthetic atmosphere. The chapter closes with some remarks on atmospheric aspects of Metapolitan anthropogenic politics. To frame the exploration, this chapter starts out with considering *how at all* the city and the urban can be conceptualized today.[1]

The city, the house, and the universe

Once upon a time, the conceptual limits of the city could with good reason be clearly demarcated. Leaning against a city's walls, a conceptual distinction between an inside of built urban structures and urban life and an outside of rural life and nature could be plausibly maintained. Such container-like conceptualizations of the city had a long-lasting influence, despite the fact that cities just as plausibly could be seen as permeable entities in constant exchange with their surroundings. Not least, urban sociology looked, long into the twentieth century, "to the city for an understanding of the city" (Pahl 1975: 234). Similarly, the idea of the city as a container-like, large house found an echo in the twentieth century in Rossi's *Architecture of the City*.

Marcel Hénaff characterized this city as *monument*, as a picture of the world, a replica of the cosmos, as the "realization on the Earth of a world that responds to the world of the gods". To build a city, therefore, was to

DOI: 10.4324/9781032655307-30

build a world. The city walls were for defence, but they mainly had religious reasons. Inside, the walls enclosed the sacred world of the city, while the outside was a world of pagan polytheism (Hénaff 2008: 33–34, 53). This city resembled a house. Entering it through the city gates was like opening and closing the doors of a house (Hénaff 2008: 173).

Similarly, designing a city and a house were parallel activities. For Leon Battista Alberti (1404–72), the universal rules of building a house were ultimately the same as the rules for building a city (Choay 1980, quoted in Queysanne 1991: 49). The city "is like a big house" and the house is like a "small city", just as well as the parts of the house are like the parts of a city (Queysanne 1991: 50, 52). Jean-Nicholas-Louis Durand (1760–1834) assured that "just as the walls, the columns etc. are the elements from which buildings compose themselves, so the buildings are the elements from which the cities compose themselves" (quoted in Queysanne 1991: 54). In the twentieth century, Aldo Rossi (1931–1997) subscribed to similar ideas. The "city is to be understood [...] as architecture" in which regard he approvingly referred to the above quote from Durand (Rossi 1992: 21, 35).

This conception of the city as a house-like monument that replicates the universe brings to mind the ontology of globes explicated by Peter Sloterdijk (Sloterdijk 1999). From Plotin to Leibniz, the authoritative ontology was one of "macro-spheres" in which the concepts of World and God were imagined as "all-enclosing spherical volumes" in which "subordinate [...] spheres [...] were embedded in a concentric way" (Sloterdijk 2004: 17). This concentric world was seen as a huge house to which humans relate as inhabitants. "Man is essentially a dwelling creature. His relationship to the world-whole is a relation of inhabitation" (Sloterdijk 2007: 231). One of the important layers in this concentric world-house was the city with its walls (cf. Sloterdijk 1999: 251–325).

The "disappearance" of the city and the urban

According to Sloterdijk, today we cannot do with such globes of nested levels. Life has become "multifocal, multiperspectival, and heterarchical", and the globe has "imploded" into foam (Sloterdijk 2004: 23, 26). Foam "households" are "introvert" and attached to each other in "co-isolation" (Sloterdijk 2004: 56). Cities, similarly, become "foam cities" (Sloterdijk 2004: 604–05). None of this gives any hint that the city should be a monument for, or mirror of a somehow harmonious universe or a clearly delimited entity at all.

The idea of dissolving city boundaries is not totally new to urban discourse. In a Danish context, planning observer Knud Bidstrup in 1963 foresaw that Denmark in a "distant future" would turn from "a small country" into a "large city". From observations of current tendencies, he predicted that Denmark would change into

> one big city – although an open city with large green areas between the densely built-up areas, but still one urban area, connected by fast communication and close communities of work and inhabited by people

> that independently of their work and profession live and think in an urban way.
>
> (Bidstrup 1963: 143–149)

Bidstrup's distant future is NOW, not only in Denmark, as demonstrated by Tom Nielsen (2023), but in large parts of the world as well. In this blurred situation, the question of how to conceptualize something that still, with good reason, can be called "city" or "urban" is particularly urgent. The city or the urban cannot be understood as "a pre-given site, space or object" anymore. "Questions of conceptualization lie at the heart of all forms of urban research" as Neil Brenner rightly says (Brenner 2013: 96); today, one could add, more than ever. The question is: *how?*

I have addressed this question elsewhere (Albertsen 2020a, b), but let me re-state the main idea here. Since about 1990, conceptualizations of the city or urbanity have been polarized between too narrow and too expansive conceptions, leading to the "disappearance" of both concepts. Either cities were conceptualized as entities with clearly identifiable limits, centre-periphery structures, centres for political power or unification of urbs (the physical territory) and civitas (the community of citizens), the inadequacy of which were obvious, hence, the disappearance or death of the city. Or urbanity was conceptually expanded into being "everywhere and in everything".

> [W]hat is not the urban? Is it the town, the village, the countryside? Maybe, but only to a limited degree. The footprints of the city are all over these places, in the form of city commuters, tourists, teleworking, the media, and the urbanization of lifestyles. The traditional divide between the city and the countryside has been perforated.
>
> (Amin & Thrift 2002: 1)

Bidstrup could have written these lines.

Similarly, the concept of planetary urbanization expanded urbanity into a "worldwide urban fabric" of spaces well beyond the traditional city cores and suburban peripheries, ranging from transoceanic shipping lanes to tourist enclaves, the world's oceans, and the atmosphere (Brenner & Schmid 2011: 12). If, however, urbanization can be everywhere and in everything, the concept "fails to refer to anything", as Andrew Sayer emphasized long time ago (Sayer 1984: 281). Or as Hénaff wrote: "[P]recisely at the moment where the world seems to us to become city, the city ceases to be a world" (Hénaff 2008: 11).

The graduated city

The disappearance of two polarized *conceptions* of city/urbanity does not, however, mean the disappearance of the *concept* itself. There is an in-between of the two disappearances, which I, leaning against geographers Jacques Lévy and Michel Lussault, propose to call *the graduated city*.

The core argument is this: *First*, a reasonable concept of city(ness) or urban(ity) cannot do without *including* traditional conceptual ingredients such as Louis Wirth's classical minimal definition of the city as "a relatively large, dense, and permanent settlement of socially heterogeneous individuals" (Wirth 1938: 8) As Lévy has stated, continuing Wirth: "[D]ensity and diversity represent a good measure of urbanness, which can be defined as what makes a city a city" (Lévy 2001: 16). Even the planetary urbanism thesis needs this; it does not forget "the analytical centrality of agglomeration" because, I would say, transformations of heterogeneous bundles of sites and entities outside the agglomerations cannot be named *urban*-izations if dense and diverse urban agglomerations do not exist. They are *urban* because they are "operational landscapes" attached to "centres of agglomerations" (Brenner 2014: 16, 18, 20).

A reasonable concept of cityness or urbanity should, then, include a conception of agglomerated density and diversity. But *second*, the inclusion should not be excluding, as this would only bring back the death of the city. How then to conceive a more expanded concept of urbanity without the concept self-exploding into everything everywhere? Louis Wirth, unwittingly, gave the clue by defining the city as *relatively* large, dense, and heterogeneous, opening the possibility that cityness is a question of degree, of more or less. Further and *third*, as soon as gradation is included within the concept of city, this may affect the components of the concept, and in different ways. Size, density, and heterogeneity do not have to co-variate. Urbanities may be large, while densities are low, without heterogeneity turning into homogeneity. Cityness may occur as *differential compositions of differential degrees* of the components of its concept.

Lévy and Lussault have categorized such compositions into urban "geotypes" of variated assemblies of density and heterogeneity. *Centralities* show high degrees of both; *sub- and peri-urban areas* have low degrees of density; *para-urban areas* have low degrees of heterogeneity; and *infra-urban areas* have low degrees of both, as it is *the limit case to non-urbanity* (Lussault 2000: 32). In contemporary large European cities, the types are dispersed among each other; they are not distributed concentrically from one, historical centre (Lussault 2007: 330–334). Infrastructural mobility networks keep this plurality of differentially composed "urban types" (my concept) together.

The graduated city includes human as well as non-human actors. The city is composed of buildings, infrastructures, mobile objects, and humans, of bricks, glass, steel, concrete, asphalt, trees, birds, foxes, open landscapes, air, weather, mobile objects and human beings. Cities contain "human and non-humans, some of which (like pigeons) are living, and others of which (like street lights) are not". They interfere with (support, constrain, divert, prevent, push, pull, enable) or neglect each other as human and non-human *actors* in urban assemblages (Farías & Bender 2009). Seen this way, cityness is not everything everywhere. It has an outside, which matches the fact that cities cover only 2% of the Earth's land surface (Pincetl 2017: 75).

Nonetheless, the concept can accommodate a multiverse of differently graded urbanities such as the metropolis, the megalopolis, the fragmented metropolis, the metacity (Shane 2011) or the big city, the station town, the suburb, the market town, or the village (Nielsen 2023: chapters 6–10), and it does not require any distinction between city and urban; hence, my use of the terms interchangeably.

The concept of the graduated city can serve as the *lowest common denominator* of the multiple urbanities of the contemporary city. In some respects, it resembles Deleuze and Guattari's concept of the philosophical concept as composed of different components, which relate variably and differently to each other at a plane of consistency or immanence (Deleuze & Guattari 1994: 15, 20). The graduated city concept is descriptive and horizontal. It does not causally explain urban processes. It does not locate urbanity in a vertical hierarchy of nested levels from the global to the household. It is not transcendental, does not find the universe in the city, and does not explain from above or from below from hidden powers. It describes a common plane of immanence (Deleuze 1981: 172, 164) at which a multiverse of urbanities can be conceptualized, theorized, hierarchized, and explained without losing common urbanity.

Metapolitan atmospheres

One contemporary form of the graduated city is the Metapolis as conceptualized and theorized by François Ascher (1995). In Europe, it arose in the post-Second World War period out of the rapidly expanding functionally zoned–"taylo-,fordo-,keynesio-,corbusian"–modernist city (Ascher 1995: 86) as formerly separate urban areas and landscapes were assembled through mobility infrastructures into "vast conurbations" consisting of "dispersed and discontinuous, heterogeneous and multipolarized" territories, where "the limits and the physical and social differences between city and country become increasingly blurred" (Ascher 1995: 58).

Aesthetic atmosphere (Böhme 2001, 1998, 1995) is a vital aspect of city life: sensuous human bodies sensing urban environments, the constellations and ecstatics of buildings, streets, squares, monuments, parks, sounds, busses, cars, bicycles, landscapes, the weather, trees, flowers, birds, and other human beings. From the point of view of aesthetic atmosphere, the Metapolis "assembles a multitude of ambient places (lieux-ambiance), [...] more or less related to each other" (Bourdin 2009: 112). In European cities, this multiplication often has shown a double movement: atmospheres were intensified in the historical centres and diversified throughout the urban types.

My hometown Aarhus offers an example. The historical centre presents an *atmosphere of age* (Albertsen 2019: 21) where one can sense the city as old without knowing much about its history. An *atmosphere of shopping* characterizes it as a commercial centre (Bisgaard 2006: 109). Simultaneously, big city life with its multiplicity of impressions and differences creates *an atmosphere of urbanism* (Wirth 1938). In recent decades, the centre was improved in different ways which intensified this atmospheric conglomerate.

In suburbia, three areas, quite similar in terms of income and upper-class settlement, turn out to be different and quite *unique* in terms of atmosphere. In the Skaade Hills south of the historical centre, a "landlord feeling" prevails; in the area called "Fedet" in Risskov towards the North a "Klondike"-like atmosphere predominates, while a small neighbourhood around Stationsgade in the same area shows a strong *feeling of community* (Raahauge 2007: 71f).

Suburbia also contains *typical* atmospheres. If we stick to habitation: the neighbourhoods of detached housing with atmospheres of privacy around the house and the lawn; the large social housing estates where atmospheres of standardized "neutrality" have been improved with various devices of homeliness; the low-density estates with an atmosphere of a village-like community, supported by the density of the enclaves, the small scale, and the relative closedness (Albertsen 2019: 12).

In the para- and infra-urban areas, a plurality of mixed built-up and landscape constellations shows up. A village is surrounded by detached private housing that faces the fields to obtain a calm landscape atmosphere. Another village has been supplied with new storey buildings that significantly alter its village atmosphere. Around the highway exits, atmospheres emerge from constellations of large box-shaped commercial buildings, traffic facilities, mobility, and open landscapes (Nielsen 2011: 46, 47; Nielsen 2009: 71).

In the suburbia of Vejlby, north of the historical centre, sharply contrasting or clashing atmospheres can be experienced. In the late 1960s, a big commercial centre in glass and concrete was planned for the area. A large old village was eradicated. The plan failed and for many years a sizeable lawn, one building in glass and concrete, and a ring road named Vejlby Centre Road were the only signs of the failed plan and the lost village. Some of the area is now built up, but the empty lawn is still there. Walking through the area, the atmospheric contrasts and clashes stick out. Here a piece of village atmosphere in the remaining parts of the lost village, there the friendly yellow bricks of a social housing area, and the red, warm, welcoming bricks of an elderly centre. In another place the geometric rationality of black, dense high-rise housing clashes atmospherically with low-rise red-brick private housing, while a short walk will lead you into a village-like "Italian" community atmosphere of a two-storey, low-density settlement (Albertsen 2022).

The Metapolitan Anthropocene

The general idea of the concept of the Anthropocene is that the Earth has entered a new geological epoch due to the hitherto unseen human impact on the planet, its climate, ecosystems and geology. Humanity has become a "geological superpower", "a force of nature" that changes the Earth itself radically through species extinctions, the depletion of fossil fuel resources, the impact of greenhouse gasses on climate change, and landscape transformations, among other phenomena (Crutzen & Stoermer 2000; Barry & Maslin 2016). The concept is a highly *contested* one, for several reasons, which I cannot go into here (see e.g., Haraway 2015). Despite the disputes, I keep the concept since it has proved a useful umbrella for interdisciplinary dialogue between the natural and the human sciences (Zalasiewicz 2017: 168). I also follow the argument that its inception can be identified as the Great Acceleration (Head et al. 2021). The Great Acceleration denotes the extraordinary growth after the Second World War in human population and the global economy, the increased use of energy (fossil fuels) and expanded industrialization, all of which have led to accelerating environmental change and rising levels of CO_2 in the atmosphere (Zalasiewicz 2017: 167).

Urbanization has increased rapidly as well. From 1950 to 2015, the number of metropoles (10 million inhabitants) rose from one (New York) to 36 (Zalasiewicz et al. 2017: 344). Cities, the Metapolis included, were a prime mover in the anthropogenic acceleration, not least due to increases in the carbon footprints of urban regions in the Northern Hemisphere (Davis 2010: 41). The "history of the Anthropocene is predominantly the history of urbanization" (Amin & Thrift 2002: 1). Mobility, dispersion, suburbanization, sprawl, and expanded infrastructures have generated anthropogenic effects. As already mentioned, cities cover only 2% of the Earth's land surface, but they consume over 75% of the material resources (Pincetl 2017: 75).

Atmosphere in the Metapolitan Anthropocene

The arrival of the Metapolis with its aesthetic atmospheric multiverse was, then, also an anthropogenic urbanization process that has contributed to global warming and climate change, affected the Earth system, and deteriorated biodiversity. Both processes, the aesthetic and the anthropogenic, influence urban living, and one may search for connections. "After all, elements as diverse as air, water and plants are ambiental factors as well as environmental resources" (Thibaud 2015: 303). In the context of the concept of atmosphere, one obvious *anthropogenic* question is the relationship between meteorological atmosphere (climate, weather) and aesthetic atmosphere.

Air and meteo-aesthetic atmosphere

Aesthetic atmosphere is often said to be experienced directly. Subject and object "melt together," as when you hear and feel the buzz of a mosquito. But how is such directness possible? The buzz would be insensible if space were not filled with air. Hence, perception presupposes air as "a *third term* (my italics) between the perceiving subject and the perceived world" (Thibaud 2015: 224). Subject and object must melt together through something (else), namely air. Paradoxically, the experience of immediacy and directness presupposes mediacy and indirectness.

Air is also a third term between meteorological and aesthetic atmosphere. In order to "transcend the opposition between the meteorological and the affective, we need to refill the atmosphere with the material stuff of air". Air mediates both the atmosphere we breathe and the atmosphere we feel (Ingold 2012: 81).

The *matter* of air is generally defined as the atmosphere, the layer of different types of gases surrounding the Earth. Air can contain water vapour, microparticles, and aerosols (air pollution). Most of these components are imperceptible to the human body unless their proportion in the atmosphere changes significantly. As a *medium*, air enables movement and perception, communication, travel and situatedness. It mediates *climate* with its territorially induced differences and seasonal cycles of repetition and stability, and it mediates *weather* as well. As climate, air is about *states and conditions*, and as weather, it is about *events and energy* (Horn 2018: 7, 9, 13).

The concept of medium can be understood by borrowing and amalgamating from Niklas Luhmann and Bruno Latour. Luhmann distinguishes between two forms of media: loosely and tightly coupled. Loosely coupled media allow mediation "without resistance" (Luhmann 1995: 167, 171), while tightly coupled media do not. Similarly, Latour differentiates between intermediaries and mediators. The former transport without transformation, while the latter transform in transportation (Latour 2015: 37). Imperceptible air functions as a loosely coupled intermediary. If somehow perceptible, it functions as a more tightly coupled mediator transforming affections and perceptions.

Meteo-aesthetic atmospheres of the Metapolitan Anthropocene

Returning now to the above description of aesthetic atmospheres of Aarhus, it obviously "brackets" the influence of climate and weather. Atmospheres are generated by human and non-human actors, but one actor is kept constant: air is implicitly assumed as a loosely coupled intermediary that does not make a difference to aesthetic atmospheres. Cancel the bracketing, and the air as a more tightly coupled mediator can end up affecting atmospheric sensibilities through different temperatures, varying air pollutions, differing "scapes" of sound and light, and different wind "scapes" in the different urban types. The seasons have huge influences as well. The different parts of the Metapolis feel aesthetically different in summertime and when covered with hoarfrost, which also highlights the importance of climatic differences for Metapoles located in different places on the Earth. Shorter term weather events also have an impact.

> Even if not every one of its variations is felt as a noticeable change in the atmosphere, everyday weather ecstasies such as strong rain, strong gusts of wind or fog are touching the conscious bodily state of being in atmospheres.
>
> (Hasse 2012: 164)

Climate change-induced weather shifts may be so strong and all-encompassing (hurricanes, cloudbursts, snowstorms, heat waves, and drought) that they dominate or annihilate the Metapolitan plurality of meteo-aesthetically felt atmospheres.

Atmospherics of Metapolitan anthropogenic politics

Atmospheres affect us and can generate *conflict*. This happened in 2017 in the small street of Volden in the historical centre of Aarhus. Almost all the houses were bought by a large urban development firm, the owner of which wanted to preserve the historical environment and atmosphere.

> When you walk through Volden with its cobblestones, charming houses and delicious shops, you get a very special feeling. The unique mixture of vibrant commercial life and captivating historic buildings creates a rare, cosy atmosphere that you do not experience in many other places. We hope that [...] this conservation project can help ensure that we also can experience this atmosphere in Volden in 5, 10 and 20 years.

The city architect and urban politicians supported this urban atmospheric conservation strategy. But the shop owners in the street worried about the rent, the environment, and the atmosphere.

> We like crooked stairs, street-art and patina. What's charming about our quarter is that we are so different and imperfect [...] but that you can kill by polishing the street [...] There is a quite special atmosphere here, because [we] are in [our] own shops, know each other and can make criss-crossing agreements. We go into the street and do something. But that kind of shops do not have backing from large capitals and cannot handle larger payments of rent.
>
> (newspaper-quotes cited in Albertsen 2020c: 118f)

Both parties wanted to keep the atmosphere, but they didn't agree on which and how. One would *polish* it, the other would keep it *unpolished*, using the language of the shop-owners. The confrontation did not lead to an open clash, but towards a slow transformation of the quarter through various changes in shop ownership and volumes of shops.

In Copenhagen, a long, fierce struggle over the non-built-up area of the "Amager Common" ran from 2017 to 2021. Against each other were pitted, on the one hand, a welfare city development project supported by the Social Democrats and the centre-right parties in the city council satisfying the needs for social housing, closeness to public transportation, and economic urban development. On the other hand, civil society movements were prioritizing biodiversity, protecting several sorts of orchids, the northern crested newt, the moor frog, and a unique piece of nature in the city. The civil society movements won the battle, and the urban development project was stopped (Pagh 2022).

Both of these cases, Volden's atmosphere of age and shopping in the historical centre of Aarhus and the Amager Common, a para-urban area in Copenhagen, are examples of atmospheric political conflict in the Metapolitan Anthropocene. The Volden case is not only obviously atmospheric, but also anthropogenic in the sense that the Anthropocene requires the use and reuse of built structures for much longer periods than present practice, which emotional attachments to the heritage of built structures and its preservation can assist (Sieverts 2017: 103). The Amager Common case is anthropogenic in the sense of being concerned with the preservation of biodiversity and the possibility of co-habitation between humans and nonhumans, of "making kin" rather than "city" (Houston et al. 2018). It is also atmospheric in the sense that emotional attachments to the peculiar "bio-atmosphere" of the Amager Common may have been a strong motivation for struggling politically for its preservation. Thus, different atmospheric attachments gesture towards different policies and ways of living in and with the Metapolitan Anthropocene.

Note

1 This chapter draws upon and develops two earlier articles: Albertsen (2020a) and Albertsen (2020b). All translations from the French, German and Danish are mine.

Bibliography

Albertsen, Niels. "Atmosfærer i Vejlby-Risskov", unpublished lecture to the Vejlby-Risskov Lokalhistoriske Forening, 15 September 2022.

Albertsen, Niels (2020a). "The Anthropogenic Metapolis: Atmosphere, Anthropocene, Urbanity" in *Arquitetura, Subjetividade e Cultura / Architecture, Subjectivity and Culture: cenários de pesquisa no Brasil e pelo mundo /scenarios and transversal paths for research*, edited by Cristiane Duarte & Ethel Pinheiro, 318–335. Rio de Janeiro/California: PROARQ, 2020.

Albertsen, Niels (2020b). "Atmosphere and the Anthropogenic Metapolis", in *Ambiances, Alloaesthesia: Senses, Inventions, Worlds. E-conference. Proceedings of the 4th International Congress on Ambiances, 2–4 December 2020*, edited by Damien Masson, Volume I, 74–79. Réseau International Ambiances, 2020.

Albertsen, Niels (2020c). "Den atmosfæriske by", in *Gentænk byen*, edited by Gertrud Jørgensen et al., 111–118. Copenhagen: Dansk Byplanlaboratorium, 2020.

Albertsen, Niels (2019 [1999]). "Urban Atmospheres", *Ambiances* [Online], Rediscovering, https://journals.openedition.org/ambiances/2433.

Amin, Ash & Nigel Thrift. *Seeing Like a City*. Cambridge, UK: Polity, 2002.

Amin, Ash & Nigel Thrift. *Cities. Reimagining the Urban.* Cambridge, UK: Polity, 2002.

Ascher, François. *Les nouveaux principes de l'urbanisme.* La Tour d'Aigues: Éditions de l'Aube, 2017.

Ascher, François. *Métapolis ou l'avenir des villes.* Paris: Editions Odile Jacob, 1995.

Barry, Andrew & Mark Maslin. "The Politics of the Anthropocene: A Dialogue", *Geo: Geography and Environment* 3(2) (2016) https://doi.org/10.1002/geo2.22.

Bidstrup, Knud. "Byen Danmark", In *Danmark under forvandling*, edited by Knud Bidstrup & Erik Kaufmann, 138–150. Copenhagen: Fremad, 1963.

Bisgaard, Ulrik. "Æstetikkens overflade og dybde – shoppingcentret i nyt lys", in *Det æstetiskes aktualitet*, edited by Ulrik Bisgaard & Carsten Friberg, 100–112. Copenhagen: Forlaget Multivers, 2006.

Bourdin, Alain. *Du bon usage de la ville*. Paris: Descartes & Cie, 2009.

Brenner, Neil. "Introduction: Urban Theory Without an Outside", in *Implosions/Explosions. Towards a Study of Planetary Urbanization*, edited by Neil Brenner, 14–30. Berlin: jovis, 2014.

Brenner, Neil. "Theses on Urbanization", *Public Culture* 25(1) (2013): 85–114.

Brenner, Neil & Christian Schmid. "Planetary Urbanization", in *Urban Constellations*, edited by Matthew Gandy, 11–14. Berlin: Jovis Verlag, 2011.

Böhme, Gernot. *Aisthetik. Vorlesungen über Ästhetik als allgemeine Wahrnehmungslehre*. Munich: Wilhelm Fink Verlag, 2001.

Böhme, Gernot. "Die Atmosphäre einer Stadt", in Gernot Böhme: *Anmutungen. Über das Atmosphärische*, 49–70. Ostfildern vor Stuttgart: edition tertium, 1998.

Böhme, Gernot. *Atmosphäre. Essays zur neuen Ästhetik*. Frankfurt a. M.: Suhrkamp Verlag, 1995.

Choay, Françoise. *La règle et la mèthode*. Paris: Seuil, 1980.

Crutzen, Paul J. & Eugene F. Stoermer, "The Anthropocene", *IGBP Newsletter* 41 (2000): 17–18.

Davis, Mike, "Who Will Build the Ark", *New Left Review*, 61 (2010): 29–46.

Deleuze, Gilles. *Spinoza – Philosophie pratique*. Paris: Éditions Minuit, 1981.

Deleuze, Gilles & Félix Guattari. *What Is Philosophy?* New York: Columbia University Press, 1994.

Farías, Ignacio & Thomas Bender. *Urban Assemblages. How Actor-Network Theory Changes Urban Studies*. London: Routledge, 2009.

Haraway, Donna, "Anthropocene, Capitalocene, Plantationocene, Chthulucene: Making Kin", *Environmental Humanities* 6 (2015): 159–165.

Hasse, Jürgen. *Atmosphären der Stadt*. Berlin: jovis Verlag, 2012.

Head, Martin J. et al. "The Great Acceleration is real and provides a quantitative basis for the proposed Anthropocene Series/Epoch", *Episodes. Journal of International Geoscience*, 2021, https://doi.org/10.18814/epiiugs/2021/021031.

Hénaff, Marcel. *La ville qui vient*. Paris: Éditions de L'Herne, 2008.

Horn, Eva, "Air as medium", *Grey Room* 73 Fall, (2018): 6–25.

Houston, Donna, Jean Hillier, Diana MacCallum, Wendy Steele & Jason Byrne, "Make Kin, Not Cities! Multispecies Entanglements and 'Becoming World' in Planning Theory", *Planning Theory*, 17(2) (2018): 190–212.

Ingold, Tim, "The Atmosphere", *Chiasmi International*, 14 (2012):75–87.

Latour, Bruno. *Reassembling the Social*. Oxford: Oxford University Press, 2015.

Lévy, Jacques (2001): "Measuring Urbanness", in *Change and Stability in Urban Europe*, edited by Harri Andersson, Gertrud Jorgensen, Dominique Joye & Wim Ostendorf, 15–25. Andershot: Ashgate Publishing Limited, 2001.

Luhmann, Niklas. *Die Kunst der Gesellschaft*. Frankfurt a. M.: Suhrkamp Verlag, 1995.

Lussault, Michel. *L'homme spatial. La construction sociale de l'espace humain*. Paris: Éditions du Seuil, 2007.

Lussault, Michel, "La ville des geographes", in *La ville et l'urbain, l'état des savoirs*, edited by Thierry Paquot, Michel Lussault & Sophie Body-Gendrot, 21–35. Paris: Éditions de la Découverte, 2000.

Nielsen, Tom. *Byen Danmark*. Copenhagen: Arkitektens forlag, 2023.
Nielsen, Tom (2011). "Nye urbane typologier", in *Grænseløse byer. Nye perspektiver for by- og landskabsarkitekturen*, edited by Thomas Juel Clemmensen, 40–53. Aarhus: Arkitektskolens Forlag, 2011.
Nielsen, Tom. *Det urbaniserede territorium. Østjylland under forandring*. Aarhus: Arkitektskolens Forlag, 2009.
Pagh, Jesper, "On the Battle for Amager Common: A Power Play of Housing Politics in Copenhagen", in *Architectures of Dismantling and Restructuring – Spaces of Danish Welfare 1970–Present*, edited by Kirsten Marie Raahauge, Deane Simpson, Martin Søberg & Katrine Lotz, 422–444. Zürich: Lars Müller Publishers, 2022.
Pahl, R. E. *Whose City? And Further Essays On Urban Society*. Harmondsworth: Penguin Books, 1975.
Pincetl, Stephanie, "Cities in the Age of the Anthropocene: Climate Change Agents and the Potential for Mitigation", *Anthropocene* 20, December (2017): 74–82.
Queysanne, Bruno. (1991): "L'échelle comme niveau: de l'architecture à la ville", in *De l'architecture à l'épistémologie. La question de l'échelle,* edited by Philippe Boudon, 47–64. Paris: Presses Universitaires de France, 1991.
Raahauge, Kirsten Marie. *En Århusantropologi*. PhD-dissertation. Hørsholm: Statens Byggeforskningsinstitut, 2007.
Rossi, Aldo. *The Architecture of the City*. Cambridge, MA & London: The MIT Press, 1992 (1966).
Sayer, Andrew, "Defining the urban", *Geojournal* 9(3), (1984): 279–285.
Shane, David Grahame. *Urban Design Since 1945: A Global Perspective*. Chichester: John Wiley & Sons, 2011.
Sieverts, Thomas, "The Principle of Heritage – Preservation and its Generalisation in the Anthropocene", *disP-The Planning Review* 53(1), (2017): 99–105.
Sloterdijk, Peter. *Der äesthetische Imperativ*. Hamburg: Philo & Philo Fine Arts, 2007.
Sloterdijk, Peter. *Sphären Band III: Schäume*. Frankfurt am Main: Suhrkamp Verlag, 2004.
Sloterdijk, Peter. *Sphären Band II: Globen*. Frankfurt am Main: Suhrkamp Verlag, 1999.
Thibaud, Jean-Paul. *En quête d'ambiances*. Genève: Metis Presses, 2015.
Wirth, Louis, "Urbanism As a Way of Life", *The American Journal of Sociology* 44(1) (1938): 1–42.
Zalasiewicz, Jan. "Die Einstiegsfrage: Wann hat das Anthropozän begonnen?", in *Das Anthropozän. Zum Stand der Dinge,* edited by Jürgen Renn & Bernd Scherer, 160–180. Berlin: Matthes & Seitz Berlin, 2017.
Zalasiewicz, Jan, Colin Waters & Mark Williams, "Les strates de la ville de l'Anthropocène", *Annales HSS* 72(2) (2017): 329–351.

22 The developer over-investment in ambiances

Laurent Devisme

Introduction: a grammar of enchantment

Anyone who takes the expressiveness of the urban phenomenon seriously (Devisme, 2013) may be struck by the significant convergence between the urban design of metropolitan city centre neighbourhoods and the hypostatized experience of a vibrant city.[1] At times, this may smack of American-style entertainment but as the shopping mall model has spread to other parts of the city, we may more generally note a linguistic inflation in the promises that city centres make. The contemporary enchantments of narrative urbanism raise the larger issue of the relationship between developers' visions, urban planning techniques, and the experience of such central spaces. We will concern ourselves here with the use of ambiance by a city's technical and political apparatuses.

In terms of building the city, our contemporary period is characterized by a tendency to de-specialization and the blurring of boundaries between the practices and disciplines for transforming urban spaces. In the last 10 years, the summer university programmes for urban planners in France have thus included notions on their agenda such as sharing, well-being, pleasure (!!), likability, hospitality, and urban psychoanalysis. The emerging practice of temporary or prefigurative urban development is rife with project management assistance to support government decisions, equipping such processes considerably, gathering words and staging them, promoting third places and, more generally, a regime that explores action. The language of developers, which is always connoted, situated, and performative (Chauvier, Devisme, 2021), is further shaped by the opinions of social scientists and artistic approaches to the city.

An aptly titled article in the monthly newspaper *Le Monde Diplomatique,* "Big cities and good feelings", looked at the life work of Richard Florida, which sought to model the city in the image of its creatives. His use of the "Hipster Index" produced by the company *Infogroup* brought up the issues with revealing and producing specific ambiances by counting the number of tattoo parlours, bicycle shops, independent cafés, craft breweries, vintage clothing boutiques, and record stores. Other consultants have used models

DOI: 10.4324/9781032655307-31

and figures that give new meaning to what used to be called "urban animation" in the 1970s. Today, a liberal grammar of urban innovation derives from a vision consisting of makers, street mechanics, and flea markets!

A new alloy of representations and embodiments seems to be forming. More than ever before, representations have become actants in the processes of decay as well as gentrification, and we can document this before pointing out the paradoxes of a narrative over-investment in central urban spaces as a whole.

Ambient stigmas and attractions: a few circumstantial observations

Let us begin by writing about circumstances as a variety of situations that appear to belong to the same family. Culled from an urbanographic journal,[2] these materials express a certain sensibility, an invitation to the outdoors which is the repeated advice of a scholar of social spaces.[3] These materials for thought (an expression used by Michel Lussault to introduce his ideas in a text on the future of public spaces, Lussault, 2001) are characterized – were it necessary to justify them yet again – by a case-based thinking (Passeron, Revel, 2005) and a pragmatic approach to the states of society (the weaving of the social fabric) that Bruno Latour outlined so valuably. The challenge of describing, characterizing, de-epistemologizing the sciences and paying attention to flows of experience instead of perilous leaps take priority in the scientific approach adopted by this inquiry. "... characterizing, or rather, raising the question of characteristics, means envisioning a situation in pragmatic terms, based on what we can think we know, but without giving this knowledge the power of a definition" (Stengers, 2009, p. 34).

What then should the impetus be for scholars? Considering representations as stakeholders in reality, as well as paying attention to the ordinarization of the language of urban planning and the discrepancies between the affective tonalities experienced and the ambiances projected, with all ensuing incongruities and oddities. This also involves paying attention to detail, to how a scholar enters, captures, and portrays reality. These three levels matter in the ethnographer's writing process (Datchary et al, 2013).

Let us specify that various metrics may be used for such an endeavour: cyclists, pedestrians, even automobiles. Motoring on the highways across the San Francisco Bay in mid-2023 thus revealed the visually imposing nature of its large advertising billboards, almost all of which were devoted to the promise of artificial intelligence. A new (or should we say augmented?) narrative from Silicon Valley clearly emerges from signs towering above the roadways about 400 self-driving cars "circulating around town", an experiment ahead of an entry into service that will undoubtedly change the tonality of our cities.

Taking salient cities seriously (Médam, 1998) means bearing witness to an era of urban development that entails a renewal of local marketing and the deployment of a fictional urban planning (Matthey, 2011). The visions of metropolitan city centres are enchanted views of ambianced cities that use images, literature, and film as markers and emblems, as well as for derisive or ironic purposes.

The exploitation and inversion of stigma: bashing, dark tourism

Poorly positioned medium-sized cities, former industrial towns in decline, and many other areas have since the end of the twentieth century been beset by social and economic difficulties, the confluence of which may generate a sense of stigma. Bashing, a term that describes a "game" or form of release consisting of collectively denigrating a person or subject also applies to these quasi-characters. With social media feeding anonymous expression, this phenomenon is booming: local areas can be denigrated, almost "slapped around", with varying degrees of ripple effects. The performative nature of such utterances (often made in the form of relatively caricatural reportages) is revealed by the scope of the reactions of denial or confirmation. Local actors invite the writer to return to see things from a different perspective, or a nationally distributed newspaper is rebuked by city officials, academics, and actors in the business sphere who complain of a biased view that evokes ambiances more typical of nineteenth-century industrial cities than of their city today. This bashing of local areas as being neglected and lifeless can in some instances be subverted and used to the place's advantage. We can see this in the city of Charleroi, for example, whose tourism department tenders the following invitation: "Visit the most incredible industrial city in Europe". On its website, the tourism company Charleroi Adventure proposes urban safaris to

> Discover the place where Magritte's mother committed suicide, the home of Raymond La Science of the Bonnot Anarchists, the 'ghost metro station', the most depressing street in all of Belgium, climb on a *terril* (waste coal pile) and visit an authentic abandoned factory.

This is a kind of dark tourism, an enchantment with decay. Between the indignation of local actors and the exploitative aestheticization of ruins,[4] we can see the spread of certain stubborn representations that are hard to shake off, for example, of the many places that are perceived as unsafe (e.g., French *banlieues*) or simply as ugly (as in "Ugly France", the unfortunate title of a cover article in the magazine *Télérama* that inspired an essay by Eric Chauvier in response).

The question of controlling a city's image and its representations is paramount. Most often this involves countering stigmas, but it can sometimes be highlighted in terms of their potential (for exploration, new perspectives, and even amazement). It may concern a group of activists in search of new experiences in the realm of *urbex*, or urban exploration.

Public actors may remain on edge in this situation, but they have also shown themselves to be proactive. We can see this clearly in the industrial port city of Saint-Nazaire, which has for several years highlighted the heritage of its port, invoking the world of the comic book character Tintin[5] and redeveloping its waterfront in a new urban style. Thus, the Place du Commando at the end of the boardwalk stands in marked contrast to its previous incarnation as a parking area. The public space is carefully tended, with large, alternating black and anthracite grey concrete slabs. It has become a place to be visited, surrounded by bars, creperies, and a restaurant. A small kiosk contains public toilets, flanked by a long bench adorned with plaques featuring engravings of plans of various boats that were built in the city. There is even a digital billboard showing images of the area, which reads: "Welcome to the Place du Commando – Restaurants, bars, and open-air terraces facing the sea – a 3km promenade along the ocean – umbrella pines and a seaside ambiance – holly oaks, cork oaks, strawberry tree... it looks like the South of France". This slip perplexes visitors; why should one want to be anything other than the Nord-Loire region? A tonality has been offered, bolstered by political messaging highlighting the construction of a "pleasurable city". However, this city's characteristics are very different, and this induces a disorientation between the city as it is and as it is being projected.

Working-class neighbourhood, dodgy ambiances, and linguistic surprises

Let's shift our attention to Grenoble. During a research programme that focused on the relationship between ambiances and stages,[6] while on a stroll through the city between the neighbourhoods of Berriat and Saint-Bruno, we happened upon an eco-district emerging among cultural facilities and successful building conversions that nevertheless failed to prevent the spray-painting of the tag "hipsters in yellow vests"![7] These are the folks captioning real estate construction programmes. Across the way is the "Petite Halle" marketplace, its ground-floor communications chock-a-block full of images and quotes; Pablo Picasso, Mark Twain, and Henry Ford all utter phrases embodying humanist capitalism, liberal meritocracy, and enlightened individualism. A stone's throw away, we found the completely different ambiance of a lively, chatty suburban area. Its signs were different: murals that had been vandalized to varying degrees, political messages, and numerous codes and over-codes. In a nod to Pierre Sansot, who taught here, the city seems incredibly chatty.

One can have similar experiences in other cities that, in the last several years, have begun to thematize their neighbourhoods (without, however, turning them into theme parks). These areas feature ambiances typical of a suburban urbanism, the overvaluation of neighbourly qualities, and the extension of the model of the "15-minute city" promoted by the urbanist-consultant Carlos Moreno which one hears about in numerous European cities. In reality, interpreting what is going on in many such "suburbs" is entirely a matter of nuance.[8] Recent doctoral theses[9] have shown the importance of the semantic frame of reference that developers use, which consists of creativity and conviviality, and which is promoted by government development offices, major transportation agencies, and consulting firms.

Ambivalences of the vibrant city

One final issue to contemplate may lead us to the creative city mentioned at the outset: a press conference in a large French city with a developer presenting a participatory approach defined as "innovative and offbeat". This is the response to the new standard of "the non-bothersome city" (or rather, an apology for the occurrence of unwanted phenomena). Numerous city development plans sing the praises of intensity while avoiding the delicate issue of urban density and height. It's true, the wording shows little inspiration. These monikers are all "translations" of the "vibrant city". The washing machine of urban planning concepts spins quickly! Two years earlier, the catchphrase was "the calm city"... a calm but vibrant city... And now, notions of the hand-sewn and of an everyday urbanity being highlighted so vigorously are not bad news by any means. The somewhat techno-centric vision of sustainable development that was put forth a decade ago appears to be ceding to an emphasis on usages, the nearby, the ordinary, the common, and atmospheres. This expresses the idea of a facilitating urbanism, and it echoes the somewhat hackneyed model of the 15-minute city.

In her research, Margaux Vigne studied the conducting of institutional experiments, showing among other things the frenetic nature of their approaches, whether they are separate from, or integrated into, urban development processes. Far from being standardized, they sometimes explicitly pick up certain subtle points of art criticism (for example, when they strive for "inclusive gentrification"), but most of the time, they raise the lively issue of how an ordinary city is rendered in words and images.

Paradoxes in the language of contemporary urbanism

Two tenacious paradoxes can be put forth with regard to ambiances as a new domain of urban action and as a narrative vector. The first concerns the need to imbue an ambiance and to get it to hold over time, even as the flow of references accelerates; the second concerns the risk of amplification, of overdoing the spatial potential and generating more disappointment than elation.

Holding steady over time

Enchantment contends with a temporal challenge: it has to "hold steady over time", even though fashions, styles, and concepts circulate and expire with increasing speed in the international flow of ideas. The intentional, public fabrication of spaces always supposes that they have been prepared and prefigured. The right tone must be given, and it must be "made to seem true". This presupposes consistency, relevance, agreement, as well as actants to bring this about, i.e., actors and their narratives, as well as devices consisting of all manners of unmovable mobiles, models, diagrams, perspectives, and views from the air. Iconic and verbal images must be put together without dissonance (or with as little of it as possible). The long timeframe of urban development presupposes a regular replenishment of its own projects. It is also a political task, which consists of rephrasing, of reminding that things are being held steady, from start to finish. Therefore, the challenge lies in getting the project or the tonality to last so as not to "break the ambiance". While many studies of tourism practices have shown the sometimes devastating effects of disappointment on arrival, this also largely holds true for major urban development programmes that are too far removed from their stated goals (without neglecting the increasingly broad scope of the criticism of major, useless projects that have been decreed, as well as of "large urban planning projects" in general).

As one example among many of this speed of semantic evolution, the plan for the Île de Nantes went in just a few years from eco-district to suburb to creative neighbourhood, not to mention the "metacentre" invoked by one of its urban planning project management teams. The concepts overlap one another like layers of a palimpsest rather than succeeding and replacing each other; they must be rendered compatible with one another, and their spatial values must be localized with ever-greater precision. This does not exclude a sense of strangeness and paradox; the street called "Le mail du front Populaire" now serves most of the buildings held to embody the creative district, along with a sky-bar, a law firm, and an architecture studio, among others. The urban sensations they generate are far removed from the populist vision inherent to such a street name.[10]

Narrative over-investment

Narrative over-investment comes to form part of the deployment of a fictional urban planning. In fact, the point of many projects seems to be to build and sustain the image of a neighbourhood. In addition to the return of a toponymy, we can see to what extent popular working-class identities are now in some way pre-empted by urban planners who transform them into veritable editorial policies. Again on the Île de Nantes, we can see the invocation of the major trope of the suburb and the (re)construction of a veritable mythology around it: the village square, the fireman's ball, the little people, and so on. We witness the association of a specific urban form with the history of populating a place, with precarious or artistic social forms. This is how a narrative over-investment is produced: the scripts for many practices are decreed and promised, and verbal and non-verbal images are associated to suggest veritable atmospheres to be experienced. In urban planning, as in other areas of professional practice, storytelling techniques have been put to work. Theory recognized the narrative aspect of planning 20 years ago (Sandercock, 2003), and this dimension has truly invaded the techno-political field since then. This has led to the over-coding of public spaces. We simply ask too much of them. This is no longer in the sense of what typical postmodern public spaces may have represented (as standard bearers for new cities, for example), rather due to a saturation of the public space with potential practices based on an urban ordinary.

The strange words of this urban factory are supposed to transport us while remaining vague. The risk of levitation arises, despite its being a foil and the fact that a sensorial urban planning in fact seeks to re-articulate history, forms, and practices.

Conclusion: descriptive strategies and semantic clarifications

These thoughts on the "ambiancing" of our cities demonstrate the interest in combining two descriptive strategies. They must of course be differentiated, as Mr. Breviglieri has asked us to do: either describe how the world builds pressure (the guaranteed city) or explore what eludes pressure and channelling in our different fields of research (Breviglieri, 2013). Without confusing these two strategies, they can be combined to document a contemporary transformation that entails both the exercising of forms of pressure and the deployment of new forms of collective coexistence, especially around a sense of shared experiences.

In addition to a genealogical effort, the analysis of development tropes and discourses enables us – according to the perspective put forth in these pages – to demand greater precision and to explore the consequences of choosing a specific option. Mathieu Berger thus explored the differences suggested by the terms of a "friendly," "inclusive," "welcoming," or "easy" city.[11] Such terms seek to lend tonality to a city, but they are dissimilar from one another. In fact, shouldn't we be talking about welcoming refugees in our urban spaces, the civic coexistence of populations in a city, and accentuating our relationship to neighbourhoods and feeling of belonging? Can we connect semantic precision to precision in political action? Precisely qualifying observable lifestyles and attitudes in certain urban spaces allows us to address the challenges of a politics of diversity; how can we install, preserve, and express this?

The "inclusive city" cannot subsume everything. The same holds true for the term "vibrant", which substantiates a notion originating with a sense of movement or a wave (vibrational). This reification of atmosphere raises the issue of creating tonalities and ambiances, for which we must heuristically differentiate the various registers of presence: impregnation, injunction, and invitation (Thibaud, 2018) are not the same, and in terms of engendering an ambiance, we clearly prefer the register of invitation, in resonance with a politics of facilitation. Going from friendly ("you have to know someone before becoming friends") to easy (allowing for strangeness and non-familiarity to be deployed) presupposes that we pay attention to, and maintain, the amenities and facilities with which we overcome the challenges inherent to this process.

Narrative over-investment can produce word-images that end up obfuscating things. They can be found in a kind of cultural essentialism that is used to personify cities: "Montreal, the welcoming city", "Hospitality,

Nantes-style", "Saint-Nazaire, a stunning city", and so on. The suitable, dual descriptive strategy entails deconstructing mythologies and understanding ordinary experiences. It identifies a process of linguistic colonization. The scholar's modest role thus consists of exploring and demonstrating that we learn words in certain contexts and that there has to be a context for the language to concord precisely – a lesson for the worlds that developers envision. In sum, between language and practice, there are more passageways than obstacles.

Notes

1 This text is a reworked version of the one that was published under the title "Vertiges de l'usage aménagiste des ambiances" in the records of the Cerisy conference *L'usage des ambiances. Une épreuve sensible des situations* (Paris: Hermann, 2021): 83–96.

2 The equivalent of Ruwen Ogien's philosophical journal or of Edgar Morin's field notes in California in 1969.

3 Initially produced for the student web-based radio station at ENSA Nantes, readers may find several episodes of the program *Invitation au Plein Air* at https://soundcloud.com/laurent-devisme/sets/chronique-invitation-au-plein

4 In this sense, we can also see how the Anthropocene is aestheticized in certain photographs that border on the problematic genre of Anthropocene landscapes that are taken generally from above as "large scale views in colour, generally without any human presence, ambiguously transforming landscapes ravaged by capitalist industry into disturbing, abstract expressionist canvases" (Noirot, 2023, 149).

5 https://www.saint-nazaire-tourisme.com/decouvrir/decouvrir-le-port/tintin-a-saint-nazaire/

6 This was the research programme funded by the ANR titled SCAENA: https://scaena.hypotheses.org/

7 An allusion to the yellow vest movement which arose in the fall of 2018 across France in protest of the rise in fuel prices and which expanded to challenge more broadly the evolution of France's social and fiscal model.

8 As Marie-Hèlène Bacqué often argues in *Retour à Roissy* whenever she encounters projects entailing an artist occupancy of a site as a form of temporary urbanism. Such initiatives show the other possible cultures that sites can host and the new forms of solidarity they may engender, but they are also pioneers of an enhancement that other economic forces will ultimately capture.

9 We can cite Margaux Vigne, "occuper les lieux, occuper les habitants" (Nantes: Univ. de Nantes, 2019), Aurélie Landon, "La fabrique de la ville à l'épreuve de l'innovation sociale" (Paris: Univ. de Paris 8, 2022), Flavia Pertuso, "l'idéologie de l'attractivité" (Nanterre: Univ. de Paris Nanterre, 2022), Vanessa Stassi, "chronique d'une métamorphose urbaine" (Grenoble: Université Grenoble Alpes, 2022), and Yoann Peres, "Les *Réinventer* : l'urbanisme négocié à l'épreuve" (Paris: Univ. de Paris Est, 2023). The features of such works often become perceptible under the auspices of mechanisms such as the calls for ideas tendered by cities.

10 In Pierre-François Lebrun's documentary film *Entre deux eaux*, we see a former shipyard worker decrying a building made from Corten steel, called the "Île Rouge", or "red island", as rust was a shipbuilder's biggest enemy.

11 See his contribution to the closing discussion at POPSU 2, September 2017, "être métropole dans un monde incertain", during the session "Vers la métropole fluide". https://www.urbanisme-puca.gouv.fr/IMG/pdf/diagonal_no_202_colloque.pdf

Bibliography

Bacqué, M. H. *Retour à Roissy. Un voyage sur le RER B.* Paris: Seuil, 2019.

Breviglieri, M. "Une brèche critique dans la « ville garantie? Espaces intercalaires et architectures d'usage", in *De la différence urbaine. Le quartier des Grottes / Genève*, edited by E. Cogato-Lanza, L. Pattaroni, M. Piraud, and B. Tirone. Geneva: Mètis-Presse, 2013, 213–236.

Chauvier, E. *Contre Télérama.* Paris, Allia, 2011.

Chauvier,E., L. Devisme. "Langage et condition urbaine. Entre donner le ton et casser l'ambiance", in *La ville et les mots*, edited by I. Chesneau. Marseilles: Parenthèses, 2021.

Datchary, C. et al. *Petit précis de méthodologie. Le sens du détail dans les sciences sociales.* Lormont: Le bord de l'eau, 2013.

Devisme, L. "Ressorts et ressources d'une sociologie de l'expérience urbaine", *Sociologie et sociétés*, Vol XLV No. 2 (2013): 21–44.

Lussault, M. "Au-delà de l'espace public. Propositions pour l'analyse générale des espaces d'actes", in *Réinventer la ville. Les espaces publics à l'ère globale, collection Espaces et cultures*, edited by C. Ghorra Gobin. Paris: L'Harmattan, 2001, 33–46.

Matthey, L. "Urbanisme fictionnel: l'action urbaine à l'heure de la société du spectacle", *Métropolitiques*, 28 octobre 2011. URL: https://www.metropolitiques.eu/Urbanisme-fictionnel-l-action.html.

Médam, A. *Villes pour un sociologue*. Paris: L'Harmattan, 1998.

Noirot, J. "Paysages critiques et luttes écologiques dans la photographie documentaire", *Paysages photographiés, revue 303*, no. 175, (2023): 146–153.

Passeron, J-C., J. Revel. *Penser par cas.* Paris: EHESS, 2005.

Sandercock, L. "Out of the closet: the importance of stories and storytelling in planning practice", *Planning theory and practice*, Vol 4/1, (2003): 11–28.

Sansot, P. Poétique de la ville. Paris: Klincksieck, 1971.

Stengers, I. *Au temps des catastrophes: résister à la barbarie qui vient.* La découverte. 2009.

Thibaud, J.P. "Donner le ton aux territoires", in *Ethnographier les sens*, edited by Colon, Paul-Louis. Paris: Editions Petra, 2013, 235–255, Anthropologiques. ‹hal-00995503›.

Thibaud, J.P. "Les puissances d'imprégnation de l'ambiance", *Communications*, Vol 102 (2018): 67–80.

Vigne, M. *Occuper les lieux / occuper les habitants. Ethnographie de deux expérimentations institutionnelles. Bruxelles, Nantes.* Doctoral thesis in urban planning, Nantes: University of Nantes, 2019.

Opening

Ambiances of thought: A contribution to the analysis of aesthetic cognition

David Zerbib

The notion of ambiance encompasses not only a sensory experience of a situation but also a cognitive condition of knowledge. Along with the empirical and phenomenological dimension of ambiance or atmosphere comes a theoretical one, which leads us to turn our attention to the form of ideas engendered by the experience of ambiance. Thus, the issue of ambiance invites us to question a specific form of thought, which presents a theoretical interest, even when it is not considered in relation to an immediate and actual ambiance or atmosphere. There is a rather large spectrum of theories that advocate for the paradoxical cognitive value of indeterminacy, from the confused ideas described by Leibniz and the ones embedded in the genesis of Aesthetics by Baumgarten to the sense of conceptual openness and vagueness analysed by semioticians and logicians, not to mention the interpretations of art and literature by contemporary philosophers like Deleuze.

In what way and in which specific cases is it fruitful to analyse the cognitive processes at stake through the lens of atmosphere or ambiance? Drawing upon the concept of an "ambiance of thought", we examine a series of examples and experiences in philosophy, literature, and the arts in which knowledge relies on a kind of conceptual indeterminacy that reflects the dynamic features of an ambiance, especially its property of *diffusion*.

We must acknowledge the paradoxical nature of the proposition to examine *ambiances of thought* as we seek to clarify our *thought about ambiances*.[1] Would it not be preferable to continue to try to define the subject of our inquiry, and to extend our thought process as rigorously as possible to understand the nature of this particular reality that is so hard to define and demarcate, one that already considerably blurs the boundaries between things, unfolds in the dimension of an in-between, and goes so far as to render the boundary between subject and object utterly fragile? Why should we add, to the theoretical challenge of this blurring of the object, the overturning effect of questioning the very thought that seeks to grasp this phenomenon? In other words, what do we have to gain philosophically by adding a chiasmus that turns us upside-down to this fog that surrounds us?

DOI: 10.4324/9781032655307-32

To answer these questions, we will begin with the hypothesis that the empirical traits of ambiance and the perceptual, relational, dynamic, and situational modalities of how we experience it give rise to a certain state of mind that may be considered on its own, independently of the experience of a specific ambiance. Ambiance might indeed refer to a consciousness that lacks a clearly defined object of thought, an awareness of *something* which is, however, not a distinct object that we can think about. It is an "ambiantal" and atmospheric state of consciousness that can only reflect on this situation, expand it, and make it resonant, but without managing or perhaps even trying to clarify, grasp, or identify the exact shape of what it is we are thinking about. Thus, once we have thematized this state of mind to reflect it back at itself, there are *ambiances of thought* that possess a kind of relative autonomy in relation to the experience of ambiance as a sensory experience. It is a matter of "relative autonomy", because this does not involve referring to the idealism of a sense of reason that understands itself, independently of any situated, empirical experience. To the contrary, it is by means of the situated experience of ambiance and its phenomenology that we can conduct our analysis, and through which we can understand that a certain kind of thinking takes place under what we could call *ambiantal conditions*. Once we apply this mode of thinking, it can follow its own speculative dynamics that resonate in us even when the conditions for its initial application are no longer current.[2] We are thus simply suggesting that this form of thought, which is genetically attached to a certain experience, may be contemplated independently of its original sensory condition. As a result, even though isolating this mode of thinking from its phenomenological dynamic may seem excessively formalist, this actually opens up the concept of ambiance and helps shine a light on its theoretical scope, especially in the field of aesthetics, by linking it to the traditional issues of "sensible knowledge", while also forming connections with the domains of psychology, the philosophy of mind, and philosophical anthropology, which emphasize the situated and ecological aspects of thought.

If we admit that human thought emerged from a somatically, ecologically, and spatially situated animal experience, and if thinking designates the fact of having and of forming mental representations, then these representations, which we can also call "ideas", are tied to cognitive matrices structured by situated experience. What ambiances ask us to consider is the form of a particular situated experience, which specifies a type of ecology and phenomenology of experience. Thus, we must look at a kind of thinking that has been forged by this type of ecology and phenomenology of experience to shed light on a relationship to a knowledge that does not consist of possessing clear and distinct representations that correspond to objects in the world but instead consists of a divestiture from objects and a blurring of representations. This situation paradoxically possesses an essential cognitive value that aesthetic thought allows us to emphasize in particular.

Knowledge under the empire of sight

Walter Benjamin demonstrated the theoretical value of *Denkbilder*, or "thought-images",[3] which can reveal the meaning of history through the flash of a visual form or in a specific detail of daily social or cultural life. This is not a question of analysing the image of a historical object to interpret the meaning that this representation gives to history, instead of grasping the historical thought immanent within this image. Nor is this form seen or thought of as an explicit discourse about this historical object. It manifests itself without even laying a claim to meaning, in a fragmentary manner, and, more often than not, as a contradictory figure, as a dialectical distillation of this meaning. For Benjamin, the thought-image referred more generally, as shown by Georges Didi-Huberman, to a "visual knowledge" that nourishes an entire range of heuristics.[4] The knowledge revealed through the Benjaminian gaze in fact defines a form of thought that arises from an experience of an image, but whose traits can be defined independently of a given image. And the most salient one would be the *fragmentary* quality of this form of thought, which thus elicits a process of *montage.*[5]

A heuristics of ambiance thus asks us to define the traits of a form of thought for which knowledge does not gush forth in the flash of an image or within the empire of sight. Other aesthetic and dynamic factors in the ambiance thwart the process of representation, as envisioned in the philosophical tradition, specifically though the metaphor of vision and light. Isn't knowing ideally seeing in full light? "Do you *see* what I mean?" is something we ask to ensure that we have been understood. We must grasp the subject of the discussion at the moment that it risks evading being grasped, which incites us to verify that the other person does indeed clearly have it within his or her sights. In this case, *comprehension* shows its etymology of *taking something with oneself,* that is, possessing the subject of the question. And this possession passes metaphorically through the vision that one projects onto it, the optical thus exhibiting a theoretically haptic dimension.

The philosophical modernity that arose from the Enlightenment largely stems from this emancipating approach to knowledge, which frees the eye from the grip of traditional, heteronomous knowledge and the dustiness of the theological and metaphysical legacies of centuries past. Among others, John Locke, in his *Essay on Human Understanding,* insists on the comparison between the mind, which distinguishes the truth, and the eyes, which perceive the light. This metaphor of the illuminating role of reason appears in various forms in this work. Is it by virtue of its ability to see clearly that humanity emerges from its state of an intellectual "minority", as argued by Kant in *What is Enlightenment*? Hans Blumenberg analysed very well the philosophical functioning of this metaphor of light in modern thought, which secularizes the medieval notion of divine light.[6] This profound alteration in the relationship to knowledge in European thought is also tied to a change in the relationship to power. According to Michel Foucault, the optical logic underlying modern thought was inseparable from a logic of control that defined a new mode of "governmentality".[7] Of course, the art and philosophy of the modern abound with forms of thought associated with forms of vision that images embody, and which do not express the dominating authority of a reason that becomes the master and possessor of nature and the world in general through its instrumentalization of the optical. From this standpoint, in Benjamin, the image represents a very peculiar model of thinking that in fact problematizes vision, seeing double in terms of the meaning it brings to light through a dialectical crisis of the meaning of history and modernity as a whole.

Beyond modernity's double vision against the backdrop of the crisis of instrumental rationality in Benjamin's work, the evolution of the heuristic problem of vision can be traced as it took shape in aesthetic experience and the forms of thought associated with it through the history of modern and contemporary art and the central problem of representation accompanying its development. Visions are double, unclear, inner, impossible, forbidden, hallucinatory, deceptive, diffracted, fragmented, decentred, trans-identified, multi-perspectival, machine-like, automatic, animist, and animal. These properties of visual experience, which problematize a relationship of representation ideally organized by a link of clear and distinct perception and knowledge between a human subject and an object, can be exemplified in a vast catalogue of works of visual art.

But if we had to take just one example, perhaps Andy Warhol's film *Empire* (1965) best illustrates the critical state of a certain model of representation that has been diagnosed many times, albeit in different ways, as well as the process of opening up to new coordinates of visual experience in which the optical relationship between subject and object is laid bare. In this film, the vision machine that is the camera captures New York's famous Empire State Building continuously, for hours, through a fixed frame. The subjectivity of the gaze that cinema glorified has here been totally devalued. As for the object, its symbolic power collapses, worn down by the non-event of what occurs in this fixed frame. For eight hours of a film showing changes in light and texture of the air that envelops the building in slow motion, the only visible agency of the transformations that appear in the image is that of the atmosphere, before night comes to extinguish the very visibility of this atmosphere for hours. The image is no longer the vector of an optical grasping, a projection towards a horizon where space unfolds in the history of its own conquest by means of the gaze. It instigates an aesthetic disengagement that considers all expressivity and narrativity as variations of a kind of atmospheric course of time. Thus, the Empire of seeing abandons its claim to control, like a surveillance camera that has been repositioned to capture the nuances of non-events, thereby making room for the experience of the sensory surroundings of things. These surroundings, in which object and subject dissolve, precisely define what we call "ambiance".

A heuristics of ambiance

"Are you within the ambiance of what I am telling you?" Our current language has not – that is, not yet! – adapted the notion of ambiance in such a question to ensure the proper transmission of an idea during a conversation. But we can "see" very clearly how this question changes the relationship to the idea in question in such an exchange, which we wouldn't know whether it was understood at all, or even that we wouldn't need to know that it would be grasped in any manner. The opposite is true: it's a matter of being grasped by what we are talking about. In effect, ambiance necessarily raises the question of a circular dimension, a projection that radiates outwards from the point at which the question is asked. It is as if it enlarged the field of vision beyond the ideal optics of linear perspective to a 360° angle to situate us *within* the question, in the middle of the meaning that is seeking to convey itself. And the parameters of its perception immediately multiply, as they cannot be reduced to the outline of a rigorously demarcated idea.

In Anton Chekhov's novella *Dreams,* a vagabond is being escorted along a road by two policemen, but this walk is not demarcated in the story by either a clear beginning or a specified end. Through the discussion between the three characters, we gradually discover the reasons for this situation, although we will not know its outcome at all. It is a deportation to Siberia that the vagabond paradoxically sees as a journey to a free land. We are stopped along a path, and the issue in question, freedom, is expressed by the author not through the goal of the journey – nor the horizon of a necessity confronted with contradictions, the overcoming of which the narrative is putting into play – instead through the presence of a constant possibility that we perceive in the vagabond's dream, even in his state of servitude. As Jacques Rancière described it:

> The muddy route in which their feet sank is exactly the same in front of them as it is behind them. And all around them, the whiteish fog forms a wall that blocks any perspective and accentuates the sensation of an immobile treading within a space that undefinably resembles itself.[8]

Against a backdrop of blocked perspectives, the detainee's dream, which he describes to his fellow travellers, seems to radiate like a diffuse glow that transcends the grip of the police escort. The dream of freedom here is "the harmony between a situation, the performance of a body, and the landscape of thought that this gives rise to in the mind".[9] Chekhov "is not a dialectician"[10] nor does he uphold a political line of thinking through characters serving as spokespeople, Rancière explains. It instead helps open up the landscape of the possible, otherwise called the "landscape of thought"[11] that situates us in a sustained relationship to what we have nevertheless lost sight of. This generates an alternative between a dialectic of lines (perspective lines, political lines, etc.) and an aesthetic of vague but constant glows. The "landscape of thought" that Rancière talks about in relation to Chekhov brings us closer to what we seek to characterize as an "ambiance of thought". "The landscape has ideas", wrote Balzac.[12] So does ambiance. Now, the specifically *ambiantal* nature of these ideas entails a difference from the idea of what constitutes a landscape.

With regard to Georg Simmel's "Philosophy of Landscape", Nia Perivolaropoulou reminds us that Simmel thought that the appearance of landscape as an art form presupposed a rupture with a consciousness characterized by "a wholeness, a unity, over and above its component elements, without being tied to their specificity or mechanistically composed of them".[13] Thinking about landscape, especially in painting, is in fact only possible for a philosopher through a specific relationship with nature that was developed by modern Western subjectivity, which, as Perivolaropoulou analyses in citing Simmel, "necessitated a tearing away from that unitary feeling of the whole of nature that was characteristic of medieval religious sentiment, for example".[14] In this regard, thinking about ambiances seems an expression of new coordinates for the subject's position in the world. These coordinates are neither those of the pre-modern unity described by Simmel nor those of the distance from nature inherent to modernity. Ambiance presupposes another ecology, another phenomenology, and another epistemology of subjectivity, meaning another conception of the conditions of knowledge for earthlings that calls into question the very definition of subjectivity.

In terms of the "coordinates" of our subjectivity, we can schematically identify the parameters of *situation*, *relation*, and *movement*. *Situation*, particularly from a spatial viewpoint, is defined according to a variable of distance and belonging: inside/outside, immersion/removal, immanence/transcendence, participation/detachment, and so on. From a structural or functional viewpoint, the parameter of *relation* varies according to the forms of connection between the subject and the elements to which the subject relates: heteronomy/autonomy, dependence/independence, connection/disconnection, and so forth. The parameter of *movement* refers to the dynamic variables of these relations: stasis/mobility, stability/transitivity, centripetal/centrifugal, and so on. In contemporary art, the environmental – and specifically the ecological question – has reconfigured our relationship to the landscape by inserting the human subject within it, and no longer at a distance from it, as in Land Art, but in ways that often lead back to self-centred forms of subjectivity.[15] With regard to certain parameters of subjectivity and their related forms of thought, it seems to us that the question of ambiance opens avenues of thinking about the vicissitudes of contemporary subjectivity as it searches for its new coordinates. As in Warhol's *Empire*, domination *of* vision and *by* vision has entered a crisis. And, as in Chekhov's *Dreams*, in premonition of an *a-teleological* time, thought opens up to its surroundings and reveals a new sense of the possible amidst a relative indeterminacy of forms and ends.

The formal traits of an ambiance

What are the traits of an ambiance that qualify this cognitive condition which we call an "ambiance of thought"? Etymologically and semantically speaking, ambiance refers to that which surrounds, embraces, or envelops beings and things.[16] Often synonymous with atmosphere, it designates the sensory precipitate in which we bathe and in which the polarities of the traditional analysis of perception (the perceived object and the perceiving subject) lose their definable contours. All knowledge of ambiance must deal with its composite, dynamic nature, as Jean-François Augoyard has shown.[17] It also challenges any attempt at definition with an unstable topology. In the spirit of a new phenomenology inspired in part by Hermann Schmitz, Tonino Griffero defines atmosphere as a "quasi-thing". It is a "je-ne-sais-quoi perceived by the felt-body in a given space, but never fully attributable to the objectual set of that space".[18] Gernot Böhme analysed atmosphere as engaging things in a movement of *ecstasy*, in the sense that their presence entails that their properties "colour" what surrounds them.[19] As an emanation of things and beings, atmosphere thus prohibits the traditional ontological separation between the properties of objects and their contingent manner of existing in space. The essential problem inherent to this notion is that it contradictorily implies both a situated experience and a mutation, or even a dissolution of the very coordinates of such as situation. The movements of resonance, emanation, diffusion, and impregnation that qualify it in fact lead to the destabilizing of the relationship between subject, object, and environment.

Thus, ambiance is what surrounds us in a way that makes us lose – or which renders less discriminate and definitive – the contours of a place, of things, and of ourselves. This morphological property appears even more pronounced than in the notion of atmosphere, which ultimately always refers to a geometrically and cosmologically well-defined form, namely that of a sphere. Any atmosphere that I experience ultimately leads back to the objective sphere within which I find myself, somewhere between the maternal uterus and the air bubble that envelops our planet.[20] This horizon does not exist for ambiance. This semantic nuance suggests that the form that can objectively and physically demarcate it does not have the ontological relevance that we see for atmosphere. Air, which constitutes the physical substratum or metaphorical vehicle of atmosphere, is merely one potential component of an ambiance. Atmosphere situates the subject within the air it breathes, while ambiance situates the subject in its surroundings without materially connoting them. The nuance is frail, but interesting. It suffices to point out that atmosphere possesses a stronger ecological connotation than ambiance, which, for this very reason, exhibits a different speculative potential. But we must dig deeper. Atmosphere metaphorizes the surroundings (for being like the air that surrounds the Earth) or metonymizes it (for being a part of the whole of the atmosphere). In his eco-phenomenology of ambiance, Bruce Bégout points out this difference between atmosphere and ambiance, the former being more metaphorical.[21] Thus, when the German philosopher Hermann Schmitz talks about an "atmosphere of emptiness", this implies a sensation of enveloping emptiness that is conceived *as* an atmosphere. For Schmitz, atmosphere is a "volume of space without a surface" that "grips us carnally", meaning in terms of sensitive flesh and not our physical body (a distinction heightened by the difference in German between the terms *Leib* and *Körper*).[22] It is the depth (instead of the surface) and its affective resonance that defines atmosphere in Schmitz's words. And for that matter, it is a material vector that constitutes the vehicle for this resonance, namely air. A gas is a state of matter that does not have its own form or volume that takes the form of its container, such as a sphere around the Earth.

Ambiance connotes a slightly different set of coordinates. In effect, it designates the place of a surrounding that merely presupposes a vague spatiality without any material characterization. Ambiance positions and situates, but according to a relative geometry that can only be determined by a point's surroundings. What surrounds us – which we will call the *Surrounding* to stress the elementary state of being surrounded more than the set of things surrounding us, i.e., our *surroundings* – is first of all neither the environment nor the milieu. The Surrounding can be specified physically, biologically, or socially as an environment or milieu, or psychologically, affectively, and aesthetically as an atmosphere or an ambiance, depending on the meaning one assigns to these notions. But ambiance undoubtedly provides a more open-ended meaning of the Surrounding. Atmosphere has its place, but it inserts itself in larger spheres in which each specific atmosphere dissolves. Ambiance exists at the relative scale of the point at which it results and is experienced. There does not appear to be any objective sphere that predetermines it, even if everything in that environment affects it. In sum, if an atmosphere is "a volume of space without a surface", then *ambiance consists of a Surrounding without a contour*.[23]

We can extract three key characteristics of ambiances from these remarks:

1. *Situationality*, which entails an experience situated in what surrounds us, in our environs, which does not mean that the coordinates for this situation are clearly given, or that the place where it is experienced is its central focal point;
2. *Relationality*, which insists on the logic of the connection, the belonging, or the resonance that characterizes the ambiance (e.g. a connection that situates us within the subject/object relationship or a connection that specifically ties us to our environs, or to ourselves);
3. *Diffusionality*, a neologism that here simply indicates that the elements that make up the ambiance have a diffuse existence and behave according to a fluid, unstable dynamic.[24]

These three characteristics refer to the parameters of situation, relation, and movement that we mentioned when talking about the coordinates for subjectivity as a way to describe various forms of experience and thought tied to the traditional, idealist model of vision, its scrambling, and the issue of landscape, which is itself transformed by the question of ambiance and atmosphere. We might specify that these parameters, in the case of ambiance, entail a form of *objectal indeterminacy*, or rather, the difficulty or impossibility of determining the ambiance in terms of a distinct, precisely delineated object. These parameters also entail that ambiance is very sensitive to the forms of *mediality,* meaning the media and milieus that give rise to its production and perception.

We may thus advance with our characterization of an "ambiance of thought". We must first of all rid ourselves of any pejorative meaning, a risk that remains inherent to the current use of the term "ambiance". An ambiance of thought does not mean a thought that is vaporous and lacking in substance, without any cognitive, logical, or critical value. In his *Vocabulaire d'Esthétique,* Etienne Souriau clearly noted this risk when analysing works. "Ambiance is, in the strict sense of the term and in general, the milieu in which a being is situated and which envelops him/her entirely", he wrote before remarking that, regardless of its meaning, the term suffered from a "fairly vague meaning", unlike the more technical term of ethos, which he preferred.[25] This is why "this imprecision allows for the use of these terms in hasty, convenient opinions", he held, to the point of warning that "for purists, the use of the word ambiance" is often "fairly poor language, even subject to some measure of ridicule".[26] We have been warned! And we are willing to bet that, if we manage to avoid this pitfall, our contribution to the Aesthetics of ambiances will be all the more useful.

From thinking about an aesthetic form to the aesthetics of a form of thinking

In terms of aesthetics, the notion of ambiance should steer us away from the sole question of the aesthetic judgement and critical analysis of art objects. Gernot Böhme insisted extensively on this matter in his phenomenology of atmosphere.[27] Talking about "ambiance of thought" suggests that we take a further step to isolate ambiance from its merely phenomenological dimension. In his philosophy of art, for example, Arthur Danto shows that the existence of art objects as such depends primarily on certain theoretical conditions. In *The Transfiguration of the Commonplace*, he called "the atmosphere of aesthetic theory" that made the ontology of art of the contemporary period historically and logically possible.[28] In his opinion, the difference between a work of art and an ordinary object that is indistinguishable from a sensory viewpoint lies in this "atmospheric" condition of the definition of art[29]. The term presupposes that the theory in question is not comparable to an explicit paradigm in which a certain evolution of artistic forms would result in a predictable manner. Generally speaking, the history of ideas depends on climates and other intellectual atmospheres that prevail at any given time or place. Thus, between deep structures, on the one hand, and surface events of a varying degree of randomness, on the other, an "ambiance" may undoubtedly qualify a kind of diffuse, intermediary intellectual conditioning that behaves less as an implacable, univalent determination than as a milieu that provokes varying effects, depending on the kind of attention we pay to them.

Far from a theoretical inconsistency, the objectal inconsistency of ambiance should direct us towards a strong meaning of the syntagm "ambiance of thought", which is not limited to grasping ambiance as an object of thought in order to designate ambiance as a quality of thought itself. The question "How to think about ambiance?", which is inseparable from the question "How do we think in the ambiance?", becomes for us "What is thought as an ambiance?", or rather, "How does ambiance think?" There is always an ambiance of thought in a broad sense, to the extent that the ambiance appears as a kind of constant or transcendental condition, provided that this term is understood here not in the Kantian sense of transcendental idealism, rather in terms of the pragmatic transcendentalism that derives from Emerson. In effect, there is ambiance the moment there is thought, as thought must be envisaged as situated. All thought operates under ambiantal conditions. Simply, we can decide whether or not to render our thought attentive to these surroundings that nourish it, which are diffused in it and in the resonance of which our belonging to the world is actualized, similarly to our relationship to ourselves, as expressed by Emerson through his concept of self-reliance.[30] Using Emerson, Thoreau, and Ives as examples, Pauline Nadrigny shows how the experience of a sound ambiance and a form of thought are linked.[31] In the flow in which the composite elements of perception are diffused, the boundaries of the thinking subject tend to become redefined, resonating more strongly as they open up to the particular sounds that surround them.

Another example of what is meant by ambiance of thought in this sense comes from the famous "5th walk" in the *Reveries of a Solitary Walker*. Rousseau describes the happiness he felt during his stay on a small island in the Lake of Bienne in Switzerland. The author, "a man who liked to circumscribe himself", discovered the opportunity on this island, isolated from the tumults of the passions of the social world, for a sensory and intellectual experience that seemed to him the ideal state of all existence. He described it specifically in terms of a moment of reverie at the lake's edge:

> The ebb and flow of this water, its continuous sound swelling incessantly into my ears and eyes, replaced the internal movements that the reverie extinguished in me and sufficed to make me feel my existence with pleasure, without having to think about it. From time to time, a faint and brief reflection would arise on the instability of the things of this world whose image the surface of the waters offered me, but soon

> these light impressions would fade away in the uniformity of the continuous movement that rocked me back and forth (...).[32]

A singular *cogito* is enunciated here that affirms existence by dispensing with Cartesian certainty, which is based on the mere fact of thinking. The sentiment provides the certainty, and the pleasure of existence is experienced without "having to think". But we would be mistaken to think that this state advocates for a radical a-rationalism. Moreover, there would be a performative contradiction in asserting "I do not think, therefore I exist". Instead, we are attributing a function to thought here that is different from grasping clear and distinct truths in the linear order of reasons. To the contrary, Rousseau is defining here the virtues of a thought of letting go, in which the things of this world that it is considering remain unstable and fluid, *around* the thinking subject and not directly in front of him or her. And this flickering thought, to which the reflection of the water literally provides a model of intellectual reflection, does not hold what it sees at a distance; it immerses itself in it and melts into its movement and rhythm. If this person who liked to circumscribe himself found an opportunity there to feel himself existing fully, this was in favour of a thought that was open to its surroundings. These comprise more than the environment of the Île Saint-Pierre; they also include the diffuse, medial region of the reverie itself. Because of its composite qualities of objectal indeterminacy, porosity, situationality, fluidity, and relationality, this Rousseauian reverie can be analysed in terms of an ambiance of thought.

An epistemology of diffuse ideas

In our interest in the form of a thought subject to ambiantal conditions and through this example of reverie, regardless of what it says about the atmosphere that reined over this small island in the Lake of Bienne that Rousseau adored, we return to the strong sense of "ambiances of thought". In this perspective, the aim is to pull ourselves even further away from an empirical and material relationship to ambiance so that we may consider ambiance in terms of a theory of knowledge that seeks to identify a significant ambiantal dimension to certain cognitive processes and conceptual forms. Said another way, we are advancing the hypothesis that thinking is to be situated within an ambiance of thought, surrounded by ideas. Every idea, every element of knowledge, every fact of consciousness, emerges against a background of intellectual noise; every exercise of reason is made in resonance with the knowledge one accumulates, or in relation to the environment of past and possible language games.

Theories of consciousness distinguish between different levels of consciousness, ranked according to their degree of self-reflexivity (according to a scale that distinguishes between "sentience", "self-consciousness", and "noetic consciousness" for instance).[33] Thus, an ambiance of thought would be a hybrid modality of consciousness that plunges the noetic level of consciousness back into the most vague and diffuse conditions of sentience. The vague and diffuse obviously appear to contradict the noetic ideal of rational knowledge.

Traditionally, a clear and distinct idea stands out against a background of obscure and confused ones. And even if we recognize the salient, luminous traits of thought more clearly than the enveloping, auroral ambiances of thought, the latter play a major role in knowledge, learning, and creation. All we have to do is reread a number of literary or philosophical works, like Rousseau's reverie, to be reminded of this. This is where we will find undefined truths that resonate within the surroundings of our certainties, and it is in the attention to this peripheral "gnoseology", which the late eighteenth century qualified as inferior, that Baumgarten, for example, gave birth to Aesthetics.[34] Working in Leibniz's footsteps, Baumgarten also adopted the typologies that opposed, in terms of ideas, clarity and obscurity, as well as distinction and confusion.[35] Truth does not move straight from darkness to the light, just as the sky does not turn directly from night to noon, Baumgarten explained.[36] This is why, when we talk about knowledge, we should refrain from looking down not only at artists, but also practitioners of aeromancy, the art of divination through the medium of air,[37] as both one and the other cultivate a form of sensory knowledge that is a prerequisite for any science emerging at the dawn of various forms of understanding. It seems to us, however, that a contemporary theory of ambiances, fed by the various strains of situated thought, all of which help to define a certain "ecology of the mind",[38] as well as Deleuze's reappraisal of obscure ideas,[39] asks us not to consider the form of ideas using a single axis of greater or lesser discernment, along which the goal of every thought is to push our sense of reason towards a full light. Is there not a dynamic of thought around this shining axis that operates in a more vibrant manner, through resonance, radiance, and emanation, in which an idea forms on the condition that one ceases to discern, and in which the ensuing dissolved certainties make room for what we should call *diffuse ideas*?

We do not hold that a diffuse idea is a confused idea. As Leibniz explained, confusion concerns an idea that lacks a sufficient amount of determinations to allow us to distinguish the thing it concerns from another thing.[40] I may clearly recognize the face of someone I passed by yesterday, but if I have to describe it, forge a concept of it, for example to be able to generate a composite sketch of it, it would be very hard for anyone, based on this idea of mine, to formally distinguish this face from another one that resembles it. That said, a diffuse idea does not engage in this type of inquiry, nor is it contrary to distinction. It instead finds its opposite in what we might call *condensation*. A diffuse idea situates us in something like the middle of a lake of thought. It actively and extensively probes the surroundings in all directions, and it achieves perfection in the intensity of its resonances, as when Rousseau thinks "I exist" in a masterful idea that is clear and *diffuse*.

We can find other occurrences of this diffuse property of ambiance in Proust, which helps us reflect on what Mauro Carbone calls "sensitive ideas" in *In Search of Lost Time*.[41] Consider the famous episode of the madeleine.[42] The narrator seeks the cause of the overwhelming sense of joy that he feels but which he does not understand. It is instead by thinking about something else that he discovers this cause, or rather, this source, as if he were listening to everything in his mind that surrounds what he is searching for. So, can we not use the notion of ambiance to describe the cognitive conditions of this search that the narrator experienced? Isn't it the field of the ambiances of thought that Proust was exploring as he described these regions of memory in which nothing can emerge that is directly labelled as a piece of information awaiting activation? The recollection means something only in the slowness and the floating of an indirect attention that diffuses and radiates, thereby creating the conditions for a new meaning to emerge.

The psychologist Eugène Minkowski evokes a more ordinary experience that can help us explore further what is an ambiance of thought, where diffuse ideas bathe. In attributing the forgetting of names to one's age, the author explains that this common experience, deplored by most people, cannot be reduced to a failure of a symbolic process, and that it is instead the result of a diffusion of the signifier, which he believes is of creative interest.[43] There are times when we "sense" a word or a name that escapes us, which is when we say that it is on the tip of our tongues. But in reality, if we hang on only to the tip of the tongue, the word will keep falling off it, and, furthermore, this experience will cease to hold any interest for us. Here is a brief, ordinary psycho-pathology experiment involving an ambiance of thought: feeling the idea in the forgetting of the name or the word.[44] How does the meaning become diffuse as the signifier dissolves?

From a logical and semiotic perspective, this ambiantal state of meaning could refer, far from any kind of psychology, to a certain type of "logic of the vague", an oxymoron for which Christiane Chauviré sees a principle arise in Peirce.[45] For that matter, it has been established that the most rigorous formal logic is itself never free of "atmospheric effects".[46] Noëlle Batt has clearly shown that we must and can approach this question of conceptual indeterminacy in a rigorous, detailed manner.[47] From this we deduce that, while ambiance is indeed a vague concept, an aesthetic of ambiances must qualify precisely the nature, qualities, and functioning of this indeterminacy, as when we propose not to equate "diffuse" with "confused", for example. In this case, the difference between these two notions is comparable to the one between waves and the vague (which are respectively *vagues* and *vague* in French), between the undulating dynamic propelled by countless forces and the complete state of that which lacks a form. If the pragmatic value of conceptual indeterminacy, of which the ambiance of thought is one modality for us, has been proven in logical and semiotic terms, it is worth noting that art and literature assign a great effectiveness to it, specifically as a dynamic process of the "semio-poetic" type in this instance. When the ordinary meaning of words dissolves in a poem, this in fact serves to give rise to a new, unprecedented meaning. So, wouldn't an ambiance of thought designate an essential condition for any creation of meaning though the semantic continuum that it allows for, thereby enabling, for example, acts of "relinking across discontinuities"?[48] Literary language is particularly well-suited to exploring the processes by which meaning emerges from a "milieu" that considers language within an ecology of meaning, as Mirella Vadean has shown in relation to the work of Marie Darrieussecq, for example.[49]

Recognizing and analysing ambiances of thought may thus contribute to theories of ambiances and atmospheres, particularly their aesthetic and cognitive dimensions. By asking us to pay attention to the property of *diffuse ideas*, they also open up avenues for analysing forms of knowledge produced by the experience of objects, works of art, or otherwise, especially those that entail an environmental or immersive situationality. An analysis of the creative process might also see the concept of the ambiance of thought as a useful tool. Just as the spatiality of ambiances and atmospheres does not entail a geometry with fixed, centred coordinates, the Surrounding in which the diffuse idea spreads similarly entails a dynamic of a decentring and a recomposition of a centre through resonance. As a condition of thought, ambiance thus also appears as an actual quality of thought. Not only does every idea appear only in the context of other ideas that make it possible, but every idea also "tinges" consciousness, like the objects that tinge space described by the new phenomenology. Far from reducing the pathic to the gnostic or from comparing a complex sensory phenomenon to a simple cognitive mechanism,[50] an ambiance of thought designates a type of paradoxical reflexivity in which knowledge refers to the formal properties of the sensory, its most diffuse dynamics at that. This reference can be analysed not only genealogically but also analogically (in the sense of a resemblance – what we would call an *analogon* in the terms of classical aesthetics), in other words, relatively autonomously from a material, ambiantal situation. An "ambiance of thought" preserves or reproduces, within the thought itself, the sensory and situated parameters that fundamentally determine it, rather than extending the field of cognition to what necessarily seems to be escaping it. An ambiance of thought thus appears to be the intellectual condition in which lived ambiances diffuse and resonate, and from which new knowledge or forms of consciousness emerge. In light of the contemporary crisis of anthropocentrism and its ecological and political symptoms, we believe that thinking in terms of ambiances – and paying attention to the ambiances of thought, within which lie still-indeterminate ideas that concern what we, as human subjects, are in the process of becoming – can only contribute to the enrichment of new subjective coordinates where, paradoxically, an increasingly situated and embodied dimension is combined with a fundamental, ex-centric dynamic.

Notes

1 We proposed this perspective during the Cerisy International Symposium "The way of ambiances. The sensory experience of situations" (France, September 2018).
2 The notion of an ambiance of thought might lead us to look at how a certain consciousness works in terms of the philosophy of mind, particularly with regard to the difference between a "phenomenal consciousness" (a basic, subjective consciousness that can be separated from any intentionality or focus on a given object, also called "sentience"), an "objectal consciousness" (which entails an intentional grasping of something), and a "reflexive consciousness" (in which consciousness grasps itself). To describe an "ambiance of thought" according to these categories, we could say that it depends on a reflexive form of consciousness, a consciousness of something that remains elusive as an object, and which leads the mind to have a fundamental, qualitative experience of its own situation.
3 Walter Benjamin, *Denkbilder,* Rolf Tiedemann (ed.), Surkhamp, 1974.
4 Georges Didi-Huberman, *The Eye of History: When Images* Take Positions, MIT Press, 2018.
5 Georges Didi-Huberman, *Devant le temps. Histoire de l'art et anachronisme des images*, Éditions de Minuit, 2000.
6 Hans Blumenberg, *The Legitimacy of the Modern Age,* trans. R. M. Wallace, MIT Press, 1983.
7 Michel Foucault, *The Government of Self and Others: Lectures at the* Collège de France, 1982–*1983*, trans. G. Burchell, Palgrave Macmilan, 2010.
8 Jacques Rancière, *Au loin la liberté. Essai sur* Tchekhov, La Fabrique, 2024, 7.
9 Jacques Rancière, *Au loin la liberté. Essai sur* Tchekhov, La Fabrique, 2024, 10.
10 Jacques Rancière, *Au loin la liberté. Essai sur* Tchekhov, La Fabrique, 2024, 10.
11 Jacques Rancière, *Au loin la liberté. Essai sur* Tchekhov, La Fabrique, 2024, 10.
12 Cited by Michel Collot, *La Pensée-paysage. Philosophie, arts, littérature,* Actes Sud/ENSP, 2011, 17.
13 Georg Simmel, "The Philosophy of the Landscape" (1913), in *Theory, Culture & Society*, 24, No. 7–8 (2007), 21. Cited by Nia Perivolaropoulou, "From Urban Ambiances to Cinema: A Few Reflections Based On Siegfried Kracauer", included in Part 4 of this volume.
14 Georg Simmel, "The Philosophy of the Landscape" (1913), in *Theory, Culture & Society*, 24, No. 7–8 (2007), 21. Cited by Nia Perivolaropoulou, "From Urban Ambiances To Cinema: A Few Reflections Based on Siegfried Kracauer", included in Part 4 of this volume.
15 David Zerbib, "Écosystème de l'œuvre d'art et matérialité du monde", *Issue. Journal of art & design,* HEAD-Geneva (May 2022): https://issue-journal.ch/focus-posts/ecosysteme-de-loeuvre-dart-et-materialite-du-monde.
16 Leo Spitzer, "Milieu and Ambiance: An Essay in Historical Semantics", *Philosophy and Phenomenological Research*, 3, No. 2, (December 1942): 169–218. See p. 198 in particular.
17 Jean-François Augoyard, "Eléments pour une théorie des ambiances architecturales et urbaines", *Les Cahiers de la Recherche Architecturale*, No. 42/43, 1998.
18 Tonino Griffero, *Quasi-Things. The Paradigm of Atmospheres,* trans. Sarah De Sanctis, SUNY Press, 2017; *Atmospheres: Aesthetics of Emotional* Spaces, Ashgate Publishing, 2014, 6.
19 Gernot Böhme, "Atmosphere, a basic concept of a new Aesthetic", in *Atmospheric architectures: The aesthetics of felt spaces*, trans. A.-Chr. Engels-Schwarzpaul, Bloomsbury Publishing, 2017, 21.
20 Peter Sloterdijk based part of his work on this fundamental form of the sphere: *Bubbles: Spheres Vol. I: Microspherology,* trans. Wieland Hoban, Semiotexte:

2011; *Globes: Spheres Vol. II: Macrospherology*. trans. Wieland Hoban, Semiotexte, 2014. To this ontological and metaphysical determination of the atmosphere by means of the sphere, the geographer Derek McCormack allows us to add the physical and aesthetic determination of the balloon as a way to think about, and experience, atmosphere: Derek McCormack, *Atmospheric Things. On the Allure of Elemental Envelopment,* Duke University Press, 2018. No paradigmatic form such as the sphere, no *ambiantal thing* or thing equivalent to the "atmospheric thing" that is the balloon exists to grasp ambiance, which in our view corroborates the morphological indeterminacy of ambiance.

21 See the contribution of Bruce Bégout in this volume, and Bruce Bégout, *Le concept d'ambiance. Essai d'éco-phénoménologie,* Seuil, 2020.

22 Hermann Schmitz, "Les sentiments comme atmosphere", trans. J.-L. Georget and P. Grosos, *Communications,* No. 102, 2018, 61. In basing ourselves on various theories of atmospheres and ambiances, we are in essence questioning the spatiality of concepts, in this case by basing ourselves on this theory of the spatialisation of affects.

23 "Un entour sans contour" in French.

24 The importance of the "diffuse" in approaching the analysis of ambiance is highlighted in Jean-Paul Thibaud, *En quête d'ambiances*, MétisPresses, 2015.

25 Etienne Souriau, *Vocabulaire d'Esthétique,* PUF, 1990, 91. *Ethos,* that is, the general psychological and affective disposition of the work. Souriau insists that ambiance refers to a human, psychological, moral, and affective environment (unlike atmosphere, which refers more to the natural environment).

26 Etienne Souriau, *Vocabulaire d'Esthétique,* p. 92.

27 Gernot Böhme, "Atmosphere as the Fundamental Concept of a New Aesthetics", *Thesis Eleven*, No. 36, 1993, 113–126.

28 Arthur Danto, *The Transfiguration of the Commonplace. A Philosophy of* Art, Harvard, 1981.

29 Even for being metaphorical, the theoretical "atmosphere" in Danto's philosophy of art crosses the empirical dissolution of the artwork in the atmospheres and ambiances which characterize our "hyper-aestheticised world" according to Yves Michaud. See his contribution in this volume and Yves Michaud, *L'art c'est bien fini, Essai sur l'hyper-esthétique et les atmosphères,* Gallimard, 2021.

30 Ralph Waldo Emerson, "Self-Reliance", in *Essays by Ralph Waldo Emerson,* ed. Edna H.L. Turpin, Charles E. Merrill Co, 1907, 79–116.

31 Pauline Nadrigny, "Sound ambiances, intellectual ambiances: Thoreau and Ives", in this volume.

32 Jean-Jacques Rousseau, *Les Rêveries du promeneur solitaire,* Flammarion, 1967, 100 (excerpt translated by Claudio Cambon). See the interpretation of this text by Philippe Simay in this volume, "Towards a therapy of ambiances".

33 "Sentience" (or "phenomenal consciousness") is the base level of feeling the living situation in which we find ourselves, and which we share with animals. At the next level is simple "consciousness", an intermediary level between "sentience" and "self-consciousness" which constitutes the third level. Ultimately, the highest point of the specificity of human cognition is "noetic consciousness", also called meta-consciousness (I am aware of having consciousness, I think that I think, I know that I know, etc.).

34 Alexander Gottlieb Baumgarten, *Esthétique* (1750), trans. (into French) J.-Y. Pranchère, L'Herne, 1988.

35 Gottfried Wilhelm Leibniz, "Meditations on Knowledge, Truth, and Ideas", in L.E. Loemker, (ed.) *Philosophical Papers and Letters*. The New Synthese Historical Library, Vol 2., Springer, 1989. https://doi.org/10.1007/978-94-010-1426-7_34.

36 Alexander Gottlieb Baumgarten, *Esthétique,* 123.

37 Alexander Gottlieb Baumgarten, *Esthétique*, 242.

38 Gregory Bateson, *Towards an Ecology of* Mind, Ballantine, 1972.
39 *Cf.* in particular: Gilles Deleuze, *The Fold: Leibniz and the Baroque*, trans. Tom Conley, Editions de Minuit, 1988.
40 Leibniz, "Meditations on Knowledge, Truth, and Ideas", p. 15.
41 Mauro Carbone, *Proust et les idées sensibles*, Vrin, 2008.
42 Marcel Proust, *Swann's Way,* in *In Search of Time Lost,* trans. Lydia Davis, Penguin, 2004.
43 Eugène Minkowski, Vers une cosmologie, Payot, 1999, 113.
44 For less ordinary cases of psychopathology, we may refer to the cases analysed in: Ludwig Binswanger, Über Ideenflucht, Garland, 1980.
45 Christiane Chauviré, *Peirce et la signification. Introduction à la logique du vague*, PUF, 1995.
46 R.S. Woodworth and S.B. Sells, "An Atmosphere Effect in Formal Syllogistic Reasoning", in *Journal of Experimental Psychology*, 18, (1935), 451–460.
47 Noëlle Batt, "Plasticité conceptuelle. Des vertus du flou, du vague, du mal délimité ou de l'indéterminé, pour l'exercice de la pensée.", in Didier Tallagrand, Jean-Paul Thibaud, Nicolas Tixier (ed.), *L'usage des ambiances. Une épreuve sensible des situations,* Hermann, 2021.
48 Gilles Deleuze, *Foucault*, trans. Seán Hand, Univ. of Minnesota, 1988. Cited in Noëlle Batt, "Plasticité conceptuelle. Des vertus du flou, du vague, du mal délimité ou de l'indéterminé, pour l'exercice de la pensée."
49 Mirella Vadean, "L'esprit comme milieu des idées. Une inspiration écologique à partir des écrits de Marie Darrieussecq", in David, Sylvain et Mirella Vadean (ed.), *La pensée écologique et l'espace littéraire*, Universitey of Quebec, Montréal: Figura, le Centre de recherche sur le texte et l'imaginaire, Cahier Figura Series, Vol. 36, 2014.
50 We owe the opportunity to make these remarks to the valuable discussions held with Jean-Paul Thibaud.

Resources

Jean-Paul Thibaud, Nicolas Tixier, David Zerbib "Ambiances. A sensitivity to ordinary situations". Revised and extended version of the article previously published in French: Didier Tallagrand, Jean-Paul Thibaud, Nicolas Tixier "L'usage des ambiances. Une épreuve sensible des situations" in Tallagrand, Didier, Thibaud, Jean-Paul, Tixier, Nicolas (eds.) *L'usage des ambiances. Une épreuve sensible des situations*, Hermann, 2021, p. 17–22.

Jean-Paul Thibaud "A brief archeology of ambiances". Previously published in French: Jean-Paul Thibaud "Petite archéologie de la notion d'ambiance", *Communications*, n°72, 2012, p. 155–174. Previously published in English: "A brief archeology of ambiances", *Unlikely. Journal for Creative Arts*, Issue 6, *Translating Ambiance*, University of Melbourne La Trobe University and Victorian College of the Arts, 2020.

Jean-François Augoyard "Sensible environment and architectural ambiances". Previously published in French: Jean-François Augoyard "L'environnement sensible et les ambiances architecturales", *L'espace Géographique*, n°4, 1995, p. 302–318.

Pascal Amphoux "The waltz of ambiances". Previously published in French: Pascal Amphoux "La valse des ambiances", *Editions du Plan Construction et Architecture*, 42, 1993, p. 83–88.

Véronique Nahoum-Grappe "The thickness of the present". Previously published in French: Véronique Nahoum-Grappe "L'épaisseur du présent. Courte réflexion ethno-phénoménologique", *Communications*, n°102, 2018, p. 233–249.

François Laplantine "Ambiances, places and links". Previously published in French: François Laplantine "Percevoir: ambiances, lieux, liens" in *Penser le sensible*, Pocket, 2018, p. 57–73.

Bruce Bégout "From Milieu to Ambiance: another conception of environment". Previously published in French: Bruce Bégout "Du milieu à l'ambiance. Réflexions philosophiques sur une autre conception de l'environnement", *Les études philosophiques*, n°142, 2022/3, p. 89–106.

Yves Michaud "Milieu, ambiances, atmosphere: a new approach on experience". Previously published in French: Yves Michaud "Milieu, ambiances, atmosphère: nouveaux regards sur l'expérience " in *L'Art c'est bien fini ». Essai sur l'hyper-esthétique et les atmosphères*, Gallimard, 2021 p. 79–97

Olivier Gaudin, Maxime Le Calvé "Ambiance crossings: a conceptual journey toward multiperspectivistic sensorial inquiries in anthropology". Revised and extended version of the article in French: Olivier Gaudin, Maxime Le Calvé "La traversée des ambiances. Regards sur les atmosphères en sciences sociales", *Communications*, n°102, 2018, p. 5–23.

Derek McCormack "Returning to atmosphere". Revised and extended version of the article previously published in French: Derek McCormack "*Atmosphere*/ambiance/ambiant" in Tallagrand, Didier, Thibaud, Jean-Paul, Tixier, Nicolas (eds.) *L'usage des ambiances. Une épreuve sensible des situations*, Hermann, 2021, p. 217–226.

Rainer Kazig, Damien Masson "Ambiance as a concept of French cultural geography". Revised version of the article previously published in French: Rainer Kazig, Damien Masson "L'ambiance comme concept de la géographie culturelle francophone", *Géographie et cultures*, no. 93–94, 2015, p. 215–232.

Tonino Griffero "Affective inflationism and atmospheric competence". Revised version of the article previously published in English: "Affective inflationism and atmospheric competence" in Griffero, Tonino (ed.) *The atmospheric 'we'. Moods and collective feelings*, Mimesis International, 2021, p. 85–103.

Philippe Simay "The therapeutic nature of ambiances". Previously published in French: Philippe Simay "Vers une thérapeutique des ambiances" in Tallagrand, Didier, Thibaud, Jean-Paul, Tixier, Nicolas (eds.) *L'usage des ambiances. Une épreuve sensible des situations*, Hermann, 2021, p. 49–75.

Anne Bossé "Becoming of the contemporary sensibility". Revised and extended version in French: Anne Bossé "Quel devenir de la sensibilité contemporaine?" in Tallagrand, Didier, Thibaud, Jean-Paul, Tixier, Nicolas (eds.) *L'usage des ambiances. Une épreuve sensible des situations*, Hermann, 2021, p. 159–163.

Pauline Nadrigny "Sound ambiances, intellectual ambiances: Thoreau and Ives". Revised and extended version in French: "Ambiances sonores, ambiances de pensée: Thoreau et Ives" in Tallagrand, Didier, Thibaud, Jean-Paul, Tixier, Nicolas (eds.) *L'usage des ambiances. Une épreuve sensible des situations*, Hermann, 2021, p. 407–418.

Nia Perivolaropoulou "Urban ambiances in movies. Siegfried Kracauer". Previously published in French: Nia Perivolaropoulou "Des ambiances urbaines au cinéma. Quelques réflexions à partir de Siegfried Kracauer" in Tallagrand, Didier, Thibaud, Jean-Paul, Tixier, Nicolas (eds.) *L'usage des ambiances. Une épreuve sensible des situations*, Hermann, 2021, p. 37–48.

Grégoire Chelkoff "For a modal conception of architectural ambiances". Previously published in French: Grégoire Chelkoff "Pour une conception modale des ambiances architecturales", *Faces, Journal d'architecture*, n°67, 2010, p.18–23.

Izabela Wieczorek "Atmospheres in formation: affective thresholds, immersive fields, and active boundaries". Previously published in French: "*Atmospheres* en formation: seuils affectifs, champs immersifs et limites actives" in Tallagrand, Didier, Thibaud, Jean-Paul, Tixier, Nicolas (eds.) *L'usage des ambiances. Une épreuve sensible des situations*, Hermann, 2021, p. 251–259.

Emmanuel Doutriaux "Affordances of the atmosphere / Architectures of the common". Previously published in French: "Affordances de l'ambiance / Architectures du commun" in Tallagrand, Didier, Thibaud, Jean-Paul, Tixier, Nicolas (eds.) *L'usage des ambiances. Une épreuve sensible des situations*, Hermann, 2021, p. 341–350.

Céline Bonicco-Donato "Pursuing the aesthetics of atmospheres, with & beyond Böhme". Original article.

Nicolas Tixier "The ambiances are the message: transects, projects and urban situations". Revised and extended version of the article previously published in French: Nicolas Tixier. "Cutting the city in half: ambiences, transects and projects" in Duarte, Cristiane Rose, Pinheiro, Ethel (eds.) *Arquitetura, Subjetividade e Cultura. cenários de pesquisa no Brasil e pelo mundo*, Proarq/Rio books, 2020.

Niels Albertsen "Atmospheres in the metapolitan anthropocene". Revised and extended version of the article previously published in English: Niels Albertsen "Atmosphere and the anthropogenic Metapolis", *Proceedings of the 4th International Congress on Ambiances, Alloaesthesia: Senses, Inventions, Worlds*, Réseau International Ambiances, 2020, e-conference, p. 74–77.

Laurent Devisme "The developer over-investment in ambiances". Revised and extended version of the article previously published in French: Laurent Devisme "Vertiges de l'usage aménagiste des ambiances" in Tallagrand, Didier, Thibaud, Jean-Paul, Tixier, Nicolas (eds.) *L'usage des ambiances. Une épreuve sensible des situations*, Hermann, 2021, p. 83–96.

David Zerbib "Ambiances of thought: a contribution to the analysis of aesthetic cognition". Revised and extended version of the article previously published in French: David Zerbib "Des ambiances de pensée" in Tallagrand, Didier, Thibaud, Jean-Paul, Tixier, Nicolas (eds.) *L'usage des ambiances. Une épreuve sensible des situations*, Hermann, 2021, p. 393–405.

Acknowledgements

I would like to thank Cédric Pichat for help in preparing files for editing.

This project was supported by:

– The Cresson team of the AAU laboratory (CNRS / ENSA Grenoble / Grenoble Alpes University)

– Ecole Supérieure Art Annecy Alpes

Index

Note: *Italic* page numbers refer to figures and page numbers followed by "n" denote endnotes.

FSC
www.fsc.org
MIX
Papier | Fördert gute Waldnutzung
FSC® C083411

Kolibri 360 GmbH
Ferdinand-Jühlke-Straße 7
99095 Erfurt, Deutschland
produktsicherheit@kolibri360.de